THE CARE AND CONSERVATION
OF GEORGIAN HOUSES

THE CARE AND CONSERVATION OF GEORGIAN HOUSES

A maintenance manual
for Edinburgh New Town

Fourth edition

Andy Davey BSc (Hons), Dip Arch, RIBA, ARIAS
Bob Heath Dip Arch, RIBA, ARIAS, FSA (Scot)
Desmond Hodges OBE, FRIAS
Mandy Ketchin BArch (Hons), RIBA, ARIAS
Roy Milne BArch (Hons), RIBA, FRIAS

Butterworth-Architecture
An imprint of Butterworth-Heinemann

Butterworth-Architecture
An imprint of Butterworth-Heinemann Ltd
Linacre House, Jordan Hill, Oxford OX2 8DP

℞ A member of the Reed Elsevier plc group

OXFORD LONDON BOSTON
MUNICH NEW DELHI SINGAPORE SYDNEY
TOKYO TORONTO WELLINGTON

First published by Paul Harris Publishing in association with the
Edinburgh New Town Conservation Committee 1978
Second edition published in paperback by the Architectural Press Ltd 1980
Third edition 1986
Reprinted by Butterworth-Architecture 1988
Reprinted 1991
Fourth edition 1995

British Library Cataloguing in Publication Data
Care and Conservation of Georgian Houses:
Maintenance Manual for Edinburgh New Town. – 4 Rev. ed
 I. Davey, Andy
 690.830288

Library of Congress Cataloguing in Publication Data
The care and conservation of Georgian houses: a maintenance manual
 for Edinburgh New Town/Andy Davey . . . [et al.]. – 4th ed.
 p.   cm.
 Includes bibliographical references and index.
 ISBN 0 7506 1860 4
 1. Historic buildings – Scotland – Edinburgh – Maintenance and
 repair. 2. Architecture, Georgian – Scotland – Edinburgh –
 Conservation and restoration. I. Davey, Andy.
 TH3361.C37   94 – 25926
 690'.83' 0288 – dc20   CIP

ISBN 0 7506 1860 4

Printed and bound in Great Britain by Bath Press

CONTENTS

**THE MORAY FEUS
AND THE WEST END**

Photo: John Dewar

FOREWORD TO THE FOURTH EDITION

By the Chairman of Edinburgh New Town Conservation Committee
The Rt Hon Lord Cameron of Lochbroom

This book is an excellent demonstration of the way in which professional skill, knowledge, enthusiasm and concern can be freely shared not merely for the benefit of Edinburgh's Georgian New Town but, as it has proved, many other historic places.

This is the third time that the authors have given of their time and expertise to revise and update the text which they first wrote for the Committee between 1976 and 1978. The decision to prepare a fourth edition was taken by the authors in 1992. The high standard of painstaking revision was then set which it has taken nearly eighteen months to fulfil. The proof lies in a text which will be of great value to all who have an interest, whether professional or lay, in the proper maintenance of fine buildings.

Advice has been given generously by a wide range of experts. Their help is warmly acknowledged. All information and the comments from many sources have been collated and edited by Rachel Dodd BArch (Hons) Dip Arch, and Fiona Rankin MA Dip Arch, who, with Adrian Boot BA (Hons), Dip Arch, also prepared new illustrations. To all who assisted in many ways, the Committee is greatly indebted.

In spite of much voluntary assistance, the cost of revision has been considerable and the Committee is very grateful to the following organizations, firms and individuals, who responded so generously to our appeal:

THE ROYAL INCORPORATION
OF ARCHITECTS IN SCOTLAND
15, RUTLAND SQUARE, EDINBURGH EH1 2BE
TELEPHONE 031-229 7545/7205

SCOTTISH CONSERVATION BUREAU

PREFACE TO THE FOURTH EDITION

Each section of this Maintenance Manual has been revised and much of the text has been rewritten; all factual information has been checked and updated as far as possible, and a new chapter on shops and shop fronts has been written and illustrated by Rachel Dodd.

The extensive advice received from many architects, consultants, contractors and craftsmen is gratefully acknowledged at the end of this book. Officials of Historic Scotland and the City of Edinburgh District Council could not have been more helpful and, in particular, the local Planning, Housing and Property Services Departments, responsible for the day to day protection of the city's fabric, gave useful comments and advice.

The revised text takes account of the ecological significance of building conservation and the tenets of the Society for the Protection of Ancient Buildings.

This new edition coincides with the publication by Historic Scotland of their *Memorandum of Guidance on Listed Buildings and Conservation Areas* which, with their new Technical Advice Notes on Lime Mortars, Plasterwork, Sash Windows, will set conservation standards in Scotland for many years to come.

February 1995

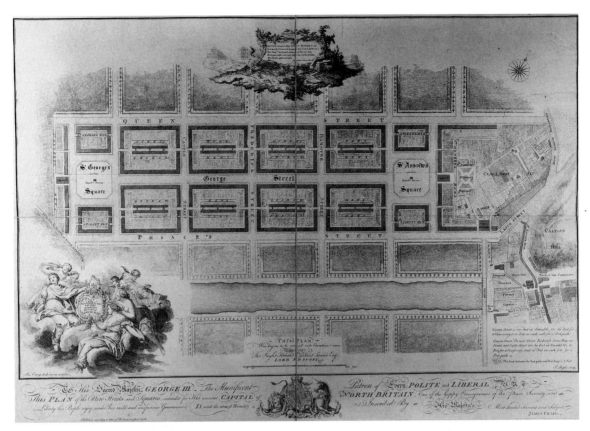

**JAMES CRAIG'S
NEW TOWN MAP**

Photo by courtesy of
Edinburgh City Libraries

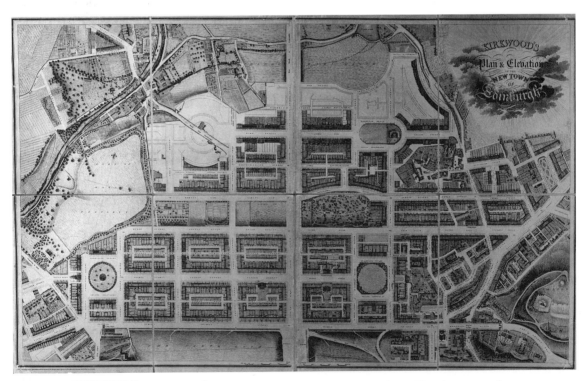

**KIRKWOODS'
NEW TOWN MAP**

Photo by courtesy of
Edinburgh City Libraries

INTRODUCTION

James Simpson BArch, FRIAS, RIBA

The New Town is among the greatest of architectural achievements and its survival largely intact is one of the great successes of the Conservation Movement. The New Town was one of the rocks which began to break up the tidal wave of destruction, profligacy and waste which has so characterized the later decades of the twentieth century. And it is in more than just the *beauty* of its Architecture – which is generally perceived to be the justification for its preservation – that the value of the New Town lies, for Architecture is about *Firmness* and *Commodity*, as well as about *Delight*.

The true greatness of the New Town as Architecture derives from the almost perfect integration of its materials and the way they are put together, the way it functions as a place in which to live and work, and its formal and romantic beauty, which is devoid of mere prettiness, but which cannot in the end, fail to appeal to the intellect as well as to the senses.

The New Town, as a complete city with its cathedral and churches, public buildings and shops, houses and tenements, streets, squares and gardens, is perhaps the ultimate achievement of the traditional building trades. It was almost entirely built by hand, with the aid of horses and carts for local transport, and sailing ships to bring timber from the Baltic and slates from the West Highlands. The principal material, stone, was from local quarries and only in the iron foundries, the brick and lime kilns and the lead and glass works was there any use of non-renewable energy, and that was derived from local coal. Nor did the building of the New Town produce any significant quantity of toxic waste or pollution, the use of lead on roofs, for tanks and pipes, and in paint, being practically the only part of the entire process which could be construed as having any harmful consequences of that sort.

To a large extent, the New Town was built of labour and skill at considerable cost in cash terms, but at minimal cost in terms of energy and resources and with only minimal consequences for the environment in the longer term. Its quality in every way is such that it is capable of lasting, with further injections of labour, skill and little else, more or less indefinitely. In effect a vast, durable and maintainable capital asset was created from very little.

Not only was the New Town energy economical to build, it is also inherently energy economical in use. The mixture of uses in a fairly densely planned but highly attractive city area means that many people live close to their work and to shops, to which they travel on foot or by public transport. The combination of houses and tenements in terraces which form well-spaced city blocks provides shelter and good exposure to the sun – the generally northern slope and the lack of differentiation between north and south elevations notwithstanding – and keeps the ratio of outside wall and roof area to internal volume to a minimum.

In the last century, when it was normal to use the internal shutters, with which all the windows of the New Town were fitted, and when the burning of coal fires made the flue-bearing party walls like great warm storage heaters, the houses were normally both warm and comfortably aired. The lack of smoking chimneys today is to the city's great benefit, but the common failure to make intelligent use of shutters as well as curtains and frequent dependence on high temperature space heating fuelled by natural gas, which makes no use of the high thermal mass of the old flue walls to provide equivalent comfort at lower temperatures and with lower heat losses, presents opportunities for the future.

The building of the New Town came at the end of an era. The Georgian period had been one of rapid development in Scotland, but the arrival of the railways and heavy industry and the rapid growth of trade in the middle and later nineteenth century changed everything, including the building trades: never again would it be possible to build so extensively, in such a mature and well-developed traditional way, so consistently and so well. Much that survived the Victorian period was blown apart by the First World War and its complex aftermath which produced the Modern Age, and more progress, more development, more population growth, more exploitation, more consumption, more destruction and more pollution than in the whole of previously recorded history.

By comparison with other cities, Edinburgh has survived this holocaust relatively well, but Jamaica Street, India Place, Leith Street, St James Square, most of George Square, and significant parts of the Old Town, the South Side and Leith were lost, and much of James Craig's First New Town has been reduced to a hollow shell. It is easy to forget the extent to which the rest of the New Town was under threat, just a few years ago; it is in reaction to this destruction and exploitation, not just of fine Architecture and useful buildings, but of natural and cultural resources of all sorts that the Conservation Movement has grown. '*Conservation is action to secure the survival, or preservation, of things of value.*' The New Town is such a thing, but so is much else and it is in this wider context of the need for *Sustainability* in all human activities, in this much more sophisticated way, with real thought for the interests of future generations, that the Conservation of the New Town of Edinburgh must now be viewed.

QUOTATIONS

'. . . our escape from the Old Town gave us an unfortunate propensity to avoid whatever distinguished the place we had fled from. Hence we were led into the blunder of long straight lines of street, divided inch by inch, and all to the same number of inches, by rectangular intersections, every house being an exact duplication of its neighbour, with a dextrous avoidance, as is from horror, of every ornament or excrescence by which the slightest break might vary the surface. What a site did nature give us for our New Town! Yet what insignificance in its plan! What poverty in all its detail!' – Lord Cockburn, *Memorials of His Time* (written 1821-30), published 1856.

'The Scottish character (and I am myself a Scot) shows an extraordinary combination of realism and reckless sentiment. The sentiment has passed into popular legend. The Scots seem to be proud of it, and no wonder. Where, but in Edinburgh, does a romantic landscape come right into the centre of the town? But it's the realism that counts, and that made eighteenth-century Scotland – a poor, remote and semi-barbarous country – a force in European civilisation. Let me name some eighteenth-century Scots in the world of ideas and science: Adam Smith, David Hume, Joseph Black and James Watt. It is a matter of historical fact that these were the men who, soon after the year 1760, changed the whole current of European thought and life. Joseph Black and James Watt discovered that heat, and, in particular, steam could be a source of power – I needn't describe how that changed the world. In *The Wealth of Nations* Adam Smith invented the study of political economy, and created a social science that lasted up to the time of Karl Marx, and beyond. Hume in his *Treatise of Human Nature* succeeded in proving that experience and reason have no necessary connection with one another. There is no such thing as rational belief. Hume, as he himself said, was of an open, social and cheerful humour, and was much beloved by the ladies in the Paris salons. I suppose they had never read that small book which has made all philosophers feel uneasy till the present day.

All these great Scotsmen lived in the grim, narrow tenements of the Old Town of Edinburgh, piled on the hill beside the Castle. But in their lifetime Scots architects and particularly the brothers Adam had produced one of the finest pieces of town planning in Europe – the New Town of Edinburgh' – Kenneth Clark, *Civilisation*, BBC, 1969.

'It was in Scotland, not in England, that the Greek revival had its greatest success and lasted longest. There seems to have been such special congruity of sentiment between Northern Europe in the first half of the nineteenth century and the ancient world. Edinburgh, which considered itself for intellectual reasons "the Athens of the North", set out after 1810 to continue in a more Athenian mood, the extension and embellishment of her New Town begun in the 1760s. The result rivals Petersburg (i.e. Leningrad) as well as Copenhagen, Berlin and Munich. Indeed, in Edinburgh, what was built between 1760 and 1860 provides still the most extensive example of a romantic classical city in the world' – Henry Russell Hitchcock, *Architecture: Nineteenth and Twentieth Centuries*, Penguin Books, 1963.

'When Playfair died in 1857 the Golden Age had been over for twenty years. In those twenty years some important building had been done, and as a result of the additions then made, the centre of Edinburgh had assumed very much the appearance which it has today, the appearance that has done so much to make Edinburgh one of the famous and beautiful cities of the world. This was the achievement of a brief, and not in all respects glorious, seventy or eighty years. But it was by and large a conscious and deliberate achievement, the result of imagination, cooperation, and hard work. There were forces which worked against it, and which began to destroy it even before it was complete. Such forces still exist. But those who worked for the creation of beauty and harmony prevailed, because they were expressing the best aspects of the life of a vigorous and on the whole healthy society. No vigorous and healthy society today will neglect or impair what was done then. It is for those who inherit the achievement of Edinburgh's classical age to understand it, to adapt it, to use it, and to enjoy it' – A. J. Youngson, *The Making of Classical Edinburgh*, 1966.

OWNERS' GUIDE

INTRODUCTION

This book is written for the householders of Edinburgh's New Town, their advisers and contractors who, as trustees of historical, beautiful and useful buildings are responsible for repair and maintenance. The contents listed in the margin are summarized in this Guide.

Provision of a new town, long advocated by Lord Provost Drummond, became a possibility when a bridge was built across the Nor' Loch in 1766. In the following year James Craig, a twenty-seven-year-old architect, won the Town Council's competition for the layout of Princes Street, George Street, Queen Street and the two squares named after Saint Andrew and (Queen) Charlotte which became the First New Town, originally entirely residential but now the business, shopping and commercial heart of Edinburgh. Successive estates to the north, east and west of this central area remain largely residential to this day.

The New Town is one of the largest and least spoilt neoclassical city developments in the world. It extends to 318 hectares and almost all the buildings are listed for their architectural or historic interest. It is an Outstanding Conservation Area. The regular pattern of streets, squares and crescents, interspersed by gardens, is laid on ground which slopes northwards to the Water of Leith. The disciplined facades of local sandstone were designed by Scottish architects, terraces of houses and flats unified by 'palace fronts' followed the superb example set by Robert Adam in Charlotte Square.

Two hundred years of exposure left the New Town in need of repairs. The 1970 Conference on '*The Conservation of Georgian Edinburgh*' was organized by the Scottish Civic Trust, the Edinburgh Architectural Association and the Civic Trust, London to highlight this decay. The Conference, supported by amenity societies and residents' associations, led to the establishment of the New Town Conservation Committee which held its inaugural meeting on 31 May 1971.

The Committee administers grants, provided, in a ratio of two to one, by the Scottish Office and The City of Edinburgh District Council. These grants are offered on a sliding scale for external repairs, with higher percentages for groups of owners undertaking comprehensive contracts. Masonry repairs account for about two thirds of total costs. The Committee has concentrated on the fringe of the New Town where much of the worst decay has been halted. The Committee's programme has been greatly helped by generous housing grants; more information about grants, and technical advice about repairs, is available at the Conservation Centre, 13a Dundas Street, where there is a reference library and exhibition space.

The development of the New Town is fully described and illustrated in *The Making of Classical Edinburgh* by A. J. Youngson and more briefly in *Georgian Edinburgh* by Ian A. Lindsay, revised by Dr David Walker. The most detailed and authoritative book so far is the Edinburgh volume of *Buildings of Scotland* by J. Gifford, C. McWilliam and D. Walker. The Committee published the *New Town Guide* (now in its second edition) by Colin McWilliam, who was also the author of *Scottish Townscape*. These publications can be read and most of them can be purchased at the Conservation Centre.

Legislation has curtailed an owner's right to alter or demolish without obtaining Listed Building Consent from the local planning authority and ultimately the Secretary of State. Most owners in the New Town accept these constraints for the common architectural good, and planning controls are reinforced by views expressed, sometimes vigorously, by individual owners and street associations or collectively through the Cockburn Association (founded in 1875) which is the Edinburgh Civic Trust.

Listed Buildings and Conservation Areas

To own a house in a terrace is to be subject to greater architectural constraints than if the house stands by itself. The combination of separate houses behind a 'palace front', brilliantly contrived by Robert Adam in Charlotte Square, was widely adapted in later extensions of the New Town. The end pavilions, and sometimes the central block of such

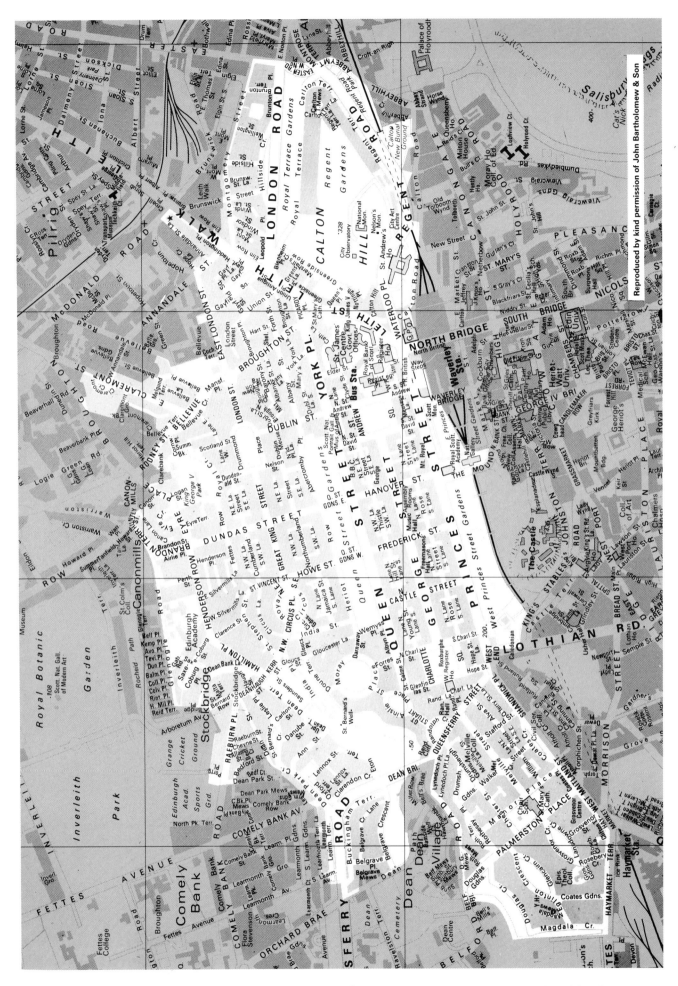

**MAP OF EDINBURGH NEW TOWN
CONSERVATION COMMITTEE BOUNDARY**

xvi

a 'palace front' in, say, Great King Street, were designed as tenements and most of the intermediate whole houses have been subdivided into flats. Owners of the few remaining whole houses still in residential use will be interested to know that the National Trust for Scotland may allow them to use its Conservation Agreement to prevent future subdivision or alteration.

Structural alterations are subject to scrutiny by the Building Control Division of the Department of Property Services (the old Dean of Guild Court), and Edinburgh people are familiar with their traditional obligations to share the cost of mutual repairs and to provide physical support for adjoining properties – obligations which can now be enforced by the Director of Property Services.

Damp is the greatest enemy of buildings. The wise owner will make sure that the property is well ventilated, that the roof does not leak and that the plumbing is in good order. Older houses are now regarded as valuable resources of shelter and craftsmanship; the post-war policy of comprehensive redevelopment was replaced by subsidized renovation, particularly for properties in housing action areas. The high cost of repairing historic buildings was also recognized and grants for this work have increased. In the New Town such grants are administered by the Conservation Committee whose policy is to offer the greatest incentives for the comprehensive repair of properties in the lower Council Tax bands.

The New Town was built with hand-wrought materials which gave the buildings a variety and a vitality which cannot be matched by machine: the tool marks on stone, the ripple in crown glass, the roughness of riven slate. If old buildings have to be demolished, such materials can sometimes be recycled. It is always better to find and use authentic old materials for repair rather than to use inferior substitutes.

Fire is still a constant danger in our homes and the importance of protecting an historic building against fire should not be underestimated, although there can be a conflict between fire regulations and the need to preserve historic features.

Historic buildings should be inspected at intervals of five years or so by an architect or surveyor who is familiar with traditional materials and methods of construction, which are very different from those used in modern buildings. Postgraduate courses in this subject are taught at several schools of architecture including Heriot-Watt University in Edinburgh, and many firms of architects now offer specialized advice to owners of historic buildings. A list of such architects is available from the Royal Incorporation of Architects in Scotland, 15 Rutland Square, Edinburgh. The Conservation Committee offers an inspection service for a small fee and will also provide lists of architects, surveyors and consulting engineers who have worked on buildings in the New Town.

Leaking roofs cause more distress and damage than any other defect and an annual or, better still, twice-yearly inspection of roofs, gutters and chimneys is strongly recommended. Several reputable slating firms in Edinburgh offer this service and one of the most valuable activities of a street association would be to organize such inspections for their members.

An architect's detailed report on the condition of a building and a quantity surveyor's estimate of the cost of repairs are essential preludes to carrying out a major conservation contract. An accurate forecast of costs is needed to support an application to the Conservation Committee for a grant and it is also the basis for a legal agreement which must be signed by all the proprietors. Experience has shown that once the need for major repairs has been accepted, it is best to leave the routine administration to a small group of proprietors who can, if necessary, employ a solicitor to help them.

In the absence of any other provision, ownership of roof spaces is dealt with under the common law, which is that the top floor proprietor owns the roof and the space between the top flat and the roof itself. Where there is more than one top flat, each top floor proprietor owns that section of the roof and roof space directly above his or her own flat. Similarly, the roof and roof space above the common stair (though frequently there is a cupola over at least part of the common stair instead of a roof space) is owned, in equal shares, by the proprietors of the common stair, who are probably (for the law is uncertain on this point) the proprietors served by the common passage and stair. The roof includes the rhone and, possibly, the downpipes if used for no other purpose. Chimney stacks are the common property of the owners of those houses having flues therein, their responsibilities being in proportion to the number of flues. Chimney cans are regarded as being part of the stack. Aerials are the movable property of the proprietors they serve, but it is

Building Regulations and Standards

Grants for Housing Improvements and Repair

Grants for Building Conservation

Salvaged Materials

Fire Safety

Survey and Report

Programme for a Conservation Project

Legal Agreements

not certain that a proprietor is entitled to place an aerial on a roof belonging to someone else, although this is often done.

Although at common law the top floor proprietor owns the roof, this does not carry with it a right to the airspace above the roof, and accordingly in an early case a top floor proprietor was prevented from adding an attic. The reason is that the airspace above the roof belongs to the owner of the solum (the ground on which the building is erected). At common law, the proprietor of the lowest floor owns the solum. The top floor proprietor can probably put roof lights into the roof that he or she owns, because these scarcely encroach into the airspace, but would almost certainly require permission from the owner of the solum before adding dormer windows.

Title deeds normally alter the common law position by adding the roof and the solum to the parts of the building to be held in common ownership. From time to time, one finds that the title of a lower flat is silent, normally by mistake, on the question of maintaining the roof, and in such cases the top floor proprietor must bear an extra share of the cost of maintaining the roof. This means that in the purchase of a top floor flat, the purchaser's solicitor must see evidence that the burden of maintenance has been properly shared and constituted in the titles of the lower flats.

Maintenance

Many of the most expensive repairs are due to neglect. Once the defect has been put right it is obviously desirable that the building should be kept in good repair. For this purpose, groups of proprietors in the New Town are encouraged to retain their committees after a conservation scheme is completed and to amend their titles to give authority to a majority of owners to have the building inspected and repaired on a regular basis. Many repairs are postponed because of expense even though postponement never saves money and usually increases the cost. When the proprietors' budget includes an allowance for future repairs and regular maintenance, the prospects for their historic building will be greatly improved. In the New Town masonry repairs are the single most expensive item and are likely to receive the most sympathetic consideration from the Conservation Committee.

Scaffolding, Hoists and Towers

The cost of scaffolding is high in proportion to the cost of minor repairs but drops to about 10 per cent or less of the cost of a comprehensive conservation contract. Non-structural external repairs to the upper storeys can sometimes be done cheaply and quickly from a mobile Simon Tower hired for a few hours, but if a working scaffold is required it is well worth using the best and safest equipment which will allow all parts of the elevation to be closely and safely examined and, if necessary, repaired. For major roof repairs, a temporary canopy, supported by the scaffolding, will protect the interior from accidental damage and flooding.

'The Conservation of Georgian Edinburgh' 1972

As Professor Youngson said at the 1970 Conference, 'In its scale and completeness it has no rival, and to let the New Town go would do more to diminish the individuality of Scotland than could perhaps be done in any other way.'

STRUCTURE

Geology and Topography

Hard Craigleith sandstone, covered by stiff boulder clay, underlies most of the New Town. The boulder clay is a good foundation, but along the northern margin it is overlain by a raised beach of sand and gravel sensitive to vibrations and changes in ground water level. Soft peat, sand and silt mark the sites of the prehistoric Canonmills, Broughton and Nor' Lochs which have long since disappeared, and the former courses of the Broughton and Greenside burns have also left pockets of unstable ground.

Foundations

The walls of eighteenth and nineteenth century buildings were raised off narrow footings. Such foundations are susceptible to differential settlement but they can be widened and strengthened, and the ground on which they rest can be consolidated by modern techniques. The owner or potential owner should not be dismayed by these hazards, proof of the structural stability of the New Town is its existence, amazingly intact, after two hundred years.

Structural Defects

All new buildings adjust themselves to their sites, and this may produce differential settlement between adjoining buildings. Cracks in the walls may have been caused by this initial movement which ceased when the structure reached equilibrium. It is important to establish the cause of cracks and, if they are found to be active, it is essential to stabilize the structure. Once the cause has been diagnosed, the cracks should be filled, particularly those on external horizontal surfaces such as cills or cornices.

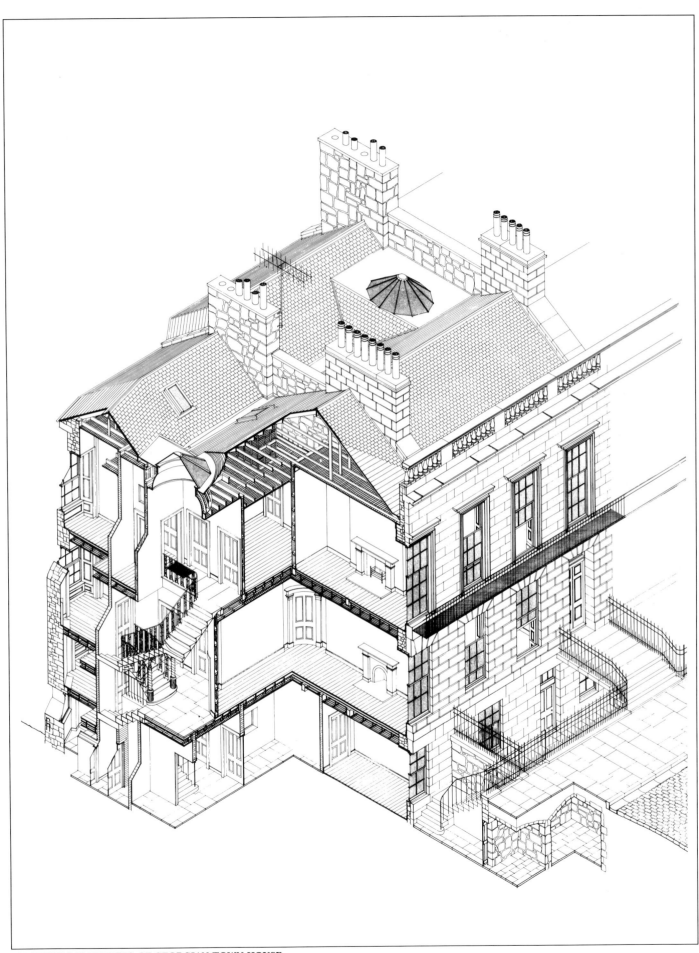

SECTIONAL ISOMETRIC OF GEORGIAN TOWN HOUSE
AT 13 CARLTON STREET   DESIGNED 1824 BY JAMES MILNE

Damp causes timber to weaken and rot, stones to disintegrate and metal to rust. Techniques for shedding and diverting water are the basis of traditional building construction. Attempts to totally waterproof an existing building can lead to problems with condensation; to be healthy, the fabric of a building must be able to breathe.

External walls in the New Town were built of sandstone – finely dressed and squared ashlar for the front and roughly dressed rubble for the back – the stones being separated by beds of lime sand mortar. The walls consist of two thicknesses of stonework bonded at intervals by 'through stones', and separated by a cavity filled with lime mortar and small stone chippings. Sandstone, being sedimentary, and consisting of layers of sand compressed over millions of years, must be placed in the wall so that the beds or laminations are pressed together by the load of the building and not face-bedded so that the laminations peel off. Stones from Edinburgh quarries, such as Craigleith which is no longer worked, are among the most durable sandstones in the country; many of the replacements now available are also very durable, but in Victorian times builders sometimes resorted to softer and cheaper stones which have since crumbled away.

In the New Town, soft stones or stones which are exfoliating, are usually only found here and there in a facade or in exposed features like chimneys. Such decayed stones should be cut out and the holes indented with new stones, just as a dentist will fill a tooth to prevent the decay from spreading. Fortunately, total refacing (which is the equivalent of having a complete set of false teeth) is seldom necessary, but it was unavoidable, for instance, in the western part of Fettes Row, on the southern side of St Stephen Street and Hamilton Place and on both sides of Clarence Street in Stockbridge. Natural stone is the best material for these repairs; any substitute will have different characteristics which will become more obvious with time. Sometimes small areas of decayed moulded stone such as cornices and window architraves are repaired with a mortar built up around reinforcement. If the correct ingredients are used, the colour of this mortar will match that of the stone at first but eventually the differences will become more obvious.

The joints in a stone wall are like the pores in one's skin; they allow moisture to evaporate from a wall, and, if the joint is sealed by dense pointing, this moisture will find its way through the stone itself bringing harmful crystals to the surface. Mortar joints should also absorb the inevitable slight seasonal movements in the wall. Lime mortar, being slightly flexible, will permit the wall to move and to breathe; strong cement mortar, being dense, will not. Non-hydraulic lime available locally, has to be mixed with a small amount of cement to make the mortar set; hydraulic lime now imported from France needs no cement and has been used successfully in the New Town. Traditional slaked lime mortar is now available again thanks to pioneering work by Historic Scotland, and it will be widely used when the traditional techniques have been mastered.

Many people associate conservation with the cleaning of buildings such as St. Paul's Cathedral in London, the buildings along the Seine in Paris, or the terraces of Bath, but they were built with limestone which can be cleaned by water alone. This cannot be done with sandstone requiring abrasives or chemicals, and these, in the wrong hands, can do irreparable damage. The grime in Edinburgh does not harm the fabric and it is better to regard the 'symphony of greys' simply as a sign of dignified age. Stone cleaning has been thoroughly investigated by Robert Gordon's University, Aberdeen and if there is a risk of permanent damage to the masonry the general rule must be *don't*. In any case, cleaning requires Listed Building Consent in Edinburgh.

Some people paint stone with oil, cement or water-based paints in a misguided attempt to smarten it up. Unfortunately, paint often has the opposite effect and it is very difficult to remove, so again the advice is *don't*.

On the other hand, moss and algae are unsightly and slippery and they should be removed, along with ivy which pushes its roots into mortar joints. Such growths may be a danger signal for damp spots which should be investigated.

One of the dampest spots in New Town property is usually the cellar under the pavement and the adjoining basement area. These are often neglected assets although it is expensive to waterproof the cellars.

Slates are laid, overlapping like a bird's feathers, to shed water and to allow some degree of ventilation in the roof space. Lead and zinc, traditionally used for gutters and flashings, must also be laid to overlap but they should be fastened to the roof in such a way that they can expand and contract without tearing when the temperature changes.

The junctions between all of these separate components are potential trouble spots requiring regular inspection. Alternative modern materials laid in one piece over the roof

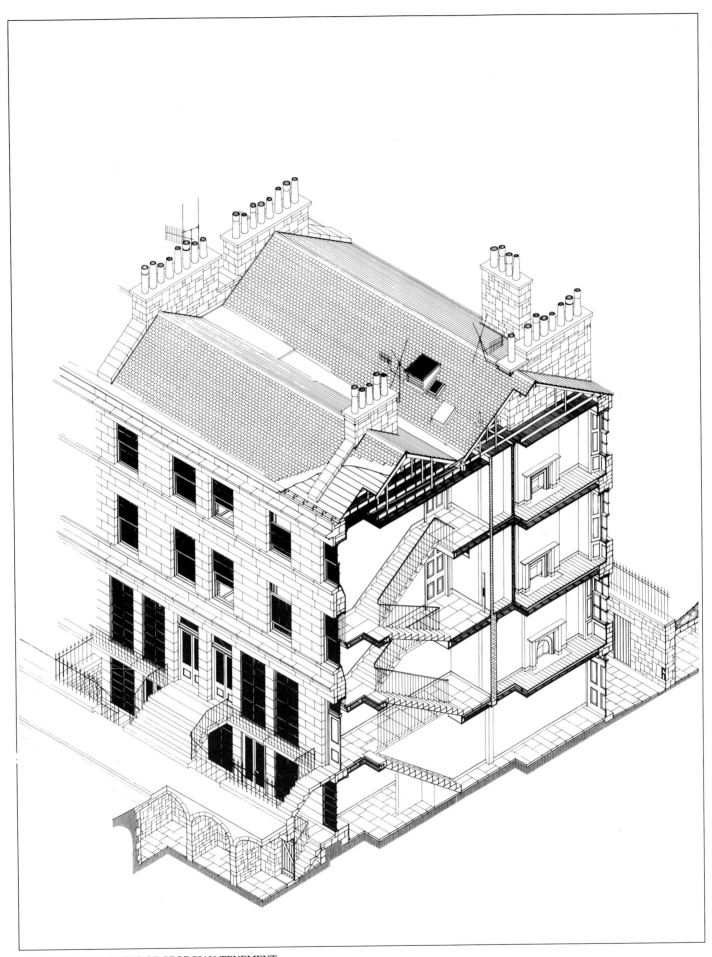

**SECTIONAL ISOMETRIC OF GEORGIAN TENEMENT
AT 52 ST STEPHEN STREET
DESIGNED CIRCA 1826 BY ROBERT BROWN**

have sometimes been substituted for the traditional ones but it is difficult to develop a material which is both flexible and strong, capable of being bonded to the roof structure and yet able to breathe. If such a continuous membrane is totally impervious, it will fail to ventilate the roof space, and moist warm air in the building may condense and create damp conditions, causing the roof timbers to decay.

Condensation may also be the unwelcome result of thermal insulation unless vapour barriers and ventilation are provided. We are more conscious of the need to save energy nowadays, and although it is not practical to insulate the walls and not worth insulating the windows, even if this did not spoil the appearance of the building, it is certainly possible to insulate the roof spaces and the basement floors at a reasonable cost.
Insulation

Slates, lead and zinc have roofed the New Town successfully for many years and it is strongly recommended that they should be retained for the future. If correctly laid, slates and lead have a very long life, although zinc, used mostly on ridges, may eventually become porous with exposure to the city atmosphere.
Roof Coverings

Repairs to roofs are inevitable but they can be minimized. Pedestrian traffic is bad for them, slates crack when they are walked on, and lead can be accidentally pierced by a hob-nailed boot or a sharp stone; permanent duck boarding and ladders should always be provided to prevent such damage. TV aerials on chimneys, with cables trailing across the roof to each entry point, require constant refixing and maintenance – at the risk of further damage to lead and slates. Aerials fixed in the loft where they are more accessible often give equally good reception, or perhaps a group of proprietors might substitute one external communal aerial instead of the usual forest on the roof.

Skilled inspections in spring and autumn, after the winter gales and when the leaves have fallen, are important to ensure that the water ways are not choked and that missing slates are replaced. Easy access is well worth providing for inspections and for emergencies.
Access to Roofs

The tops of the walls, the wallheads, are most vulnerable. If a slate is missing, water can run down to this lowest point where roof and walls meet, saturating and decaying the timber wallplate and the feet of the rafters which rest on it. Some of this water may come from leaks higher up the roof but defective lead-lined gutters at the front walls and in the centre valleys are more likely to be the cause, especially if they become cracked or choked with leaves and debris at the outlets. These outlets and the tops of rainwater pipes are favourite nesting places for birds. The backs of pipes are often forgotten during external repainting, and are therefore particularly susceptible to rusting and eventual cracking.
Gutters and Downpipes

Much of the New Town is built on ground which slopes to the north and the roofs of lower houses are clearly visible from higher parts of the city. This exposure makes it even more important to retain the texture and colour of the slate and lead and to scrutinize every planning application which threatens to alter the world famous skyline. The silhouette of many buildings in Edinburgh is enlivened by dormer windows, which are a Scottish feature, and in the New Town the austere neoclassical facades are relieved by wallhead and gable chimneys in the vernacular tradition. Such architectural ornaments mostly have a practical purpose; admittedly wallhead balustrades are purely decorative, but they are all visually important and none can be removed without diminishing the original conception and the present reality of the 'Athens of the North'.
Parapets and Balustrades

Exposed features which pierce the roof covering need careful and regular attention. Chimneys in particular have to withstand rain, frost and wind from without and flue gases from within, as well as having to support a miscellaneous collection of TV aerials. It is recommended that decayed chimneys should be rebuilt in stone. Dormer windows may not have been part of the original design but most of those in the New Town are almost as old as the buildings themselves; new ones, if allowed by the planning authority, should conform to a scale appropriate to the age of the building.
Chimneys and Flues

Dormer Windows

The stairs in the centre of New Town houses are usually lit by a cupola or rooflight often with decorative plaster at the base. The junction between the cupola and the surrounding roof is a difficult one to waterproof and this requires special attention.
Cupolas and Rooflights

The internal structure of the house should not require maintenance unless extra loading or timber decay has weakened it. An understanding of the construction of floors, walls and stairs is needed before any alterations are made, for instance not all timber stud partitions can be assumed to be non-load bearing.
Internal Walls
Floors
Stairs

Total destruction by fire is fortunately almost unknown in the New Town but the possibility raises the question of insurance. Insurance companies recommend that policies should be based on the cost of total replacement. Such a cost in the New Town would be astronomical, but it is a matter of concern that, for partial damage, the insurance

BEFORE RESTORATION   Photo: Alastair Hunter

AFTER RESTORATION   xxiii   Photo: Stewart Guthrie

20-22 FETTES ROW

companies may use the ratio of value covered/replacement cost when assessing compensation. The only practical suggestion for proprietors is to insure for twice the market value of the property and to practise rigorous fire prevention. The planning authority has no power to insist on the total replacement of the existing building even if it is listed. But they may require that replacement roofs and elevations should match the design and materials of the original or adjoining buildings. The co-proprietors in a tenement might consider the advantage of insuring with the same company and linking house insurance to regular inspection and maintenance.

FITTINGS

The design of windows, doors, railings and balconies gives a subtle liveliness to the austere stone terraces of the New Town. For example, the sparkle of the small panes of crown glass, the elegant astragal mouldings, and the well-made panelling of the shutters all add up to one of the most practical and durable of windows – the double-hung sash. **Sash Windows / Sash Window Re-cording**

Attempts to improve on the sash window are fraught with problems; the best advice is *don't try*. In Edinburgh many people think erroneously that conservation is just another word for putting back the astragals; the wide variety of mouldings on these glazing bars mark different periods in the development of the New Town and there are also many fanlight patterns in different parts of Georgian Edinburgh. Because these details are important, it is worth ensuring that the mouldings of modern replacement windows exactly match the original design (often to be found on a humble rear window) and are not just stock items from a timber merchant. It is better to restore astragals to old sashes because irreplaceable eighteenth and nineteenth century timber is of a much higher quality than modern softwood and every effort should be made to retain it by patching, splicing and reinforcing with metal brackets. **Astragals and Window Mouldings / Fanlights**

Sound and draught proofing are both problems which can be partly overcome by making full use of shutters and curtains. Secondary windows make it impossible to use the shutters, and the average sash members are not deep enough to take double glazing without spoiling the appearance of the windows.

The servicing of New Town buildings has seen major changes in recent years as standards of comfort have improved. New wiring, plumbing, heating and ventilation work has not always been sympathetic to the interiors and there are a number of pitfalls to be avoided. **Services**

Symmetry was the golden rule for the architects who designed the New Town and, because of the variety of house plans, they put in dummy windows to create a balanced elevation. It was symmetry and not window taxation which created so many blank stone panels and fixed glazed sashes – aesthetic but inaccessible. Occasionally these panels can be unblocked and fitted with sliding sashes, but usually they conceal fireplaces or flues, and must be retained and maintained. **Dummy Windows**

Attention to detail is even more important when restoring and maintaining entrance doors. The brass numerals and Edinburgh handles are early features of the New Town and they are still available either second hand or as reproduction items. Damaged doors can usually be repaired but if this is impossible, they can be replaced by similar or second hand doors – the ENTCC has a list of dealers. The earlier doors had six panels and the later (Victorian) doors had four panels with different types of mouldings. Fresh paint will help to preserve external doors and a variety of colours can make the street more lively and attractive. The choice of colour is subject to planning permission but this is unlikely to be withheld if the colour is chosen from the wide range recommended by the Planning Department. **Door Furniture / Doors**

Many sash windows have lost their original brass fittings. Lifts, rings and catches are all available in cast brass which is the proper material to use. Cast iron weights, top sash mountings and knot holders can be obtained, as can simplex fittings (otherwise known as open butt hinges) which allow the bottom sash to be swung open; again the ENTCC has a list of suppliers. **Window Ironmongery**

It is important that prominent features such as porticos and entrance steps should be well maintained and that repairs should be carried out with natural materials even though this may be expensive. There is a superb example of modern stone carving on the new porticos at 18 and 22 Henderson Row. **Porticos**

After experimenting in the New Town with precast concrete and artificial stone, it is now recommended that only natural stone should be used for steps leading up to the front entrance doors and that applied finishes such as tiles, stone veneers or asphalt should be avoided.

Steps, Platts and Arches

There is no substitute for ornamental cast iron for railings and balconies in the New Town. Cast iron is less likely to rust than mild steel, which is sometimes wrongly used to repair broken railings. Copies of the many decorative iron patterns can still be cast at the foundries in the Edinburgh area. Because there is a strong tradition of metalworking in the city it is still possible to have ironwork repaired to a high standard.

Railings
Balconies

The City and Regional Councils working with the street associations and proprietors have greatly enhanced the appearance of several streets in the New Town by replacing concrete lamp standards with cast iron railing mounted fittings whose lights form necklaces around Charlotte and Rutland Squares, Regent, Royal and Carlton Terraces, Northumberland Street, Ann Street, Heriot Row and the Moray Estate. This improvement has also spread to other streets which are not too wide and which do not carry too much traffic such as Fettes Row, Royal Crescent, Royal Circus, India Street, Gloucester Place and Danube Street, where the residents have paid for the new standards to be cast and the Lighting Department of Lothian Regional Council has fitted them to the railings complete with the light fitting.

Lamp Standards

To keep well-detailed ironwork or woodwork in good condition is not difficult. Cleaning and repainting at regular intervals is all that is required but the specification for this paintwork is important and proprietors should know about types of paint and numbers of coats. This is the trade which needs the closest supervision and strict adherence to the chosen specification.

External Paintwork

Joinery is another finishing trade whose work everyone sees and touches; doors, windows and shutters – the moving parts of a house – handrails and handles which we feel; decorative architraves, skirtings and dado rails; all should be of the best materials and craftsmanship.

Internal Joinery

Principal rooms in the New Town are enriched by decorative plasterwork which should be redecorated as seldom as possible, with distemper or water-based paints, not emulsion, which clogs and obscures the delicate details. Even when a disaster like dry rot or flooding damages the plasterwork or loosens its key, do not despair, it is possible for skilled plasterers to reinforce damaged sections and to replace missing decorations. Only recently have we learned about the designs and fashions in decoration which changed while the New Town grew from its beginning in the 1770s with Robert Adam's design for No. 8 Queen Street to Playfair's National Gallery in 1854. Ian Gow, of the National Monuments Record of Scotland, author of *The Scottish Interior*, has summarized his historical and expert knowledge of New Town interiors specially for this book.

Plasterwork

Interior Decoration

Internal redecoration is a familiar job for most householders; special care is required for historic buildings and the work is not always straightforward. It is well worth trying to reproduce authentic colour schemes to suit the age of the building and many of the authentic materials can still be found.

Internal Paint and Varnish

Wallcoverings

The main decorative feature and the focal point of most reception rooms in the New Town is the fireplace, which although tempting to thieves, and made redundant by the Clean Air Act, is still an essential part of the historic character of the building and should be retained and treasured if only as part of the ventilation system.

Fireplaces

Many of the useful corner shops still survive at basement or ground floor level in the New Town; they are part of the neighbourhood and they serve the residents well; they should be treasured and supported and whenever possible their owners should be encouraged to restore their original simple and distinctive character.

Shops and Shop Fronts

Cartoons: Graphic Partners

A GUIDELINE TO THE GEORGIAN PERIOD

1714	Accession of George I
	Population of Edinburgh 25,000
1724	Lord Provost Drummond proposes 'extensions to the north'
1725	James Court, James Brownhill
1728	Robert Adam born
	Music Society of Edinburgh formed
1738	Infirmary (now demolished), William Adam
1739	*Scots Magazine* founded
1740	James Craig born
1746	Defeat of second Jacobite rising
1752	Proposals for carrying on certain Public Works in the City
1753	Act of Parliament for improving streets and access, and new public buildings
	Royal Exchange (now City Chambers) John Adam
1759	Draining Nor' Loch started
1763	Northern access planned
	North Bridge (now replaced), W. Mylne
	Layout of George Square, James Brown
1766	Competition for Plans of a New Town won by James Craig
1767	Act for Extension of City
	Population of Edinburgh 50,000
	First New Town
1772	Act for Lighting and Cleansing
1774	Register House started, Robert Adam
1775	St James Square, James Craig
1777	James Gillespie Graham born
	Robert Reid born
1781	Work started on Mound
1784	Assembly Rooms, John Henderson
	Thomas Hamilton born
1785	Act for South Bridge and College (University) of Edinburgh
1789	University of Edinburgh, Robert Adam
	William Henry Playfair born
1790	Forth and Clyde Canal completed
1791	Charlotte Square, Robert Adam
	Gayfield Place, James Begg
1799	York Place
1800	Picardy scheme, Robert Burn
1801	Population of Edinburgh 67,000
1802	Second New Town, Robert Reid and William Sibbald (City Architect)
1812	Competition for layout of Calton Hill
1813	West End layout, James Gillespie Graham (elevations by various architects)
1815	Eastern access planned. Calton Bridge scheme, Archibald Elliot
	West Extension planned
1816	Act against building on South side of Princes Street
1818	Gas lamps introduced
1819	Calton Scheme, William Playfair
	Rutland Square, Archibald Elliot (not built until 1830/40, John Tait)
1822	Royal Institution (now RSA), William Playfair
	Union Canal completed
	Moray Estate scheme, James Gillespie Graham
1823	Royal Circus, William Henry Playfair
1824	St Bernard's Crescent, James Milne
1825	Blacket Place, James Gillespie Graham
1826	Royal High School, Thomas Hamilton
	Act approved St Leonards, Dalkeith railway
1829	Dean Bridge, Thomas Telford
1830	Death of George IV
	Population of Edinburgh 136,000

From *The Conservation of Georgian Edinburgh*, Edinburgh University Press, 1972.

LISTED BUILDINGS AND CONSERVATION AREAS

Old buildings are treasured for their beauty and their historic interest. As long as they fulfil a purpose, their fabric is likely to be well maintained, but if an historic building ceases to have a useful or appropriate function it is liable to be neglected, abused and ultimately demolished; ignorance and indifference also put old buildings at risk. Protection therefore depends on control over the use, alteration and demolition of historic buildings, on a general appreciation of their importance, and on an efficient policy of maintenance and repair.

The Town and Country Planning (Scotland) Act 1972 provides protection by the listing of buildings and other structures of special architectural or historic interest, while the Town and Country Amenities Act 1974 provides protection by the designation of Conservation Areas. The exceptional degree of public participation built into British planning legislation is a reflection of a general interest in conservation, which has greatly increased over the past twenty years.

A useful introductory guide to the town and country planning system which operates in Scotland is given in the booklet, *Planning in Scotland: A Basic Guide.*

LISTED BUILDINGS IN SCOTLAND

The list compiled by the Secretary of State places protected buildings in three categories: categories A and B contain buildings of national and local importance respectively, category C(s) contains buildings of modest merit; all receive statutory protection under the Planning Acts. The object of listing is to avoid destruction of buildings of special architectural or historic interest, or alterations which would adversely affect their character. Alteration and demolition are subject to consent by the local planning authority and the Secretary of State.

About 3 per cent of all buildings in the UK are listed; in Scotland in 1993 there were 38,391 listed items (2824 A, 23,972 B and 11,595C(s)); an item may be a terrace or tenement containing a number of individual properties. In Edinburgh the total was 3326 (567 A, 2262 B, and 497 C(s)); about half of these items, amounting to about 11,000 separate properties, are in the New Town and of these, 75 per cent are still in residential use, an unusually high proportion for an inner-city area in the UK.

Listing applies to everything within the curtilage of the property including the interior, and consent is required before making any change which will affect the architectural character of a listed building (it is considered that curtilage extends to include communal amenity gardens which lie beyond what is now a publicly maintained street). Many of the finest buildings are conceived in their entirety, architects such as Robert Adam designed furniture, fireplaces and fittings as well as the external shell of the building, and the whole design would be impoverished by the removal of any element. For instance, the character of a listed building will be at risk if the proportions of rooms are altered by sub-division, if a staircase is removed, or if separate properties are interconnected (as often happens when terraced houses become offices); detailed character will suffer if fireplaces, flagged floors, original panelling, plasterwork, doors, joinery and so on, are removed or defaced. The Planning Department is particularly determined to preserve the interiors of the remaining original Georgian flats and houses throughout the New Town.

The architectural character of the exterior will suffer if original features such as doors, windows, roofs and chimneys are altered, or if decorative features such as cornices or balustrades are removed. This applies particularly to neoclassical terraces in the New Town, because Georgian architecture relies on simple shapes and subtle surface textures within an overall classical discipline. The colour and detailing of any part of a classical building should be assessed in relation to the whole elevation. In the New Town this means that the appearance of any one property must be judged in relation to the whole terrace. For these reasons, Listed Building Consent is required in the New Town for any alteration which affects the character of a listed building. This applies to the interior as well as the exterior and to all built features within the original property boundary such as garden walls, mews buildings and outhouses (if erected before 1 July 1948). Other alterations which require Listed Building Consent include the alteration or cleaning of masonry, the painting of walls and signs and the erection of roller shutters, alarm boxes and satellite dishes.

Repairs to listed buildings do not require consent provided that they exactly match the construction, colour, texture, profile and materials of the original. In Edinburgh, the Planning Committee of the District Council has

CHEYNE STREET BEFORE RESTORATION   2

adopted consistent and detailed policies on the alteration, sub-division, repair and maintenance of listed buildings. The policy documents and illustrated summaries of each aspect may be consulted at the Planning Department.

PROCEDURE

Descriptive lists of buildings of special architectural or historic interest are compiled by Historic Scotland on behalf of the Secretary of State and may be seen at the offices of district councils. The councils, under Section 52(5) of the Town and Country Planning (Scotland) Act 1972, must notify proprietors that their property has been listed or removed from the lists. Copies of lists with amendments can be seen at the National and Central Libraries and also at the office of the Royal Commission on the Ancient and Historical Monuments of Scotland, and for Edinburgh, at the office of the ENTCC, where a copy of Historic Scotland's *Memorandum of Guidance on Listed Buildings and Conservation Areas* may also be consulted. Ideally, the title deeds of the property should record the listing, but this is not compulsory.

Owners who wish to alter, extend or demolish any part of a listed building must apply to their local planning authority for Listed Building Consent. In Edinburgh it is best to consult the City's Planning Department before starting any work, however trivial, which may affect the character of the building. (Although under review, Listed Building Consent is not at present required for works to buildings in ecclesiastical use but planning permission may be required.) Applications for Listed Building Consent and planning permission are advertised by the local planning authority in the local press, which in Edinburgh means the *Edinburgh Gazette* and the *Evening News*, and notices are displayed on or near the building in question. This allows comments to be received from individuals, street associations and the main amenity bodies, namely the Scottish Civic Trust, the Architectural Heritage Society of Scotland, the Cockburn Association and the Cockburn Trust. All comments must be taken into account when applications are considered by planning authorities. If consent for a particular proposal is refused, the applicant has a right of appeal to the Secretary of State for Scotland.

If the planning authority intends to grant consent for alteration, the Secretary of State may call the case in for his own decision; however, this affects only buildings in categories A and B. Final decisions on applications concerning category C(s) buildings were delegated to planning authorities in January 1988, except where demolition is involved. All demolition cases, including those of unlisted buildings in a Conservation Area, must be referred to the Secretary of State. Where consent for demolition is granted, three months must be allowed for the Royal Commission on the Ancient and Historical Monuments of Scotland to take photographs and to make record drawings of the building.

It is a statutory offence to alter or demolish any part of a listed building without official consent, unless it can be proved that the work was urgently required in the interests of safety or health, or for the preservation of the building, and that the planning authority was so advised at the earliest opportunity. Under the 1972 Act, planning authorities can serve enforcement notices when unauthorized work has been carried out on occupied listed buildings, and there are legal penalties if the owner fails to comply with the notice. The planning authority is also empowered to recover expenses from the owner. The planning authority may give temporary protection to unlisted buildings by means of a Building Preservation Notice. This effectively lists the building for six months during which time the Secretary of State must decide whether or not to list the building permanently.

If the owner of an unoccupied listed building (or part of a building) allows its condition to deteriorate, the planning authority or the Secretary of State, under Schedule 9 of the Housing and Planning Act 1986, may carry out urgent works to ensure its preservation, and may require the owner to pay for the works under Section 97 of the 1972 Act. The authorities must ensure that the emergency work is the minimum required for the preservation of the building.

Under Section 104 the planning authority or the Secretary of State can acquire compulsorily any listed building which is not being properly maintained. It should be stressed, however, that compulsory purchase proceedings cannot begin unless a repairs notice has been served on the owner under Section 105 at least two months previously.

The Secretary of State can extend similar powers to protect unlisted buildings in Conservation Areas.

CONSERVATION AREAS

The concept of protecting historic areas rather than individual buildings stems from the Civic Amenities Act 1967. Conservation Areas are designated in Scotland by the planning authorities under the Town and Country Amenities Act 1974, Section 2(1). Designation ensures statutory control of demolition and places a duty on the planning authority to consider whether development protects or enhances the character of the Conservation Area. This effectively means tighter control over development.

As with Listed Building Consent procedures, applications for development within a Conservation Area must be advertised to allow the planning authority to take public comments into account. Consent is required for the demolition of any building over 115 cubic metres in capacity, or any part of such a building, within a Conservation

Area, although there are a number of exceptions. Consent is also required for the demolition of gates, walls and fences higher than 1 metre to the road or 2 metres elsewhere. Trees are also protected as though they were subject to Tree Preservation Orders and they may not be topped or lopped without consent; six weeks' notice must be given of the intention to carry out works to a tree. Cleaning and painting of unlisted buildings in Conservation Areas now requires formal planning consent under Class 9 of the Town and Country Planning [General Permitted Development] (Scotland) Order 1992. Alterations to windows of unlisted buildings in Conservation Areas are also restricted. Further advice on these matters should be sought from the Conservation Section of the Planning Department.

Since 27 July 1984 the application of an Article 4 Direction applies to all of the New Town, removing permitted development rights for specific classes of development. This means that planning permission is required for classes 1, 3, 6, and 7 which are permitted development under the Town and Country Planning [General Permitted Development] (Scotland) Order 1992. Class 1 controls the enlargement, improvement or other alterations of dwelling houses. Class 3 controls other developments within the curtilage of dwelling houses. Class 6 controls the erection of satellite dishes while Class 7 controls the erection of gates, fences or walls. These refinements are important to preserve the uniform design of terraces and their settings.

The 1992 Act also increased conservation controls by enacting that Class 2 (alterations to roofs), Class 4 (formation of hardstandings), Class 5 (storage of oil) and Class 9 (stonecleaning and exterior painting) are not considered to be permitted development in Conservation Areas. Planning permission is therefore always required for these types of development.

It should be stressed that as part of the District Council's local plan process, the number of classes falling within the Article 4 Direction may be extended in the future. Further advice should be sought from the Historic Buildings Section of the Planning Department.

Government grants are offered for the enhancement of areas of outstanding architectural or historic interest on the advice of Historic Scotland. The 318 hectares of the New Town, one of the largest single Outstanding Conservation Areas in the United Kingdom, benefits from grants from this source and also from the City of Edinburgh District Council. These grants are administered by the ENTCC.

CHEYNE STREET AFTER RESTORATION   5

BUILDING REGULATIONS AND STANDARDS

INTRODUCTION

Over the last hundred years building legislation has become increasingly complicated as new methods of construction, more complex services and higher standards of safety have been introduced. The principal Acts and Regulations are as follows:

- *The Building Standards (Scotland) Regulations 1990 to 1993*, and the supporting *Technical Standards:* These are the main regulations affecting the construction and alteration of buildings. They comprise detailed requirements for the performance and durability of materials, structural strength and stability, structural fire precautions, the provision of means of escape from fire, the construction of chimneys and the use of heat-producing appliances, the prevention of dampness, the resistance to the transmission of sound and heat, ventilation, drainage and sanitary appliances, electrical installations, housing standards, refuse disposal, the construction of stairs and, most recently, facilities for disabled persons. (*Note:* The system of regulations is different in England, Wales and Northern Ireland.);
- *The Building (Self-certification of Structural Design) (Scotland) Regulations 1992;*
- *The Building (Procedure) (Scotland) Regulations 1981 to 1991;*
- *The Building (Forms) (Scotland) Regulations 1991 to 1992;*
- *The Building Operations (Scotland) Regulations 1975;*
- *Building Standards (Relaxation by Local Authorities) (Scotland) Regulations 1991;*
- *Building (Scotland) Acts 1959 and 1970* as amended.

Combined, these Acts and Regulations set out the whole procedure for controlling construction works, the requirements for Building Warrant, the processes of relaxing the Building Standards in certain instances, and the issuing of Certificates of Completion.

Construction can also be required and controlled by the following Acts:

- *The Workplace (Health, Safety and Welfare) Regulations 1992* and the supporting *Code of Practice:* These Regulations cover many aspects of health, safety and welfare in the workplace. (They do not apply to construction sites, for which separate regulations, the *Health and Safety: Construction (Design and Management) Regulations 1994*, come into force during 1995.
- *Fire Precautions Act 1971:* This Act sets out the general requirements for Fire Certificates and their enforcement.
- *Health and Safety at Work etc. Act 1974:* This Act includes general requirements for buildings used as workplaces.

From the seventeenth century building construction in Edinburgh was subject to increasing control by the Dean of Guild Court. The very extensive powers of the Court were defined when it was reconstituted in 1879, and since then all drawings submitted for Warrant (see below) have been stored in the city archives, where they can be examined. The archives contain a few drawings made before 1879 for parts of the New Town. There are also drawings made after 1879 for alterations to earlier buildings. This is a source well worth checking before carrying out a measured survey. A facility for photocopying is available in the archives.

All new buildings, alterations and extensions have to comply with current regulations and requirements. Local authorities can also enforce the regulations where there is a change of use or under circumstances where public health or safety is concerned. A detailed paper (1985) on the application of Building Regulations to existing buildings may be consulted at the ENTCC's office.

BUILDING REGULATIONS

A building warrant must be obtained for the construction, alteration, extension, demolition or the change of use of a building. It does not exempt the owner from getting planning permission or Listed Building Consent.

To obtain a warrant, application forms and detailed drawings have to be submitted to the Building Control Division of the local authority (in Edinburgh the office forms part of the Department of Property Services). As well as issuing warrants, the Building Control Division inspects work in progress and issues a certificate on satisfactory completion.

Some minor repairs may be carried out without one, but a warrant is certainly required for structural work and for major repairs even where like is to be replaced with like. Alterations incorporated into a repair may also require a warrant if they affect the extent to which a building complies with any of the regulations. It should also be borne in mind that repairs involving demolition of structural elements also require a warrant. It is always wise to check with the Building Control Division before carrying out any repairs – in practice, the extent of the repair often determines whether or not a warrant is required.

Where alterations to an existing building are concerned, the Regulations apply only to those parts which are being altered or extended. The local authority may refuse to grant a warrant if, as a result of the alteration or extension, a building ceases to comply with the Regulations.

The Building Regulations categorize buildings by the nature of their use. Changing the use of a building will require a warrant if the new use brings it into a group to which different or more onerous regulations apply. Planning permission is usually required for change of use.

Schedule 2 of the Building Regulations lists fixtures for lighting, heating, ventilation, drainage, and so on, which may be replaced, and various fixtures such as gas fires, TV aerials and replacement windows which may be installed without a warrant.

Since the Building Regulations are designed primarily to control the quality of new construction, they may not always be appropriate for work to older buildings, and may even conflict with legislation protecting the character of listed buildings. Under current building legislation it is possible to apply to the local authority for a Relaxation of the Regulations, and, if this application is refused, to appeal to the Secretary of State against the refusal. This appeal system takes time and it is wise to discuss any problems with the Building Control Division and the Scottish Office as soon as they arise.

BRITISH STANDARDS, CODES OF PRACTICE AND BRE DIGESTS

The Technical Standards which support the Building Standards (Scotland) Regulations contain the specifications which are 'deemed to satisfy' the requirements of the Regulations. Many of these specifications are based on the British Standards (BS) and Codes of Practice (CP) produced by the British Standards Institute.

Non-mandatory but useful official information on building construction is available in the monthly Digests issued by the Building Research Establishment (BRE), whose advisory service, for a fee, will examine and report on specific building problems.

The Agrément Board issues certificates relating to the performance of newly developed building products and techniques which have not yet undergone sufficient testing to merit inclusion in the list of British Standards. The existence of an Agrément Certificate is no guarantee that there will be no harmful long-term effects.

REPAIRS NOTICES

In Edinburgh, the Director of Property Services has the duty of ensuring that buildings in need of repair are put into good order or, if beyond repair, are demolished. The exercise of these duties in the past has meant the loss of some historic but neglected buildings, but has also had the effect of forcing owners to maintain the fabric of many important buildings in the city. Repairs to listed buildings, even under Statutory Notices, must comply with Planning Regulations. Typical repairs, such as rebuilding defective masonry, mending leaking gutters and clearing drains and pipes, usually relate to parts of a building in common ownership; single terraced houses seldom receive Statutory Notices.

The issue of a Statutory Notice for repairs under Section 87(1) of the Civic Government (Scotland) Act 1982 follows either a routine inspection from street level by officials of the Property Services Department or a request by one or several owners. The latter may be particularly advantageous if the voluntary cooperation of all other owners cannot be obtained. The Notice requires the owner, or owners in the case of a tenement, to rectify the specified defects so that the building is brought into a reasonable state of repair with regard to its age, type and location. Action is normally required within twenty-eight days of issue of the Statutory notice, although in practice owners will seldom be held to this and Building Control will allow time for the appointment of consultants and the preparation of tender documents. Under Section 106 of the 1982 Act, appeals to the Sheriff against any requirement of the Notice, or the cost if executed by the Department, must be made within fourteen days of issue, by Summary Application to the Sheriff Clerk, 82 Lauriston Place, Edinburgh.

Should the owners fail to execute the works specified in the Notice, the matter would be considered at a meeting of the Joint Building Control/Repairs Sub-Committee of the District Council and, unless the owners could show at the meeting that they have agreed or are likely to agree to have the necessary works carried out, the Sub-Committee is free to exercise powers under Section 99 of the Act and authorize the Director of Property Services to execute the works with the costs being met by the owner or owners.

These repairs are carried out by private contractors under the direction of the Director of Property Services, who charges an administration fee for this work. The liability for the cost of the repairs is divided equally among the proprietors. These costs, which are recoverable through the Courts, are payable to the District Council. The Finance Department may accept payment by instalments or, in the case of hardship, may provide a loan which becomes a burden to be discharged when the property is sold.

Repairs under Statutory Notices are in the nature of 'first aid' rather than 'radical surgery'; the Director of Property Services is empowered to deal with the immediate problem but not to embark on long-term repair or renovation. For this reason, and because proprietors cannot control the quality of contracts placed by the District Council, it is strongly recommended that repairs should be directly commissioned by the owners. Repairs by the District Council do not qualify for grants from the New Town Conservation Committee.

Under Section 13 of the Building (Scotland) Act 1959, the Director of Property Services can evacuate a building which is in a dangerous condition and can insist on its immediate demolition or repair. This power is exercised only as a last resort.

GRANTS FOR HOUSING IMPROVEMENT AND REPAIR

Under Part XIII of the Housing (Scotland) Act 1987 the local authority can, with government assistance, provide grants to improve or repair a house. The booklet *Improve Your Home with a Grant*, issued by the Scottish Office Environment Department, explains which grants are available and how they can be obtained. Advice about the local application of these grants is available from the Housing Department of the City of Edinburgh District Council.

The Council is obliged to pay grants to provide the following standard amenities in sub-standard dwelling houses:

- A fixed bath or shower
- A wash-hand basin   } with hot and cold water supply
- A sink
- A water closet

The cost limit of standard amenities is £3010 but an additional amount of up to £3450 may be available to cover works of repair and replacement which result from the provision of one or more of the standard amenities. This gives a maximum allowable cost of £6460 which may qualify for a 50 per cent grant of up to £3230. This grant is available even when the rateable value (in 1989) of the house exceeds the limit for the area. Grants for improvements are paid at the Council's discretion to bring houses up to the Tolerable Standard (i.e. the *minimum* requirements a house must meet), but in Housing Action Areas the Council is obliged to give grant aid. The Council also has discretion to offer Repairs Grants to prolong the life of a house, to remove lead plumbing, and to provide mandatory fire escapes in houses which are in multiple occupation.

A house meets the Tolerable Standard only if it:

- Is structurally stable;
- Is substantially free from rising or penetrating damp;
- Has satisfactory provision for natural and artificial lighting, for ventilation and for heating;
- Has an adequate piped supply of wholesome water available within the house;
- Has a sink provided with a satisfactory supply of both hot and cold water within the house;
- Has a water closet available for the exclusive use of the occupants of the house and suitably located within the house;
- Has an effective system for the drainage and disposal of foul and surface water;
- Has satisfactory facilities for the cooking of food within the house;
- Has satisfactory access to all external doors and outbuildings.

To bring a house up to the Tolerable Standard, grants of up to 75 per cent of the cost of improvements may be available for individual houses or flats in Edinburgh with a rateable value (in 1989) of £1135 or less; if an improvement order is served the rateable value limit does not apply and grant is mandatory. If the house meets the Tolerable Standard and has a fixed bath or shower, or when other buildings are converted into houses, the grant cannot exceed 50 per cent of the maximum approved expense limit.

The current maximum allowable cost for discretionary Improvement Grants is £12,600 (including VAT and professional fees) covering plumbing, electrical services, damp-proofing and structural defects. (Depending on the nature of the works involved, the Council has the power, with the approval of the Secretary of State, to increase the maximum approved expense limit.)

These grants do not cover the installation of central heating, improvements to houses in good repair, or houses built or converted after 15 June 1964, although the latter restriction can be waived by the Secretary of State on application by the local authority. Not more than half of the allowable cost may relate to repairs or replacements directly associated with the improvement works. Any increased costs for external repairs on listed buildings in the New Town can be partly offset by a grant from the Edinburgh New Town Conservation Committee. The ENTCC does not support the sub-division of the few remaining complete terraced houses in the New Town, but flats within sub-divided houses which had a rateable value (in 1989) of £1135 or less can qualify for Housing Improvement Grants. It should be understood that these grants are discretionary and are subject to the District Council's policy at any given time. The Scottish Office hopes to introduce new legislation by 1995.

Edinburgh has many tenements in multiple ownership, and generally all flats share the cost of common repairs to roofs and chimneys. Such common repairs require positive cooperation between proprietors whose obligations are usually set out in the title deeds of each flat. Edinburgh has benefited in recent years from the liberal provision of grants for common repairs and although the proportion has now returned to 50 per cent of £5500, or £7800 if the property is a pre-1914 tenement, per owner, the total assistance still potentially available to a tenement with ten or more owners is substantial. The Housing Repair Grants are taken into account by the ENTCC when considering applications. Business premises in an otherwise residential tenement may also qualify for a Repairs Grant.

If the Council requires owners in tenements to provide a fire escape, the cost of the work up to a limit of £9315 may qualify for 20 per cent grant or up to 90 per cent in cases of hardship.

Lead plumbing is regarded as a health hazard and the cost of replacing lead pipes and tanks with approved materials may qualify for a Repairs Grant of up to a maximum of 90 per cent of £5500, regardless of the rateable value of the property. The amount of grant a person can receive will depend on his or her financial resources.

Grants may be available from the Housing Department for environmental improvements to gardens and the surroundings of houses including the repair of railings. This Environmental Grant is intended to improve groups of houses rather than single dwellings. At present, Environmental Improvement Schemes are limited to Housing Action Areas.

Housing associations, financed by Scottish Homes, may also qualify for Improvement, Repair and Environmental Grants. These associations provide a non-profit-making property management and development service, midway between private and public ownership, which can be well suited to administer repairs to tenements in parts of the New Town.

Further information is available from the City of Edinburgh District Council Housing Department and the Scottish Federation of Housing Associations.

GRANTS FOR BUILDING CONSERVATION

INTRODUCTION

Paragraph 41 of Historic Scotland's booklet *Scotland's Listed Buildings* states: 'The fact that a building is listed does not give its owner automatic entitlement to a grant.' Despite this discouraging advice there are several sources of finance for the repair of historic buildings.

PUBLIC FUNDS

The Secretary of State for Scotland, acting on the advice of the Historic Buildings Council for Scotland, may approve grants for the repair and maintenance of buildings listed for their architectural or historic interest and for buildings in 'Outstanding' Conservation Areas. Conservation grants from this source are intended only to help any shortfall which the applicants are able to demonstrate cannot be met by any other means. Further information is given in Historic Scotland's *Memorandum on Listed Buildings and Conservation Areas*.

The local authority, through the Planning Committee, has wider grant-making powers for assisting the repair and maintenance of more modest listed buildings and buildings in non-outstanding Conservation Areas and it also assists with repairs to buildings which are a vital part of the townscape.

Both Historic Scotland and the City of Edinburgh District Council contribute to the funds administered by the ENTCC, to which the Historic Buildings Council for Scotland has delegated its powers to recommend grants for repairs to buildings in the New Town, other than churches and interiors.

The Committee uses the Council Tax band of the property as an indication of the proprietor's ability to pay for the repairs, and percentages are also graded in relation to the scope of the repairs.

It is the policy of the Committee to encourage groups of proprietors (i.e. the owners of all the flats in a tenement or of all the houses in a terrace) to undertake comprehensive external repairs, and to offer grants of higher percentages for this purpose.

A proportion of the discretionary Improvement Grants awarded by the local authority, under the Housing (Scotland) Act 1987, may be spent on repairing the external fabric. If the house is of architectural or historic importance, local authorities have the power, with the approval of the Secretary of State for Scotland, to increase their discretionary grants.

PRIVATE FUNDS

Some charitable foundations, such as the Pilgrim Trust, the Gulbenkian Trust and the Carnegie Trusts, may also assist the repair of notable buildings with loans or grants. The terms of reference vary with each trust and are likely to limit assistance to buildings used for a specific purpose, e.g. education, art or literature. Further details can be found in the *Directory of Grant Making Trusts*.

The concept of a revolving fund for the purchase, restoration and resale of notable buildings, which was pioneered so successfully in the Little Houses Scheme by the National Trust for Scotland, is the basis of a £2 million Architectural Heritage Fund administered by the Civic Trust in London. The Cockburn Association (the Edinburgh Civic Trust) has established a Conservation Trust, using revolving finance, and there are now Building Preservation Trusts in Lothian and many other Scottish regions. Housing associations, within the limitations imposed by their constitutions, are an excellent vehicle for the restoration, rehabilitation and future maintenance of dwellings, particularly in inner-city areas.

TYPICAL CONDITIONS APPLICABLE TO MOST GRANTS OFFERED BY THE ENTCC

1 The work shall comply with drawings/specification on which tenders were based.
2 Prior approval in writing must be obtained for any departure from the agreed scheme.
3 The grant may be reduced or recovered if the eligible costs prove to be less than estimated, or if the work is not carried out satisfactorily.

4   Grant will not be paid on VAT charged to VAT registered proprietors. Proprietors must confirm their VAT status in writing to the Director.

5   The work shall be completed to the satisfaction of the Director, and/or the Historic Buildings Council's architect, who may inspect the work in progress at any time.

6   It has been established in the Courts that the proprietor is responsible for insuring his or her property during repairs against loss and damage by fire, lightning, explosion, storm, tempest, flood, bursting or overflowing of water tanks etc. The Director requires to see evidence of such insurance cover before authorizing payments of grants for work to existing buildings (Clause 22c of the Standard Form of Contract, 1980 edition latest revision).

7   Proprietors shall notify the Director of any repairs which qualify for insurance compensation.

8   The period allowed in contracts for interim payments to contractors shall be 28 days after the issue of Certificates, but cheques for the proprietors' share, if payable through the Committee, should be issued within 14 days.

9a *For tenements in multiple ownership:* The proprietors shall amend their titles to the properties so as to provide for their annual inspection and to empower a majority of owners to order mutual repairs, and shall send a copy of an appropriate Deed of Conditions to the Committee within three months of the date of the letter of offer.

9b. *For whole houses in single ownership:* Within twelve months of the date of the Final Certificate and annually thereafter (unless specifically agreed otherwise by the Committee), the roofs, chimneys, roof spaces and valley gutters, walls and common stairs shall be inspected by the Director or his representative, and brief reports shall be available for each proprietor. Repairs, other than routine maintenance, identified in these annual reports can be recommended for supplementary grants.

10 If a property which has benefited from a grant from the Committee is sold or sub-divided within five years of the grant offer, the Committee may in certain circumstances require the owner to repay his or her share of the grant in proportion to the number of years which have elapsed since the work was completed.

11 Architects shall certify to the Director that they have examined the interior of each room in each property before the contract commences with a view to establishing that the structural timbers are not affected by dry rot or other weaknesses.

12 Photographs showing the condition of the property before and after restoration shall be provided free of charge for use by the Committee. The photographs shall measure not less than 210 mm x 297 mm, shall be in black and white and be printed on glossy paper.

13 All comprehensive contracts and major repairs shall be supervised and certified by a registered architect.

14 The Committee may publicize the fact that a grant has been made and may attach a permanent sign to the exterior of the building to record the restoration.

15 All mouldings in stone and timber should match the original design and full-sized details shall be approved by the Director before construction.

16 The mortar to be used for repointing and bedding of stone shall consist of hydraulic lime and sand or another mortar specifically approved by the Director.

17 Adequate and secure temporary protection must be provided during major roof repairs involving re-slating and new centre gutters.

18 Adequate permanent ladders, access hatches, walkways and duck boards shall be provided so that future inspections and repairs can be made as easily as possible.

19 All repairs to ornamental ironwork shall include removal of all rust, priming with an approved primer applied immediately after cleaning the metal, two undercoats and one finishing coat of gloss black paint.

20 The Committee may require an assurance that all residential properties concerned are up to the Tolerable Standard as defined in the Housing (Scotland) Act 1974.

21 The offer may be withdrawn if work does not start within *six months* of the date of the letter of offer.

22 The agent shall make the proprietors aware of these conditions and shall confirm in writing to this Committee within 28 days that the proprietors have accepted these conditions and shall confirm the VAT status of each proprietor. The agent shall confirm within three months that the proprietors have adopted Clause 9a or 9b. The agent shall forward to this Committee evidence of the insurance cover required under Clause 6.

MAINTENANCE

'Stave off decay by daily care', was one of William Morris's favourite maxims. The basis of conservation is good housekeeping. When a spectacular restoration project is complete and an historic building has at last reached 'the plateau of good repair' it is profligate, if not immoral, to allow it to relapse into decay.

The only way to avoid dilapidation is by the unglamorous chore of routine maintenance. We accept such a discipline for cars, which generally have a life span of less than ten years (regular servicing is recorded in their log books and older cars have to pass an annual test). Our houses, which unlike cars may increase in value, deserve at least as much attention and if a building is also part of the national heritage it is even more important that it should be well maintained. It is recommended that, on average, one per cent of the market value of a property should be set aside each year to cover all maintenance and decoration costs. Once a building has been put into good order, it should be regularly inspected to make sure that no defects which might damage the fabric of the building reappear. Failure to maintain an historic building may prejudice any application for grant assistance from Historic Scotland, and in addition, the local authority has planning powers which can be used to carry out emergency works in cases of dereliction of unoccupied listed buildings, and can serve repair notices or compulsory purchase orders on any listed buildings which are neglected by their owners (see LISTED BUILDINGS AND CONSERVATION AREAS).

Most buildings in the New Town were designed and are used as residential tenements containing between six and twelve heritable properties. The titles of these properties usually refer to the obligations placed on each owner to contribute to 'common repairs' and many disputes have resulted from the interpretation of such titles.

Failure to agree on a repair often leads to neglect and decay of the tenement, particularly at roof level, resulting ultimately in the issue of a Statutory Notice to enforce works which may become more costly as a result of delays (see BUILDING REGULATIONS AND STANDARDS). The ENTCC requires that those accepting grants for comprehensive repairs amend their titles to give power to a majority of proprietors in a tenement to undertake regular inspections and maintenance. Further information is given in LEGAL AGREEMENTS.

PREVENTIVE CONSERVATION

Risks and maintenance costs can be reduced if the owner takes the following inexpensive precautions:

1. Lag all pipes and tanks in roof spaces or where exposed to frost.
2. Remove ivy and plants which can weaken mortar, hold damp, damage stone or conceal decay.
3. Remove sources of damp such as soil heaped against the walls. Provide site drainage where necessary around the building.
4. Provide easy access and good lighting in roof spaces, as well as permanent roof ladders and, in extreme cases, safety wires and harnesses.
5. Keep a constant moderate temperature throughout the building, avoid spasmodic high temperatures near joinery, and never use portable room heaters burning paraffin or bottled gas; they cause excessive condensation because their product of combustion is water.
6. Ventilate rooms, roof spaces and ducts, keeping flues clean and open; relative humidity should be at about 50 per cent.
7. Protect important fabrics and contents against ultraviolet radiation from strong sunlight and artificial light.
8. Improve electrical insulation with earth leakage trip and miniature circuit breakers.
9. Ensure that all gas pipes are free from corrosion and are properly jointed, ventilated and insulated to comply with the latest regulations.
10. Ensure that all rainwater conductors and drains are clear, that all joints are sound and that all bends can be rodded.

MAINTENANCE PROGRAMME

The following list is not exhaustive but it is appropriate for the average house in the New Town:

1. *As required:*
   - Inspect roofs, gutters and rainwater pipes (after every storm);
   - Eliminate vermin.

2 *Autumn and spring:*
- Inspect roofs;
- Remove leaves and other debris;
- Replace loose, slipped, cracked or missing slates;
- Repair leaking gutters if necessary;
- Check rainwater disposal and drains;
- Inspect external walls for signs of damp, trace the cause and make good the defect;
- Remove vegetation from the building;
- Wipe gloss paintwork with a mixture of water and white spirit in equal quantities with a squirt of liquid soap (not soaps containing acid; consult a conservator before cleaning valuable materials);
- Spring clean rooms in rotation.

3 *Annual inspection:*
- Inspect roofs (slater);
- Sweep chimney if in regular use (chimney sweep);
- Inspect/rod drains (plumber);
- Check heating system, header tank and water circulation, inspect boiler, clean ducts and sweep flue if required (heating engineer);
- Inspect fire extinguishers (supplier);
- Repair and repaint south-facing woodwork if necessary (joiner and painter).

4 *Every five years:*
- Clean out roof spaces;
- Inspect and report on condition of building (architect);
- Prepare schedule of repairs in order of urgency for next five years (architect);
- Inspect and test electrical installation (electrician);
- Repaint all external woodwork;
- Inspect all external ironwork and repaint as necessary.

INSPECTION

The ENTCC offers an annual maintenance inspection service with a brief report on the condition of the roof and external walls of listed buildings in the New Town. Many general and roofing contractors, heating engineers and electricians offer an annual maintenance service at a reasonable cost. This service is recommended, but if proprietors have already appointed an architect or surveyor they should take his or her advice before appointing contractors so that the inspections can be coordinated. Proprietors in a tenement are advised to appoint a management committee to organize inspections and repairs.

All reports of inspections and descriptions of repairs should be filed either by one of the proprietors or by the consultant or solicitor. These documents are the basis of a log book, which will be useful in the future.

SALVAGED MATERIALS

INTRODUCTION

Architectural character is a precious and fragile quality the loss of which diminishes a building and weakens the case for its protection. For instance, the loss of cornices, fireplaces or panelled doors degrades the architectural integrity of a room; where there are proposals to remove original features such as these, formal consent will be required from the Department of Planning. The rescue of items from buildings about to be demolished makes good economical and ecological sense. However, some people are concerned that the use of reclaimed materials may prevent a new generation of craftsmen from developing the traditional skills required to work with traditional materials.

Removal of historic materials from the site should be the last resort, undertaken only with Listed Building Consent; such materials should be photographed *in situ* and recorded, labelled and catalogued before removal so that, with a known provenance, they will be likely to find a new home appropriate in date, style and ideally by the same architect.

All who deal in second-hand goods will be aware of the risks of handling stolen property and the architectural salvage market is increasingly organized and self-policing. However, the demand for salvaged materials can lead to theft, the subject of English Heritage's *Theft of Architectural Features*, and the source of any salvaged materials should always be questioned before purchase.

Failure to use compatible materials which closely match the original can result in:

- A marked change in the architectural character of the building as in some tenements where traditional slates have been replaced by concrete tiles, stone replaced by brick, or sashes by UPVC windows;
- A shorter life for all or part of the building because substitutes may not last as long as the original materials;
- Decay of the original fabric when incompatible materials are introduced.

SOURCES

Fortunately, as the principles of conservation have become more widely accepted, fewer traditional buildings are being demolished and so there is less opportunity to salvage material. In the New Town, Listed Building Consent is required from the planning authority before original features can be removed. There are, however, three principal sources of materials:

1   *Within the building itself:* A principle of good conservation is that all existing materials and components should be retained wherever possible, and where repair is necessary this should be done on a like-for-like basis. (See later for further information on specific materials.)

2   *Direct from other building sites:* It is comparatively simple to recover components and fittings from buildings which are about to be demolished. Contractors may occasionally allow individuals on to a site to salvage particular items before demolition starts, although this should never be done without prior approval (not least because of the hazards of entering a demolition site). The contractor will normally salvage valuable features and offer them for sale. Buildings undergoing extensive alterations may also provide a useful source, and again negotiation with the building contractor, the architect or the owner is essential. One should beware of dealing with those who strip all the original features from one building to embellish another.

    When stripping out original features or even demolishing all or part of a building, it is recommended that all salvaged materials become the property of the building owner rather than the contractor. In this way greater control can be exercised over sympathetic reuse. Alternatively, if the owner agrees, the architect can instruct that all suitable materials be delivered to the nearest local authority salvage store, if there is one. The City of Edinburgh District Council Department of Planning recommends that salvaged materials should be retained and reused in the building from which they have been taken. If this is not possible, the materials should be offered or sold to a local commercial salvage yard, unless they are of particular historical importance, in which case they should be offered to the Edinburgh People's Museum.

3   *From salvage stores:* The *Salvo Directory* lists dealers in architectural antiques, reclaimed building materials, fine architectural items and statuary, in Britain, Ireland and even a few in Belgium and France. The 1994 edition lists eight firms in Lothian, nineteen in Strathclyde, three in Central Region, five in Fife, seven in Tayside and

seven in Grampian. The ENTCC can advise on other firms in the Edinburgh area which are not listed in the directory. A successful salvage firm relies on a wide range of materials, good publicity, easy retrieval, accurate records and rapid turnover. Some of the largest stocks are to be found in London and Bath, where materials from all over the country have been collected and are available to any purchasers, including those from overseas.

MATERIAL AND COMPONENTS

Slates

West Highland slates are only available second-hand and stocks are held by slaters and builders' merchants. For further information on suitable available types of slate see ROOF COVERINGS.

Even when full re-slating is necessary, it is usually the nails and not the slates themselves which are defective and a large proportion, generally 70-80 per cent of the original slates, can be used again if redressed and reholed.

Lead

Both sand-cast lead and milled lead sheet (which was initially available from Glasgow and became widely available by the 1820s) were used originally in the New Town. When roof and gutter coverings or flashings have to be stripped out and renewed, the old lead has some commercial scrap value. Customers should ask for a credit item in a contractor's estimate when removal of lead is involved. Alternatively, old lead sheet can usefully be reused as loose sheets for sacrificial flashings (see FLASHINGS).

Stone

Sound, good quality stone from downtakings should be reused wherever possible even though redressing is usually necessary. Care needs to be taken during demolition to ensure that the stones are not damaged. Redressing to new dimensions, and even reworking to remove local erosion and other defects, is easily accomplished with modern tools, and although wet stone will often be softer to work than dry, the legend that old stone without its quarry sap is difficult to dress can be discounted now that modern power tools are available. When using second-hand stone it is important to copy the original method of construction (see MASONRY WALLS).

Old lime can be easily cleaned off stone but hard cement may be more of a problem. Stone from chimney stacks will probably contain crystalized acidic salts from flue gases and should not be reused. Flags, cobbles and setts should all be salvaged for reuse.

Carpentry

The softwoods used originally in the New Town will almost certainly have a greater structural strength and a higher resistance to fungal attack than modern fast-crop timbers. Original joists and rafters should therefore be retained wherever possible, even when it means that new ends have to be spliced on. If possible, repairs should be carried out in new timber. However, if suitable new timber is not available, recycled timber can be bought from one of the firms specializing in second-hand timber. In the past, redundant warehouses have provided a good stock of second-hand timbers such as pitch pine, but the supply is decreasing as more buildings are converted to new uses instead of being demolished. As a general precaution, all salvaged timber brought in from elsewhere should be carefully inspected and, if necessary, treated with fungicide. Sometimes old nails make timber difficult to reuse.

Joinery and glazing

There is a substantial market in second-hand panelled doors because these are easy to retrieve, store and select. The mouldings and configuration of panels vary enormously, and these need to be carefully matched with those on existing doors.

Window sashes are less easily reused in new locations because, despite the apparent uniformity of New Town street elevations, window opening sizes vary and it can be surprisingly difficult to find one to fit. Sash windows can almost always be repaired and, because the timber is of such good quality, it is usually well worth spending the money on repair, even if this is more expensive than renewal. Derelict sashes should be carefully laid aside so that any original crown or cylinder glass can be saved.

Other joinery such as architraves and skirtings can also be usefully and economically reused, although careful measurement and survey is necessary to ensure that mouldings are appropriate and that everything fits. The same applies to window linings and panelling.

Brasswork

There is a well-established trade in Edinburgh in second-hand brasswork, from handles, numerals and bell pulls to rim locks and sash lifts. All are worth saving even if their respective doors and windows are beyond repair.

Sanitary fittings

Victorian and Edwardian fittings, usually of rather grand design, may be more suited to the New Town than those in modern catalogues. Second-hand fittings, including taps, are relatively scarce but can occasionally be picked up from salvage yards, plumbers or through classified advertisements.

Chimneypieces

The market in chimneypieces is a flourishing part of the antique trade; they are available in a number of shops and also through advertisements. Unfortunately their high resale value has resulted in the illicit plundering of empty properties, and precautions should be taken to minimize the risk of loss when property is left empty or advertised for sale. The chimneypieces and doorcases should always be individually photographed and preferably protected within a stoutly constructed timber box.

The temptation to install an over-elaborate chimneypiece should be resisted. Second-hand grates, hobs and other accoutrements of the hearth are also widely available.

Ironwork

Decorative ironwork such as railings, finials, balconies and standards are eminently suitable for successful salvage and reuse, and can usually be adapted to suit new locations.

General

Bricks, wrought iron, decorative plaster, chimneypots, cupolas, stairs and handrails are all worthy salvage items.

FIRE SAFETY

Damage or loss by fire is a constant threat to historic buildings. When the buildings in the New Town were first occupied the owners were no doubt fully aware of the fire risks associated with the use of a naked flame for cooking, lighting and heating. Today we are much less aware of the fire-raising potential around us, but we also have much more efficient safety systems.

Fire safety laws have a history almost as long as building itself. Indeed, some traditional design features and materials are less flammable than those being used today!

By law, the owner of any building open to the public is responsible for providing adequate fire protection for the occupants. However, fire safety legislation does not apply to unaltered, original, New Town houses, although its principles should be followed. Any proposed alterations should be discussed with the Listed Building Section of the Planning Department prior to starting on site, and Listed Building Consent is required for any works which may affect the character of a listed building.

This chapter recommends some commonsense fire precautions, and refers to measures needed to comply with the fire regulations when upgrading or adapting a building.

LEGISLATION

The Building Standards (Scotland) Regulations 1990 and the supporting Technical Standards apply to the fire protection of the structure of all new buildings and where alterations are being made in existing buildings. The Technical Standards are not retrospective and can only be enforced when the use of the building is to be changed or when the structure is to be modified. In any case, Planning and Building Control Departments should be consulted.

Whenever alterations or change of use applications are being considered for historic buildings, the fire authorities will initially require that the premises comply fully with the relevant regulations. The onus is on the owner thereafter to seek a relaxation, from the Building Control Department, of those standards which would compromise listed features. In many cases there are deemed-to-satisfy standards which may be acceptable to both parties.

The Fire Precautions Act 1971 sets standards for fire protection, which apply to certain categories of existing buildings. The Act is concerned only with the protection of life in the event of a fire; it does not apply to the protection of the building fabric. The Act is designed to cover all buildings in one of the following categories of use, which do not include private residential use:

- Recreation, entertainment or instruction, or for any club, society or association;
- Teaching, training or research;
- Institutions providing treatment or care;
- Any purpose involving the provision of sleeping accommodation (e.g. hostel, hotel);
- Any use involving access to the building by members of the public, whether on payment or otherwise;
- The use of the premises as a place of work.

However, at present only hotels and boarding houses, factories, offices, shops and railway premises have been designated under the Act.

The Act sets standards for:

- Means of escape and their safe and effective use;
- Means of firefighting;
- Means of giving warning in case of fire;
- Maintenance of all the above facilities.

The Act requires that the fire authorities inspect the premises and grant a Fire Certificate if the required standards are achieved. If not, the fire officer will prepare a schedule of work required and the premises must be upgraded. The application of the Act in an historic building should strike a balance between protection and preservation. The measures should represent the minimum necessary to ensure the protection of the building's occupants while avoiding, where possible, alterations to the architectural character of the building. This may include some compromise between active and passive fire protection measures.

During 1994 new regulations for fire safety at all workplaces came into force under the Work Place Directive of the European Commission.

REDUCING THE RISK OF FIRE

A reminder of how fires may start in historic buildings is a salutary and useful basis for understanding how to avoid them. The Building Research Station publishes useful Information Papers and the local fire prevention officer can give further advice free of charge. The main causes are:

1 *Wilful fire-raising:* Where appropriate, a high standard of security should be maintained to prevent unauthorized entry into the building.

2 *Electrical equipment:* Any electrical installation is potentially a fire hazard, especially if it is more than twenty-five years old. Installations should be regularly inspected and tested by a qualified electrician. Fuses should be of the correct rating; long flexes and multiple use of sockets should be avoided.

3 *Heating installations, particularly open fires:* Open unguarded fires, damaged flues and structural timber too close to fireplaces are hazardous and should be avoided. Chimneys in daily use should be swept regularly. Sparkguards complying with BS 3248 should always be used with open fires; damaged flues should be repaired with fireclay liners (see CHIMNEYS AND FLUES), and fires should not be allowed to burn directly on the hearth but should be contained in a fire-basket, to prevent the hearthstone from becoming too hot and charring timbers below. Central heating boilers and furnaces should be regularly serviced and combustible materials and fuel stored well away from fires, heaters and lights.

4 *Cooking:* The fire brigade recommends that a fire blanket be kept in every kitchen to enable the smothering of cooking fires.

5 *Smoking:* Carelessly discarded cigarettes and matches are one of the main causes of fire, particularly in living accommodation. Whenever practicable, smoking should be banned, or at least restricted to specific areas.

6 *Combustible materials and bad housekeeping:* Every effort should be made to prevent the accumulation of rubbish in cellars, roof spaces and spare rooms. Wood preservatives and insecticidal treatments are often flammable and special precautions should be taken during and after application. Regular removal of waste materials (including ashtrays!), correct storage of flammable materials and the proper tending of fires all contribute towards fire safety and are all part of good housekeeping.

7 *Lightning:* Lightning protection to BS 6651 should be provided, especially on buildings with projecting features such as tall spires.

8 *Workmen carrying out maintenance or alterations:* A large number of serious fires in old buildings have been caused by the careless actions of building contractors. Examples include:
   - Timber ignited by a blowlamp while removing paint;
   - Timber ignited by a portable stove;
   - Timber fire surrounds ignited by burning rubbish;
   - Flammable vapours from adhesives ignited by a pilot light;
   - Waste at the bottom of a shaft ignited by sparks from oxyacetylene cutting equipment;
   - Roof timbers set alight by workmen while burning/welding during the relaying of lead;
   - Roof timbers ignited by faulty electric lead to a lamp;
   - Timbers set alight due to flammable vapours from solvents in pesticides or preservatives sprayed to treat insects or decay.

Owners should be fully aware of the vulnerability of historic buildings to damage by fire (for example, Windsor Castle) during repair, alteration or maintenance work and should take all possible precautions while work is in progress. Before work starts on site there should be initial consultations with the architect, surveyor, planning authorities and insurer. This should include clear allocation of responsibilities for issue and receipt of hot-work permits which should be required for all operations which produce heat or require the use of naked flames. The control and avoidance of hot-working (blowlamps, welding torches and so on) is recommended. For example, non-flammable solvent-type paint stripper should be used in preference to blowlamps or electric hot-air blowers.

Conditions applicable to a hot-work permit should include:

- All apparatus to be operated only by skilled tradesmen;
- Smoking by hot-work operatives, in enclosed areas, especially in roof spaces and other danger areas, to be forbidden;
- All litter, rubbish and combustible material in the vicinity of any work to be removed or protected;
- Insulating shields to be used, where applicable, to protect the areas surrounding the work against heat and sparks;
- Flame-producing equipment not to be used on or near containers of flammable liquids or compressed gases;

- Working with a blowlamp is to cease at least one hour before the end of the working day, and during the following two hours, regular and thorough inspections to be made for smouldering fires;
- Checks and limitations on the burning of waste materials (never in fireplaces!);
- An adequate number of fire extinguishers of the correct type (at least two), to be close at hand.

Another new set of regulations is likely to be drafted under the Temporary and Mobile Work Places Directive of the European Commission. These are expected to apply to all building works.

UPGRADING

Upgrading may become necessary in order to comply with fire regulations when an existing building is altered or its use is changed. The following are only broad guidelines for upgrading an historic building. In practice, an architect should always be involved, working closely with the Planning and Building Control Departments, the fire authorities, and Historic Scotland. Any alterations needed to comply with fire regulations should, if possible, be reversible.

Escape routes

The generous proportions of most New Town buildings will usually mean that, given a little thought, adequate escape routes can be devised. The length and construction of all escape routes must comply with the Building Standards (Scotland) Regulations 1990. External fire escapes should be avoided where possible. If adequate escape provisions cannot be devised, Building Control may accept a system of smoke or heat detection, to reduce the length of time between ignition and discovery of the fire, and/or the amendment of the use or disposition of the area of concern. It may be possible for an escape route to lead through to an adjoining property if the 'receiving' owner is prepared to enter into a proper agreement.

Fire brigade access

It is important that staircases and roof hatches are wide enough for fire officers wearing breathing apparatus. It is also a good idea to have a set of master keys available on the premises in case of fire.

Structure

Both the Fire Precautions Act and Building Regulations require structural elements such as floors, walls and roofs to be fire resistant, in terms of:

- Stability – resistance to collapse or excessive deflection;
- Integrity – resistance to passage of flames and hot gases;
- Insulation – resistance to excessive temperature rise on the exposed side.

The materials and construction originally used in the New Town often have very good fire-resisting properties, which go a long way towards satisfying these requirements. Methods of increasing the fire resistance of walls and floors can be found in BRE Digests Nos 208 and 230, though the effect of any additional materials on structural loading must be checked. Alternatively, it may be acceptable to install a fire-detection system which will ensure that the occupants are alerted to the presence of fire before the structural element fails.

Fire-resistant doors and assemblies

Self-closing fire doors protect escape routes by delaying the spread of fire and smoke. Existing timber doors can be upgraded in several ways:

- Covering with intumescent coatings which swell up to form a fire-resistant char;
- The fixing of thin fire-proof panels (so long as no damage is done to the original door).

Any proposals to upgrade an existing door must be agreed with Building Control. If none of the above solutions is practical, new fire-resistant panelled doors can be specially made. It is important to remember that the frame, overdoor light, hinges, door furniture and sealing and closing system will require upgrading at the same time as the door. Details of the upgrading are discussed in DOORS.

Services

All new or altered domestic premises are now required by Building Control to have a smoke alarm of the mains-operated type. The protection of electrical systems is mentioned in SERVICES. Special attention must be given to the careful positioning of all heat and smoke detectors, break-glass alarms, warning signs and control panels. Detection systems should be unobtrusively sited, with cable runs hidden where possible; they should comply with

BS 5839. A new system now exists which is radio operated, requiring no wiring, and therefore causing minimal damage to structure or decor. In rooms of great importance, possibly those with highly decorative plasterwork, consideration should be given to the use of the Vesda system; the detectors are isolated from the room itself and only a small hole is required through which air is drawn for sampling. Sprinklers are difficult to hide and should only be used in historic buildings when unavoidable; they require an associated drainage system. Water can often do more damage than the fire itself.

In buildings designated under the Fire Precautions Act, emergency lighting to BS 5266 must be provided; certain types can be incorporated within the existing light fittings, especially when a back-up power supply is being used. It is important that the correct fittings are installed; consult the local fire authority for advice.

Firefighting equipment

A fire blanket in the kitchen should be sufficient for domestic premises. Means for fighting a fire must be provided in all the categories of building designated under the Fire Precautions Act, i.e. hotels and boarding houses, factories, offices, shops and railway premises. The following factors should be taken into account:

- There should be at least one fire extinguisher per storey, which should be suitable for the type of fire risk (for example, water for carbonaceous fires or carbon dioxide for electrical fires);
- Ideally, all the extinguishers in a building should have the same method of operation;
- All extinguishers must comply with BS 5423 in terms of operation and performance;
- They should be light enough for one person to lift;
- They should be clearly labelled;
- Extinguishers should be located adjacent to the area of risk, be clearly visible, located at the same position on each floor, and preferably wall mounted. The location should be discussed and agreed with the fire officer;
- Regular maintenance must be arranged;
- All extinguishers must comply with BS 5306 for the purpose of siting and maintenance;
- Fire hydrants and dry risers should be clearly marked and their location discussed and agreed with the fire officer.

RECORDS

A measured survey with photographs and an inventory of fittings and finishes will be invaluable in the event of fire, and can be the basis for the restoration of a listed building. All documentation should be kept in a fire-resisting cabinet on the premises and a copy held elsewhere, preferably the National Monuments Record of Scotland.

PROGRAMME FOR A CONSERVATION PROJECT

INTRODUCTION

This chapter, which is written with building owners in mind, describes the sequence of events during a typical comprehensive repairs project and discusses some of the problems and pitfalls, particularly those affecting tenements in multiple ownership. The chapter emphasizes the importance of early and professional assessment of the building, and the advantage of delegating authority to a few proprietors elected to ensure that repairs are carried out satisfactorily.

Ideally, every historic building should be inspected annually and the maintenance schedule should be updated after each quinquennial inspection; with such a system the building should soon reach a plateau of good repair, and thereafter future maintenance, although unavoidable, should be minimal. However, all too often maintenance is overlooked and major repairs are precipitated by the unexpected discovery of dry rot, leaking roofs or spalling stonework. In Edinburgh proprietors can be compelled by Statutory Notices to undertake such repairs (see BUILDING REGULATIONS AND STANDARDS). It is in the national interest that listed buildings which have been repaired with public funds should be regularly inspected and maintained thereafter. See MAINTENANCE for specific recommendations about annual schedules and details of ENTCC's own inspection service.

REPAIR OF HOUSES

Most owners of terraced houses in the New Town are conscious of the need to inspect and maintain their buildings. Unlike proprietors of tenemental property, they do not require the consent of adjoining owners to carry out repairs unless the defects are mutual, but neither can they share the expenses for anything other than mutual work. In any case, they will be well advised to appoint an architect to report on the condition of the house, estimate the cost of repairs, apply for grants, recommend contractors, obtain tenders and supervise repairs. The ENTCC can make an initial inspection of the building and will advise proprietors about procedures.

REPAIR OF TENEMENTS AND OTHER BUILDINGS IN MULTIPLE OWNERSHIP

It can be difficult enough for an individual house owner to get repairs organized on a proper footing but for a tenement block in multiple ownership the obstacles may seem insurmountable. In fact eighty-seven comprehensive repair schemes have been started since 1972 under the guidance of ENTCC. Most of these involved groups of owners, and have been completed to a high standard.

In Edinburgh it is a requirement written into many title deeds and upheld in common Scots law that owners share the cost of common repairs. This is reinforced by Statutory Notices, now issued by the District Council under the Civic Government (Scotland) Act 1982. If it were not for this mixture of individual obligation and municipal control, many buildings in central Scotland might have long since collapsed.

Statutory Notices are issued by the Department of Property Services (Building Control Division) of Edinburgh District Council to ensure that a building is put into a safe and sound, though not necessarily original, condition. If the owners do not comply with the Notice, limited repairs will be instructed by the District Council and the cost, plus an administrative fee, will be charged to each owner in equal shares. It is far better for the proprietors, acting together, to have the building repaired, not just to the minimum standard but in a permanent manner which will restore the building to a sound condition and replace missing features where appropriate.

It is unlikely that everyone in a tenement will have equal resources or an equal desire to invest in a major repair project and few titles make provision for majority rule. The Deed of Conditions described in LEGAL AGREEMENTS attempts to overcome this problem. If the defect is serious and cooperation is totally absent the problem must be referred to Building Control. However, an owner cannot be forced to do more than the minimum repair required by the Statutory Notice, e.g. to contribute to the extra cost of reslating a whole roof if patching will keep it watertight for a year, even if the more comprehensive repair is better long-term value, or to restore missing architectural features such as astragals and ornamental stonework, even if this is aesthetically desirable.

The proprietors must agree in advance, and preferably in writing, how they will apportion the costs for repairs, fees, VAT and expenses. The ENTCC recommends that costs should be divided in proportion to the relative values of the properties, using Council Tax bands for simplicity. Alternatively, chimney repairs are sometimes costed on the number of flues per flat and relative floor areas can also be used. The title deeds will probably set out the proportion of cost for common repairs to roof, chimneys and drains to be paid by each proprietor. Whichever system is adopted, it should apply to the whole external fabric of the building and to all common elements.

The grants offered by ENTCC go a long way towards meeting the extra cost of external repairs to a listed tenement, and there are supplementary grants for proprietors who qualify for rebates on their Council Tax. The Housing Department of the City of Edinburgh District Council should also be consulted about grants (see GRANTS FOR HOUSING IMPROVEMENT AND REPAIR and GRANTS FOR BUILDING CONSERVATION).

FIRST COMMITMENT

All such matters are best discussed at the first meeting of the proprietors, convened by a concerned proprietor or by the ENTCC, at which all residential and commercial proprietors should be represented. The committee room at the Conservation Centre is available for meetings during office hours and on weekday evenings, when the Director or Assistant Director can usually attend. If requested, the ENTCC will inspect the building beforehand, so that the proprietors can be advised on the need for repairs, the correct procedure and eligibility for grants.

If the proprietors agree that some common repairs are unavoidable they must then choose between minimum short-term patching without a grant or more durable repairs to a higher standard which may be eligible for a grant. It is usually impossible to make this choice without having some idea of the costs and professional advice is needed for that purpose. The appointment of an architect and consultants, the election of a small representative committee and an agreement to share the initial fees for a detailed survey, report and probable cost are the first, and often the most important, commitments which the proprietors must make.

APPOINTMENT OF PROFESSIONAL CONSULTANTS

With listed buildings it is recommended that all but very minor repairs should be carried out under the professional guidance of a consultant specializing in the conservation of historic buildings; even if the cause of the problem seems to be obvious, it is better to have professional advice before employing a contractor. The ENTCC and the Historic Buildings Council for Scotland make it a condition of grant aid that professional consultants are appointed (see GRANTS FOR BUILDING CONSERVATION). The architect or building surveyor will advise on the extent of necessary repairs, assist in obtaining grants, specify the required work in detail, obtain competitive tenders, inspect the work on site and certify each payment due to the contractors.

Proprietors should satisfy themselves about the professional competence of a consultant; they should not make an appointment until they have examined the architect's or building surveyor's work on similar buildings and consulted their previous clients. The Royal Incorporation of Architects in Scotland will give, free of charge, general advice on the appointment of an architect, and will provide, on request, an impartial list of three or four firms. The Royal Institution of Chartered Surveyors (Scottish Branch) will give similar advice about the appointment of building surveyors and quantity surveyors. The ENTCC keeps a record of architects and surveyors who have particular experience in the New Town.

Fees should be negotiated before the appointment is confirmed in writing; the consultant should set out the condition of engagement explaining in some detail the services to be provided and their costs. Fees can be based on a hourly rate, and this often applies to preliminary advice or surveys, a lump sum, or calculated as a percentage of the estimated or actual contract sum.

Payment by instalments should reflect the need for more frequent inspections by the consultant during repairs than is needed for a new building which requires more time to be spent on design. The RIBA/RIAS 1990 Scale for Historic Buildings (basic services) Class 2 is used by the ENTCC when calculating grants on fees. Allowable expenses should not exceed 1 per cent of the contract sum. On large contracts consultants may recommend that a Clerk of Works should be employed to inspect the work in progress. This has advantages but it does not absolve the consultant from responsibility and it is not eligible for a grant from the ENTCC.

The architect will also advise on the need for other consultants, e.g. a structural engineer or a quantity surveyor, to be appointed by the proprietors. Normally such consultants are employed by the client and their fees are charged separately on an agreed basis. All fees charged by the architect, surveyor, engineer and other consultants should be negotiated and agreed before the appointments are confirmed in writing. The choice of architect, building surveyor and engineer is the most important decision which the proprietors must take.

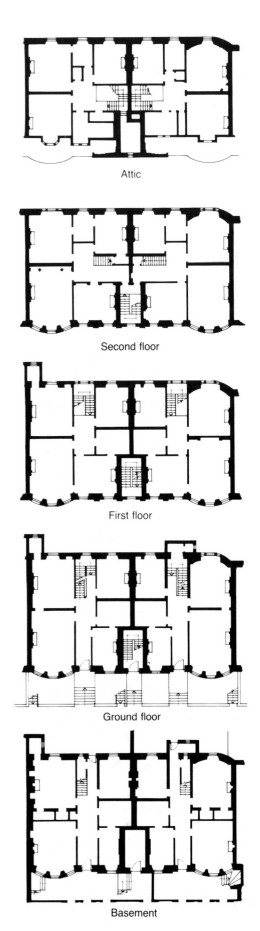

Attic

Second floor

First floor

Ground floor

Basement

**39-43 CASTLE STREET, BUILT 1793
(SIR WALTER SCOTT LIVED AT NO. 39)**

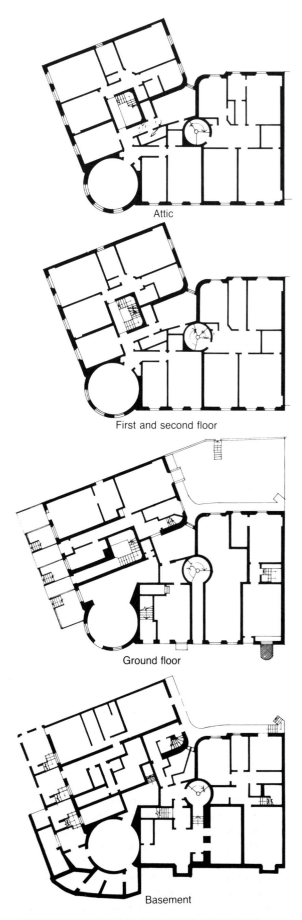

Attic

First and second floor

Ground floor

Basement

**1-7 BARONY STREET/ 34-40 BROUGHTON STREET
BUILT 1829-34**

Most consultants will provide a formal letter of appointment to be signed by the proprietors. The proprietors should insist that costs which exceed approved estimates should not be accepted without their written authority.

INSPECTION AND REPORT WITH OUTLINE PROPOSALS

After checking the findings of any previous inspections the architect should thoroughly inspect the entire building fabric (see SURVEY AND REPORT). The report presented to the proprietors should refer to the history of the building and its architectural quality, making recommendation for repairs in order of priority such as *immediate*, *urgent* or *desirable*. It is important that this inspection should be as thorough as possible; the inconvenience and disruption caused by opening up parts of the building are more than offset by the benefit of an accurate assessment. The architect will be required, at tender stage as a condition of grant, to certify to the ENTCC that every room has been examined for signs of dry rot or structural movement. Fees for this inspection are usually based on the time taken, but some architects are prepared to include this in the overall percentage fee for the whole project. Fees for the preliminary report become eligible for ENTCC grant when the contract starts.

SECOND COMMITMENT

At the second formal meeting the proprietors or their representatives will receive the report, with recommendations from the architect and an approximate estimate of costs from the quantity surveyor. The costs, including fees and VAT, should then be apportioned by a method agreed by the proprietors and, if each share is reduced by the grants already negotiated in principle, each proprietor should have a reliable estimate of his or her net contribution. If the proprietors accept these approximate costs, the architect and quantity surveyor should be instructed to prepare tender documents; such an instruction also commits the proprietors to pay substantial fees, which may become due at this stage.

The proprietors should then confirm their commitment by signing a legal agreement, authorizing their committee to instruct the architects and to accept tenders if they are within the approximate estimate (see LEGAL AGREEMENTS). A solicitor may be required to adapt the form of agreement and, unless one of the proprietors agrees to be an active honorary secretary/treasurer, the solicitor may also be asked to provide these services. Legal fees for the preparation of the agreement only are eligible for an ENTCC grant.

A joint deposit account should be opened with a bank or a building society and, unless a solicitor is engaged as secretary/treasurer, the signatures of at least two members of the proprietors' committee should be required to operate the account.

It is best that all contributions should be lodged before the contract starts but at least one fifth of each proprietor's total contribution should be lodged when the architect is instructed to prepare tender documents, one fifth when the repair contract is signed and the remaining three fifths lodged in regular monthly instalments during the contract period.

PREPARATION OF TENDER DOCUMENTS AND SELECTION OF CONTRACTORS

Following approval in principle by ENTCC, and the signing by the proprietors of the legal agreement, the architect or building surveyor can proceed to the preparation of drawings and specifications which describe in detail the extent of the repairs and the standards of materials and workmanship required, calling in a structural consultant if necessary. He or she will also advise on the need, where appropriate, to obtain Listed Building Consent and a building warrant, and will apply for these on behalf of the owners. He should advise on the need for temporary protection during the contract, especially if there are to be extensive repairs to the roof and chimneys.

For large contracts, a bill of quantities – an extremely detailed description and measurement of all the materials and activities required to carry out the work – will be prepared by the quantity surveyor as the basis for obtaining truly comparable tenders. Alternatively, for smaller contracts, a less detailed document can be produced, either by the quantity surveyor or by the architect; this may reduce the professional fees but it gives less control over costs during the course of the work.

Architects and consultants are advised to consult the Director of the ENTCC at an early stage because draft tender documents should be approved by the ENTCC before being sent to contractors.

It is usually advisable to obtain at least three tenders, but rarely necessary to have more than six. Open tendering, whereby any interested contractor can submit a price, is not recommended, particularly for work involving historic buildings where specialized workmanship is required. There is always the danger that a contractor who produces the lowest tender will also produce the lowest quality work, but careful selection can ensure that all contractors invited to tender are capable of producing work of the required standard. The architect will advise on suitable contractors and either he or the quantity surveyor will invite tenders on the owners' behalf.

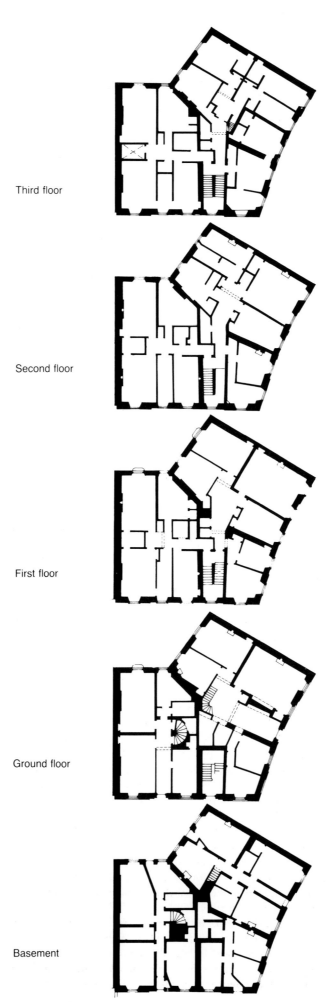

Third floor

Second floor

First floor

Ground floor

Basement

TYPICAL TENEMENT PAVILION, 36 DRUMMOND PLACE/1-3 LONDON STREET

THIRD COMMITMENT

The proprietors' committee should already have authority to accept tenders which are within the probable cost prepared by the quantity surveyor; tenders in excess of the probable cost will require further approval by the proprietors and it may be necessary to omit some of the work or to reduce the standard of materials and workmanship in order to make savings. Any omission or reduction in standards requires prior approval by the proprietors and, if the work is to be grant aided, the ENTCC and the local authority will need to approve any alterations to drawings or reductions in specifications before the contract is let.

BUILDING CONTRACTS

Except for minor work of a jobbing nature, all building and repair projects should be carried out under the umbrella of a formal building contract – an agreement between the proprietors and the contractor. There are various standard forms in current use and the architect will advise on the most appropriate (he or she will have specified the type of contract in the tender document). At present it is likely to be the latest editions of either the JCT Standard Form of Building Contract or the SBCC Building Contract for Minor Building Works. Other contracts, however, are available.

The contract defines at some length the rights and responsibilities of each party and the role played by the consultant (who acts as quasi-arbiter between the owner and contractor, even though he or she is employed by the former), and it lays down procedures for settling disputes, should they arise.

INSURANCE

All building contracts include a specific requirement for the builder to indemnify the owner against injury to people and damage caused by his or her own negligence, omission, and so on. In addition, because of the nature of repair works in terraces and tenements, it is sensible for the contractor to insure against damage to adjacent property.

It is important to establish responsibility for the protection of an existing building during repair; the contract holds the contractor liable for any injury or damage due to negligence, but it is the building owner's duty, as part of the contract, to insure the building and the work against damage by fire, lightning, storm, flood, and the like. Normally the owner will already have the house insured against these perils, but the insurance company should be told of the impending contract; it may be necessary to pay a small additional premium, and failure to do so could jeopardize a successful claim. In a tenement, individual policies held by each proprietor should allow for this increased risk, but to avoid lengthy disputes between insurance companies it is often preferable for the architect to instruct the contractor to take out a joint 'block' insurance in the names of proprietors and contractors.

It is better to avoid the risk of flooding by having a temporary canopy to protect the roof during repair; such temporary roofs, although expensive, are increasingly common on larger contracts and they are recommended by the ENTCC (see SCAFFOLDING, HOISTS AND TOWERS).

TIMESCALE

Major repair works cannot be organized speedily, and a conservation project can take anything from one to five years from start to finish, depending on the scope of the work. It may take some months for the proprietors to organize themselves, and in present economic circumstances it may be anything up to three or four years before grants are available. At the time of writing there is a long waiting list for local authority housing grants.

If a delay seems likely it may be necessary to bring forward repairs to prevent continuing water penetration, structural movement or the spread of dry rot. In such cases, provided that the ENTCC are consulted in advance, the cost of this work will usually be grant aided retrospectively when the main bulk of the work is carried out. The production of a Statutory Notice is convincing evidence of the urgency of such work.

The architect will have specified in the tender documents a date for commencement and a date for completion. In most contracts there is an optional provision for claiming Liquidated and Ascertained Damages, i.e. a pre-set sum payable per week by the contractor to the owner if the contract is not completed within the specified time. The sum must be a genuine estimate of the owner's likely loss. This is usually not appropriate for residential property, because the owners can seldom make a convincing case for loss of income and, if it is included in the contract, a damages clause may inflate the tender prices unnecessarily.

There are many factors which can conspire to delay the work such as bad weather, late delivery of materials (stone from quarries is a regular culprit), unexpected additional work, and so on. Delays can be infuriating and inconvenient but, sadly, they are all too frequent and not necessarily anyone's fault.

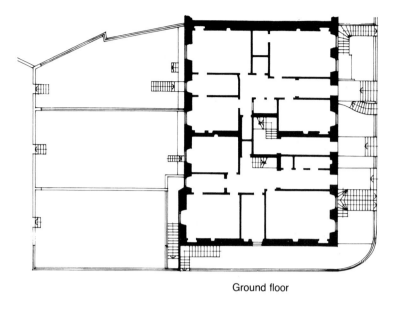

Ground floor

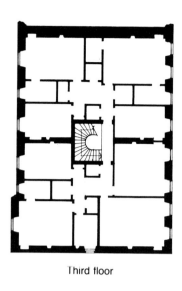

Third floor

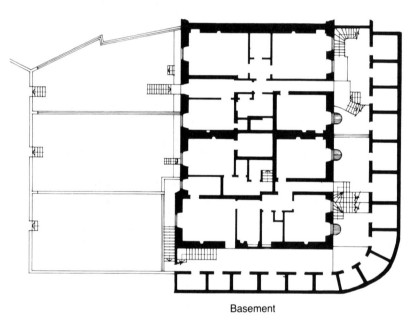

Basement

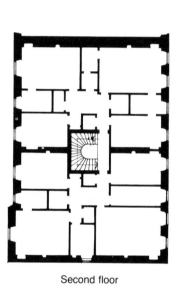

Second floor

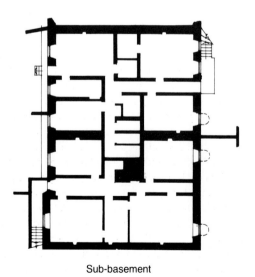

Sub-basement

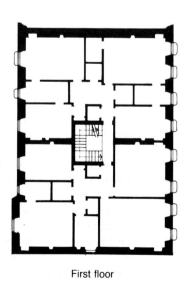

First floor

9-13 DUNDONALD STREET

28

COSTS AND PAYMENTS

All costs presented to proprietors should include estimated expenses, professional fees and VAT on top of the actual contract value. VAT is currently payable on all fees and on all repair and maintenance work and it is unlikely that this burden will be eased in the near future. The ENTCC does not pay grant on VAT charged to VAT-registered proprietors who should make sure that each registered proprietor obtains an individual receipt for his or her share of the VAT. All variations to the contract, and particularly those which substantially increase the cost, should be reported to the proprietors for their approval before being instructed.

During the course of the work the architect will issue monthly interim certificates authorizing a payment by the owner to the contractor. The owner has a duty under the terms of the contract to pay the builder within fourteen days of the date of issue of the certificate. If the owner feels that he or she needs an extension of this period this should be discussed with the architect at an early stage so that it can be clearly stated in the tender documents. Each certificate will show a deduction of 5 per cent of the value of the completed work. This is known as the Retention Fund and is held back as a safeguard against unsatisfactory materials or workmanship.

As well as monthly payments to the contractor, the owner will also need to pay fees due to the architect and other consultants on a regular basis as the work proceeds.

Where an ENTCC grant is involved the normal procedure of payments passing directly from owner to contractor does not apply, and the owner, or the proprietors' representative or treasurer, forwards each certificate, together with their share of the payment, to the ENTCC, who (through the District Council) add their proportion of grant and send both cheques to the contractor. Further grant payments cannot be requisitioned unless all previous certificates have been honoured. Twenty-eight days should be allowed for clearance of certificates and the contract should be worded accordingly.

WORK IN PROGRESS

The architect should hold formal monthly site meetings with the contractor and make regular site visits to inspect the work in progress. Separate monthly meetings between architect and owners are also advisable, as is the regular issue to proprietors of a written progress report and a cost report. Copies of minutes and progress reports should be sent to the ENTCC. With grant-aided schemes, the Director of the ENTCC or Historic Scotland's own architect may inspect the work at any time.

During the course of the work the architect will need to issue various instructions to the contractor and copies of these should be sent to the ENTCC. Mostly these will amplify existing information for the contractor and may have no cost implication. Small items of additional work will be paid for out of the contingency fund which will have been included in the contract sum, again with no increase in overall expenditure. If significant extra work, resulting in additional cost, is found to be necessary it should be discussed and agreed in advance with the proprietors and with the ENTCC.

To avoid ambiguity and any possibility of conflict it is important that owners do not give direct instructions to the contractor; all requests, instructions and comments must be issued by the architect. Similarly, where a number of proprietors are involved, all communication with the architect or building surveyor should be via their secretary or representative.

PRACTICAL PROBLEMS

Although the architect will spell out in the tender documents the need for adequate protection against the weather, and so on, things can and do go unfortunately wrong! A wise proprietor should be mentally and physically prepared for the inconvenience and disturbances which can plague him or her during the course of the contract.

Scaffolding may inconvenience people entering and leaving the building; it will also cut out a considerable amount of light from rooms, and it may be necessary to keep curtains permanently drawn for privacy while people are working directly outside. Scaffolding also makes upper storeys easier to burgle, so all windows should be kept locked and shuttered when rooms are empty. However careful a contractor may be, some dirt and debris will fall onto gardens, basement areas and pavements, and some dust inside the building is unavoidable. The noise of people working, particularly drilling or cutting out damaged stone, can also be quite disturbing.

Damage caused by water penetration will usually be paid for by the contractor or, if the weather can be proved to have been exceptional, through the proprietor's own house insurance (see earlier). Although inconvenient, leaks of this kind should not cause undue concern, provided that the cause of the leak is diagnosed and rectified quickly. Proprietors should inform the architect immediately and in writing if water damages their property. Occasionally something worse than a minor leak can occur – for instance, if there is unusually inclement weather when the building is at its most vulnerable. Architects and contractors should do all in their power to prevent leaks and it is certainly wise to move valuable possessions away from main working areas.

The ENTCC may offer grants towards the extra cost of a temporary roof during repairs, but in the case of major roof repair or dry rot eradication it may be necessary to vacate rooms entirely. This happened at 2-5 East Broughton Place and 73-79 Broughton Street where the roof and upper floors were totally reconstructed.

GUARANTEES

When the work is to all intents and purposes finished, the architect will issue a Practical Completion Certificate and will, at the same time, release half of the 5 per cent retention. The remaining 2.5 per cent retention is withheld during the Defects Liability Period, usually a year, after which the architect will inspect the work and instruct the contractor to make good any defective work free of charge. When this has been satisfactorily completed and the final account has been agreed with the contractor, the architect will issue a Final Certificate and release the remaining retention.

The best way of ensuring a sound job is to employ a good architect and a good contractor. Guarantees are not offered in the building trade except by firms offering a guarantee for eradication of dry rot and damp. Such a guarantee is limited to the areas actually treated and is only as good as the integrity and financial standing of the firm which issues it. If, after the Final Certificate has been issued and paid, it transpires that the contractor did not follow the specification or that the architect or other consultants acted negligently, the owner has recourse to common law.

OUTLINE PROGRAMME FOR A COMPREHENSIVE CONSERVATION CONTRACT FINANCED BY THE PROPRIETORS AND THE ENTCC

Proprietors	Architect/building surveyor and quantity surveyor	ENTCC	Local authority	Historic Scotland	Contractors
(1) Need for repairs noted					
(2) Approach ENTCC for grant		(3) Inspect property and commence meeting with proprietors and/or representatives. Grant system explained			
(4) Appoint small committee and agree to pay share of fees for survey		(5) ENTCC notified			
(6) Committee appoints architect or building surveyor and quantity surveyor to make a preliminary report with probable costs of repairs	(7) Consult ENTCC about scope of report				(9) Contractor to open up suspected areas
	(8) Consult engineer if structural faults suspected				
	(10) Survey report and estimate presented to proprietors				
(11) Legal agreement signed, executive committee appointed, share of costs lodged in special bank account, architects instructed to obtain tenders		(12) Grant approved in principle subject to tenders			

Proprietors	Architect/building surveyor and quantity surveyor	ENTCC	Local authority	Historic Scotland	Contractors
	(13) Tender documents prepared. Consents sought from local authority, ENTCC consulted				
	(14) Listed Building Consent and warrant applied for if required				
	(15) Tenders invited from contractors nominated and approved by architect and proprietors				(16) Tenders submitted
		(17) Grant recommended on lowest tender + fees + VAT		(18) Grant approved by Secretary of State	
		(19) Grant offered with conditions to proprietors			
(20) Grant conditions accepted in writing. Tender accepted and contract signed. Insurances noted	(21) 'Before' photos sent to ENTCC				(22) Plant and scaffolding on site. Prepare programme. Order materials
	(23) Programme of work agreed with contractors and sub-contractors. Minutes of each site meeting issued by architect to each contractor, to the proprietors' committee, and to ENTCC				
	(24) Issue interim certificate based on quantity surveyor's valuation of work done less retention. Copies to proprietors' committee and ENTCC				(25) Forward certificate to proprietors with request for payment and VAT

Proprietors	Architect/building surveyor and quantity surveyor	ENTCC	Local authority	Historic Scotland	Contractors
(26) Cheque payable to contractor for proprietors' share of certificate sent to ENTCC		(27) Proprietors' cheque with requisition for grant forwarded to Finance Dept. of City of Edinburgh District Council (CEDC)	(28) Grant and proprietors' cheque paid to contractor by CEDC	(29) SDD share of grant reimbursed to CEDC	
	(30) Request final inspection by ENTCC				
		(31) Final inspection			
	(32) Issue of Certificate of Practical Completion				(33) Agree final account
(34) Final meeting with architect	(35) Issue Final Certificate				
	(36) 'After' photographs sent to ENTCC				
(37) Make final payment		(38) Pay final grant			
(39) Agree system of future maintenance		(40) Fix plaque to building			

LEGAL AGREEMENTS

In New Town tenements subject to multiple ownership, many minor conservation schemes and jobbing repairs have been completed successfully without any formal agreement between proprietors. However, for all schemes which involve major expenditure, grant aid from public bodies such as the Edinburgh New Town Conservation Committee (ENTCC), and the appointment of professional consultants, a formal agreement signed and binding on all the proprietors is essential. The notes and style of the simple agreement which follow have been widely used in schemes supported by the ENTCC and have generally been found to serve satisfactorily.

NOTES ON THE EDINBURGH NEW TOWN CONSERVATION COMMITTEE'S STYLE OF AGREEMENT FOR USE BY PROPRIETORS

The blank draft of a simple form of agreement which follows is intended for use by proprietors or their lawyers in conservation schemes assisted by the ENTCC. It is intended that the agreement should be signed once the architect has prepared an inspection report and the quantity surveyor has prepared a probable cost, but before tender documents have been prepared. The agreement authorizes a committee of the proprietors to instruct the architect, collect money from the proprietors and the ENTCC and eventually pay the contractor. The full legal rights of the proprietors, architect, quantity surveyor and contractor are preserved. The main purpose of the agreement is to ensure that the architect and contractor can deal with one body only – the proprietors' committee, normally represented by the secretary – instead of all the proprietors as individuals. It also provides that the proprietors should place their shares of the cost of the work in a bank or building society account, either in full before the work starts or by instalments as the work proceeds, so that the contractor has some assurance of payment.

The following steps should be completed before the agreement is signed:

1 Proprietors agree to commission an architect (or building surveyor, in which case the term 'Building Surveyor' should be used instead of 'Architect' throughout the agreement) to prepare an inspection report and a quantity surveyor to prepare a probable cost of the work. The proprietors may appoint a provisional committee or a spokesman to authorize this work and to collect and pay the agreed shares of the costs involved.

2 Proprietors apply to the ENTCC, through the Director, for a grant towards professional fees for the inspection report and probable cost. This grant is offered if it seems to the ENTCC that the repairs are likely to start and that funds are likely to be available in the near future. The grant is paid when the contract starts.

3 Inspection report and probable costs are presented to the proprietors and approved.

4 Professional fees are paid and the ENTCC is informed.

5 Proprietors agree to appoint a committee of three to five representatives, including chairman, treasurer and secretary.

6 Proprietors obtain the consent of adjoining proprietors if work involving mutual property is intended.

7 Proprietors prepare a draft agreement (or instruct their lawyer to do so), and decide upon details such as where the bank or building society account will be opened and how the cost of the work will be apportioned. Cost need not be apportioned according to title deeds if otherwise agreed.

8 Proprietors make formal application to the ENTCC for grant assistance and have the signing copy of the agreement prepared for signature.

9 Signed agreement is sent to the ENTCC for noting and return.

The ENTCC's grant can either be paid into the bank or building society after all of the contractor's bills and professional fees have been settled by the proprietors, or it can be paid by instalments direct to the contractor and consultants, provided that the proprietors pay their share of each instalment first and that receipts are available for the full amount of previous instalments.

The proprietors should consider the advantages of retaining a committee responsible for the regular inspection and routine maintenance of the property after the scheme has been completed. A clause to this effect can be added to the agreement. It is a condition of ENTCC grant for comprehensive schemes that proprietors sign a Deed of

Conditions, amend their titles to provide for the annual inspection of their properties and empower a majority of the proprietors to order mutual repairs.

FORM OF AGREEMENT

AGREEMENT between ....

(FULL NAMES OF ALL THE PROPRIETORS)..

....

(hereinafter called 'the proprietors')

EDINBURGH,..(DATE)..

CONSIDERING THAT the proprietors have appointed...(NAME &

.... ADDRESS)

as their Architect and..(NAME &

.... ADDRESS)

as their Quantity Surveyor and have agreed that certain work should be undertaken at their property at

....

....

....(ADDRESS)

all in Edinburgh, as described in the Architect's inspection report dated...
and the Quantity Surveyor's probable cost dated...................................., part of which is contained in the Schedule annexed and signed as relative hereto, and have further agreed that a Committee of proprietors should have authority to instruct the Architect, collect such sums as may be required from the proprietors, the Edinburgh New Town Conservation Committee (hereinafter called 'the ENTCC'), or any other source, and ultimately pay the Contractor and others: NOW THEREFORE it is agreed between the proprietors as follows:

1. A Committee is hereby appointed, consisting of the following proprietors:

.... (CHAIRMAN)

.... (TREASURER)

....(SECRETARY)

....

....

(hereinafter called 'the Committee')

Any member disposing of his or her interest in the property may, or if so requested by all the other proprietors shall, resign from the Committee. On the death, incapacity or resignation of a member the remaining members of the Committee may, or if so requested as aforesaid shall, appoint another proprietor to fill the vacancy. The Committee shall keep all the proprietors informed from time to time of the progress of the work and shall exist until the Architect has certified that all the work has been completed, all charges, fees and expenses have been paid and all the proprietors agree that the Committee should be dissolved.

2. The ENTCC having agreed in principle that a grant of £..................... should be recommended towards the probable cost of the work plus VAT, professional fees and expenses (OR AS THE CASE MAY BE) and the proprietors each having signed the Schedule annexed hereto, the Committee is hereby authorized to instruct the Architect, through the Secretary, to prepare tender documents and accept the best tenders from those obtained, provided that the total cost of the proposed work shall not exceed the gross cost stated in the said Schedule and that the recommended grant is confirmed by the Secretary of State. The proprietors agree that such instruction shall bind each one of them to pay his or her share of the cost of the work carried out in consequence, plus professional fees and expenses, up to the amount stated in the said Schedule. After such acceptance, the Committee alone shall deal with the Architect or Contractor in any question involving variation from or addition to the Contract, but otherwise it is understood that the Architect or Contractor may contact the proprietors direct if necessary, for example to make arrangements for access to premises. No addition to the Contract shall be authorized by the Committee without the consent of a majority of the proprietors (notice of the proposed addition

having been given to all the proprietors) and any variation or addition which could increase the cost of the work shall require the consent of all the proprietors. In a vote, each flat shall be entitled to one vote only even though it may have more than one proprietor. Members of the Committee shall be entitled to vote as individual proprietors, and proprietors shall be entitled to appoint proxies to vote on their behalf.

3. The proprietors having agreed with the Architect that the cost of the work shall be apportioned between them equally (OR AS THE CASE MAY BE), each proprietor agrees to pay the Committee, or such agents as may be appointed by the Committee, such sums as the Architect may certify, after taking all possible grants into consideration, as being adequate to meet his or her share of the cost of the work, professional fees and expenses, and it is agreed that the Committee may refuse to allow the work to start until every proprietor has made such payment, either in whole or in part as the Committee shall decide is necessary. The Committee shall place all sums received on Deposit Account with

....

....(NAME OF BANK OR

BUILDING SOCIETY) at...(ADDRESS)
and shall keep accounts which shall be open to inspection by any financially interested parties (including the Contractor) on demand. In the event of the sums paid being insufficient, each proprietor agrees to pay the Committee such further sums as the Architect may certify as being necessary, and any surplus left after payment of professional fees and reasonable expenses shall be refunded in proportion to each proprietor's contribution when the Architect has certified that all work has been completed. If the work is for any reason abandoned or cancelled, the proprietors shall be bound to pay the professional fees and reasonable expenses of the Architect, Contractor and others incurred to the date of and as a result of such abandonment or cancellation.

4. The Committee shall pay the Contractor such sums as the Architect shall certify as being payable after inspection of the work carried out, and payments to any other person or body, other than professional fees and expenses, shall only be made with the consent of a majority of the proprietors. Payments to the contractor shall be made immediately on receipt of the Architect's certificates, usually within fourteen days as required by the Building Contract.

5. It is especially agreed that the Committee shall have no responsibility of any kind relating to the satisfactory completion of the work instructed, and nothing in this Agreement shall prevent the Contractor from recovering sums due direct from the proprietors by process of law if necessary or deny the proprietors any remedy against the Contractor, Architect, Quantity Surveyor or any other person which would have been available to them had they instructed the Architect themselves. Each proprietor agrees to free and relieve the Committee and the other proprietors of any claim whatsoever in respect of work undertaken on his or her own property.

6. Each proprietor hereby undertakes that the rights and obligations hereby created shall, in the event of his or her death before the work is completed and paid for, be due to and a burden upon his or her executors and successors. Further, any proprietor disposing of his or her interest in the property before all his or her liabilities under this Agreement have been discharged shall assign the said liabilities to the new owner and exhibit to the Committee the new owner's signature to an Addendum to this Agreement to that effect, failing which he or she shall remain personally bound.

7. The proprietors' consent to the registration of this Agreement for preservation and execution.

(ALL PROPRIETORS SIGN, writing the words 'Adopted as holograph' before their signatures)

....

....

....

....

....

....

....

....

SCHEDULE showing gross cost, including professional fees and expenses, of recommended work estimated by the Quantity Surveyor on the basis of a tender dated .... apportioned between proprietors according to shares (OR AS THE CASE MAY BE).

PROPERTIES	PROPRIETORS	SHARES	GROSS COSTS
....	....	....	....
....	....	....	....
....	....	....	....
....	....	....	....
....	....	....	....
....	....	....	....
....	....	....	....
....	....	....	....

Total gross cost of work £...........................

ALL PROPRIETORS SIGN, writing the words 'Adopted as holograph' before their signatures)

....

....

....

....

....

....

....

....

NOTE: The ENTCC's style of Agreement has been prepared for use in the New Town. It may be inappropriate for use elsewhere and in particular where Scots Law does not apply, in which circumstances legal advice should certainly be sought.

SCAFFOLDING, HOISTS AND TOWERS

INTRODUCTION

Full scaffolding is required for comprehensive external repairs, reinforced scaffolding for structural support and a mobile hoist (or cherry picker) or light scaffolding for limited repairs or inspections. Towers can provide access to upper floors and roofs. Cherry pickers require road access and parking space beside the building and advance warning must be given to the Department of Highways of Lothian Regional Council. The nature and location of the work will influence the scaffold design. Roof or chimney repairs may require only a tower for access and a hoist for lifting materials; a temporary roof will be necessary if extensive roof work is to be carried out and, if there is a risk of debris falling onto the pavement, a projecting fan of boards (and possibly debris netting) will be required. For further information refer to the following publications:

- *Construction (Working Places) Regulations, 1966;*
- *Health and Safety: Construction (Design and Management) Regulations, 1994;*
- *Health and Safety at Work – Guidance Note GS 15: General access, scaffolds;*
- *Health and Safety at Work – Guidance Note GS 10: Roof works, prevention of falls.*

For the purposes of inspection, maintenance or repair, the Health and Safety at Work Act 1974 requires access that is safe and free from risks to health. Various interpretations can be put on what constitutes safety, but failure to take adequate safety precautions can result in the prosecution of the person responsible. Employers must ensure that the working conditions of employees are safe. Employees should not take risks and put themselves in danger and self-employed people must take responsibility for their own safety and not put others at risk.

SCAFFOLDING

Secure, well-braced scaffolding, fully boarded to give access to each part of the work and with ladders of the correct length, each firmly clamped at the top, provides the best conditions for efficient work. Some contractors may be prepared to do minor work with temporary harnesses and ropes without scaffolding but a substantial anchorage point must be available for such work. Easy and safe access facilitates thorough inspection and good workmanship.

In Edinburgh, under the Roads (Scotland) Act 1984, permission has to be obtained from the Department of Highways of Lothian Regional Council before erecting any scaffolding on a street frontage. The Department of Property Services of the City of Edinburgh District Council should also be consulted about their requirements because a building warrant may be required in some instances.

Scaffolding should be left to specialist sub-contractors but the design, erection and maintenance must be properly coordinated by the architect, contractor and structural engineer. Scaffolding must comply with BS 5973, which gives comprehensive guidance on the construction of access, working and special scaffolds, although there are many proprietary systems which exceed these requirements. To ensure maximum safety and strength, scaffolding should always be erected and maintained according to manufacturers' instructions and, if it is hired, instructions should always be provided and followed closely. In no circumstances should a scaffold be altered without the designer's consent, nor should access scaffolds be used as platforms for the storage of heavy materials such as stone and slate. The Health and Safety Executive require that an F91 book be kept on site and completed after each inspection of the scaffold, for review by their officers. The length and design of battening, ladder position, toe board and other protection should all be such as to allow the work to proceed without danger to the user or the general public. For example, where scaffolding is built off a pavement suitable precautions must be taken to protect pedestrians against injury from scaffold poles.

Loadbearing scaffolds are generally designed by an engineer for a specific purpose, (for example, the propping of a facade during replacement of decayed lintels or unstable stonework). A building warrant will be required and each situation should be discussed with the Department of Property Services.

There are various proprietary types of chutes which help to remove rubbish safely and efficiently from the upper storeys of buildings. These are the responsibility of the contractor and must be adequately secured to the scaffolding.

Scaffolding can give burglars easy access to property. Removing the lowest ladders every night and ensuring that window catches are secure are elementary precautions which should be taken. Thieves can be deterred by burglar alarms installed on the lowest lift and by floodlights on the scaffolding.

Special care should be taken, when erecting or dismantling scaffolding, to avoid damage to any building but this applies particularly to listed buildings, because irreparable damage can be done. Window glass, projecting ledges, cills, cornices, balconies, steps and cast iron are particularly vulnerable to accidental damage. The contract should impose severe penalties on the scaffolding contractor for any damage caused. All carved or ornamental stonework should be protected with timber covers. The ends of putlogs (horizontal scaffold poles) should not touch the surface of the wall and should be covered with plastic caps. The load from the scaffolding should be distributed evenly through timber sole plates at basement or pavement level, first checking that the supporting area can carry all expected loadings, including the self-weight of the scaffolding, without cracking stone slabs underneath. Prior to dismantling a scaffold, the contractor should check that all debris has been removed from the boards.

Scaffolding must be tied to the building. The number, type and position of ties will be determined by the nature and extent of the work. The most common method used in the New Town is a combination of reveal ties and masonry anchors (drilled-in fixings). Reveal tie fixings are achieved by placing telescopic extending tubes across the window openings and wedging them against the window ingoes (reveals). The British Standard allows up to 50 per cent of the fixings to be reveal ties, but every attempt should be made to keep the numbers of this fixing type to a minimum because they may exert excessive loads on the wall. Masonry anchors normally take the form of stainless steel rings, screwed into stainless steel expanding bolts set into the masonry. It is imperative that these anchors are used in accordance with the manufacturer's instructions. At least 10 per cent of all such ties must be load tested and a copy of the test results noted in the hand-over certificate given to the contractor. Masonry anchors can be fitted inconspicuously to all buildings but should never be fitted to lintels, ingoes or cills. When the scaffold is dismantled and the rings have been removed, the bolts should be covered up, preferably with a soft lime mortar or, alternatively, with proprietary plastic plugs which allow fixings to be reused at a future date. In both cases bolts must be recessed back 10 mm from the stone face.

If a building is unoccupied it may be possible to tie scaffolding through open windows by positively secured internal fixings. This method can only be used effectively if the buildings are derelict, and is not recommended for general use.

TOWERS

Towers require as much design as major scaffolds, and this should be carried out by a scaffolding expert or an engineer. They are often used for small-scale repair work or redecoration when fast, easy access is required. For safety reasons, the height of scaffolding towers should not be more than three times the minimum base width (or three-and-a-half times when used indoors). Mobile towers may be used where the ground is level and the tower well secured.

HOISTS

Manual, mechanical or electrically operated hoists are necessary for lifting materials and building equipment; manual pulleys are suitable only for small loads. Whatever type is adopted, the scaffolding should be strengthened to take the extra load and suitably tied into the building. Any loads being hoisted by a jib crane must be controlled by rope from the ground to avoid accidental damage to the load or to the building.

COSTS

The cost of scaffolding on conservation contracts in the New Town in 1993 averaged 5–6 per cent of the total contract sum. The erection and dismantling charge is normally greater than the hire charge alone. This means that purchase, rather than hire, is only justifiable when the scaffolding is likely to remain in position for a long time. It is important, because of its cost, to make good use of the scaffolding while it is in position by carrying out necessary repairs to all areas accessible from it.

TEMPORARY ROOFS

A temporary roof should be considered where repairs to slating, gutters and chimneys are extensive or prolonged. The main object is to protect the fabric of the building from rain but the design of the temporary roof covering will depend on the specific contract and on the configuration of the building.

Usually temporary roofs take the form of metal tube trusses spanning between front and rear scaffolding, with or without intermediate supports. These trusses are then covered simply by reinforced plastic sheeting or by plywood decking covered with mineral felt.

The need for a temporary roof must be considered before tenders are invited because the design of the supporting scaffolding may be affected, and contractors taking account of the extra protection offered may reduce their prices. In 1993, a temporary roof cost between 4 and 7 per cent of the contract sum, but against this sum must be weighed the benefits of avoiding the damage and misery caused by flooding, the full protection of the historic fabric and the benefit to the contractor of uninterrupted work even in adverse weather conditions.

The design of temporary roofs is highly specialized and should be undertaken by a specialist scaffolding contractor or structural engineer, in consultation with the general contractor and the architect. Consideration must be given to the general method of disposal of rain from the temporary roof, which may not be considered by the scaffolder, whose remit is to achieve a structurally sound, wind- and water-resistant temporary cover. There are often shortcomings in a temporary roof at the wallheads and around the chimneys. Props in the central gutters around cupolas and other such locations can inadvertently lead to the blocking up of existing rainwater outlets and so a clear span across the whole roof is preferred.

SURVEY AND REPORT

INTRODUCTION

Historic buildings have to be observed in considerable detail over a period of time to be fully understood. This chapter shows how this requirement can be met by combining careful research with a thorough and systematic inspection of the fabric.

A written inspection report is an essential preliminary to any programme of repairs (and is the basis for an application for conservation grants). It should contain a description of the architecture and the history of the building, a detailed description of the fabric and its condition, recommendations for repairs, preliminary cost estimates and advice on future maintenance, together with annotated sketches and photographs. If it is clear from the outset that a measured survey will become necessary during repair and restoration work the building should be measured during the inspection. An inspection report is not a specification and should not be used as the basis for tenders from contractors.

PREPARATION

Background information

Any useful information about the building should be collected and studied before the inspection is carried out. Sketch drawings, photographs, records and accounts of past alterations, geological, historical, statutory, legal and other information from public sources, architects, lawyers, proprietors and occupants may all contribute to a proper understanding of the building.

Published sources of information on the New Town include:

- Lindsay, I. G., *Georgian Edinburgh*,1948 (Walker, D., revision 1973);
- Royal Commission on Ancient Monuments Scotland, *The City of Edinburgh Inventory*, 1951;
- Youngson, A. J., *The Making of Classical Edinburgh*, 1966 (reprinted 1988);
- Gifford, J., McWilliam C., and Walker D., *The Buildings of Scotland, Edinburgh*, 1984.

These are available in the ENTCC library, as are Historic Scotland's *List of Buildings of Architectural or Historic Interest* and the Edinburgh Architectural Association's photographic survey of the New Town (1970-74). The National Monuments Record of Scotland is a primary source of information on Scottish architectural history, as is the Scottish Records Office and the City Archives which include records of the former Dean of Guild Court in Edinburgh. A detailed summary of available documentary sources is given in the article, 'Source Materials for the Study of Scottish Architectural History' (John Dunbar, *AHSS Journal* No. 12, 1985). Cases of settlement can be studied in the report of the Institute of Geological Sciences (now called the British Geological Survey), *Investigation in Building Subsidence in the New Town of Edinburgh*, 1973, available at the ENTCC library.

Title deeds are usually the best starting point for historical research on a particular building, and these, together with Dean of Guild records, street directories and census returns, were used to investigate the ENTCC's own premises; a paper, in the ENTCC library (by Andrew Kerr WS, February 1979), traces the commercial sub-division of 13-13c Dundas Street, which was originally a town house.

Preliminary assessment

Having studied this background information and after a preliminary examination of the fabric, the architect can use a number of techniques to assess the original design, construction and subsequent development of the building. For an historical analysis of important buildings, the guidance of a professional architectural historian will often be essential, as will the skills of specialist conservators for historic paint, wallpapers, textiles, sculpture, and so on.

Equipment

Overalls and stout shoes	Penknife	Safety helmet	Torch
Clip board and notepad	Spirit level	Measuring tape or rod	Ladder
Pocket tape recorder and tapes	Binoculars	Screwdriver	Boroscope
Camera with flash and film	Plumb bob	Sectional inclinometer	

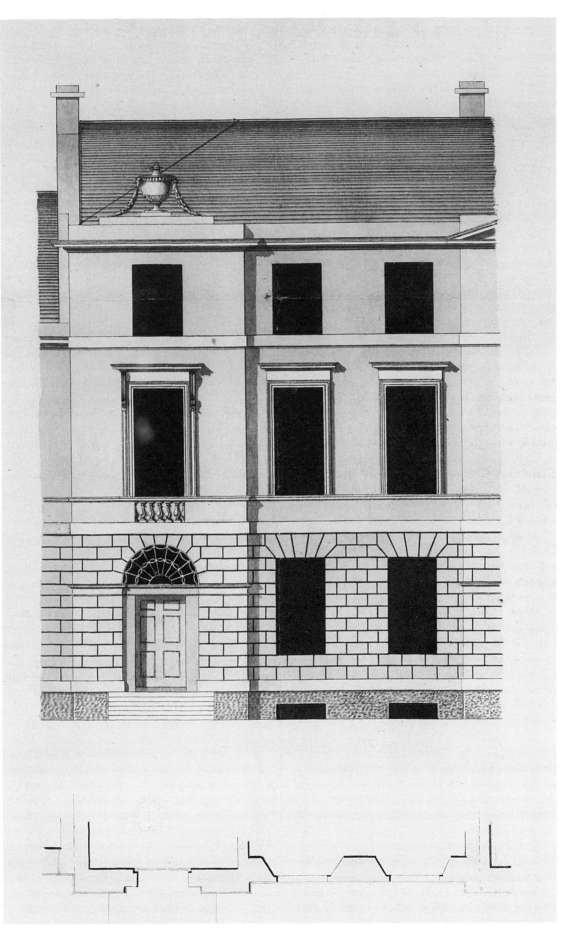

HOUSE IN HERIOT ROW (1802-1808)
DRAWING BY W SIBBALD
ARCHITECT TO THE CITY COUNCIL

41

Photo: Alastair Hunter

Arrangements

Access to all parts of the building should be arranged in advance, particularly when it is necessary to have tradespeople in attendance. Minor maintenance, such as the cleaning of gutters, can sometimes be carried out at the same time as the inspection. Roof inspections or surveys of derelict buildings can be hazardous and for safety should be carried out by at least two people. Activity in large empty buildings can arouse suspicions and it is wise to notify the police before starting a survey. Both the police and the Department of Highways must be notified if a mobile hoist is to be used for closer inspection. In Edinburgh a permit must be obtained from the Department of Highways if a hoist is to be used on public land, because traffic and pedestrian flow or parking bays can be affected (see SCAFFOLDING, HOISTS AND TOWERS).

THE INSPECTION

A standard and logical route should be followed through and around the building. Materials, construction and condition should all be noted and special reference made to defects. Care should be taken to think about the building as a whole and to consider whether individual defects are symptoms of less apparent but potentially more serious problems. External Survey Fabric forms are available from the ENTCC.

Particular attention should be paid to signs of structural movement, dampness, water penetration and condensation, timber decay and insect attack. All accessible parts of the building should be examined, not forgetting roof spaces, basement areas, garden and retaining walls, and railings. Note should be taken of the condition of stonework, pointing, roof coverings, gutters and flashings, downpipes, joinery, paintwork and services.

Never try to be too ambitious; do not inspect dangerous areas like outer roof pitches without adequate security. Closer inspection from vantage points such as roofs and upper windows can be better than viewing through binoculars from street level, but may require ladders, a safety harness, a mobile hoist or temporary scaffolding.

Fitted carpets, furniture or fittings may limit inspections and suspicious signs should be checked by opening up (during the inspection or at a later date). In any case, the report should state the limitations of the inspection, and the architect should disclaim liability for defects in inaccessible places which were not examined, recommending further investigation where appropriate.

The tendency for architects and surveyors to leave the inspection of timber or stonework to specialist contractors is to be discouraged. A report from such a firm does not absolve the architect from professional responsibility and a personal inspection of roof timbers, exposed joists and wooden lintels should always be made. A professional examination for structural defects and dry rot is a requirement of ENTCC grant-aided schemes, and the architect should recommend that a consultant structural engineer investigate the causes of any serious structural movement.

MEASURED SURVEY

Sketch survey drawings may sometimes be useful, but they should be clearly marked as being not to scale. Any measured survey drawings should, above all, be accurate. If crucial dimensions are not to be forgotten, the measuring should be done systematically; plans should be recorded with running dimensions, taken in a clockwise direction, noting the diagonals, wall thicknesses, through sizes, heights and joist directions, centres and spans. When drawing elevations, photogrammetry can be very useful for large and complicated structures. For more simple buildings, rectified photographs, traced or developed on film paper, annotated and dyeline printed, can supplement elevational drawings.

Parts of the building which will have to be demolished and rebuilt should first be accurately measured and photographed. For instance, when chimney stacks are severely decayed or inappropriately altered, it is essential to record course heights of stacks relative to the roof ridge and to each other. The cope profile should be carefully measured and authenticated in relation to the building and its neighbours. Detail drawings of these and other features requiring renewal should be prepared.

THE REPORT

Before the report is written, an overall assessment of the fabric should be made, conclusions drawn and the repair strategy worked out in draft. A quantity surveyor is normally consulted about the approximate costs of the recommended repairs. The report itself is a working document, and should be clearly and concisely written.

The contents of the report might be as follows:

1 *Introduction:*
   - The purpose of the report;
   - The scope and limitation of the report: this section should explain how the survey was carried out, listing those areas which could not be inspected, and stating its limitations;
   - Summary of conclusions: this should be an assessment of the condition with a brief list of recommended repairs and estimated costs, including fees and VAT.

11 HERIOT ROW IN 1993
43
Photo: The Royal Commission on the Ancient
and Historical Monuments of Scotland

2 *Background information:* An architectural and historical assessment of the building evaluating the original and any later features; additions may have their own artistic or architectural merits, and they are part of the history of the building. Reference should be made to the geology of the area and any adjoining properties. The uses of all parts of the building and site should be stated, and any previous or historical uses, while any unusual aspects of title should be mentioned. Details should be given if a building is listed or within a Conservation Area, if it is within a Housing Action Area, or if it accommodates any non-conforming use (in planning terms).

3 *Detailed account of the fabric:* This should be a completely factual record of what is observed and should form the main section of the report upon which the recommendations are based. Construction and finishes should be described and the condition of individual elements and any particular defects noted. The written sequence should follow that of the inspection, normally the roof, the external walls and lastly the interior (room by room) and services.

4 *Detailed recommendations for repairs:* The aim of a conservation scheme is to stabilize the condition of an historic building and to put it into good order, while balancing economic repair, the conservation of the existing fabric (whether original or secondary) and the enhancement of its appearance. Bearing this in mind, the repairs should be placed in four categories:
- Immediate work: required immediately;
- Urgent work: required within one year;
- Essential work: required within one to five years;
- Desirable work: required for enhancement but not structurally necessary at present.

A draft programme of work with estimated costs for each phase will help the owner(s) to assess their long-term commitments. Much skill goes into balancing various factors, such as the sensible use of scaffolding, the requirements of grant-aiding bodies, the grouping of work into logical building sequences and the estimated life of existing finishes.

5 *Recommendations for future maintenance:* This part of the report should outline the normal routine work which needs to be carried out in order to maintain the building on a plateau of good repair. It should be the foundation for a maintenance log book, making recommendations for work in annual and quinquennial cycles, with an estimated cost per annum. Recommendations for improved access, safety and lighting should be made, so that future inspections will be easier, safer and more thorough.

6 *Appendices:* These should be clearly indexed and, where necessary, annotated.

7 *Acknowledgements and sources of information:* It is important that others reading the document should know the source of the information.

GEOLOGY AND TOPOGRAPHY

The construction of the New Town involved the building of vaulted cellars on the original ground surface, followed by excavation for foundations for the houses with any surplus material being dumped in the space which would become the road. For this reason, most roads in the New Town are some three metres or more above the original ground surface and the natural land form is often obscured.

Many roads constructed along the sides of slopes were built up with a steep embankment on the downhill side, notably Princes Street, Queen Street, Douglas Crescent, Lennox Street, Royal Terrace and Regent Terrace. Other roads are built on embankments across hollows such as The Mound and Clarence Street while others form ramps as in Blenheim Place. Unrecorded made-up ground is undoubtedly widespread.

The original ground surface of the New Town was mainly shaped by ice flowing from west-south-west to east-north-east imparting a strong topographic grain of drumlins (elongated ridges) with tapering crag-and-tail features on the high ground and deep hollows scoured out in the low ground. Later transgression by the sea to 40 m O.D., the flooding of hollows by lochs, and the cutting of valleys by rivers and streams all produced level flats or gentle slopes.

Central and southern parts of the New Town lie on drumlins of glacial till, also known as boulder clay. This deposit is a firm, well-consolidated, silty clay full of pebbles, cobbles and boulders which give good foundations. In these areas bedrock lies within a few metres of the surface and has frequently been reached by foundations, as in the Princes Street/George Street area.

By contrast, the northern part of the New Town is underlain by raised beach deposits, soft, unconsolidated silt, clay, sand or gravel, resting on the glacial till, which give poorer foundation conditions. In this area bedrock can be at depths of 5 m, 10 m or more. The southern edge of this deposit has been obscured by building but is around the 35 m contour.

Princes Street Gardens mark the site of the Nor' Loch, a deep hollow gouged out by ice and filled to a depth of 50 m or more by lake alluvium of boulders, gravel, sand, silt, clay and peat. The original flat lake bed has been obscured by embankments and cuttings. Similar loch sites and deposits occur at Silvermills and Bellevue.

The flat areas alongside the Water of Leith are formed by river alluvium consisting of unconsolidated gravel, sand, silt and clay; these areas have a high water table.

Raised beach deposits, river and lake alluvium are porous and low-lying, and the latter two are liable to be saturated with a high water table. Where the glacial till forms an impervious blanket, the water table is not usually a problem.

Bedrock is mostly concealed by surface deposits except on the Calton Hill and in the incised gorge of the Water of Leith, where it is exposed. The rocks underlying the New Town are all Carboniferous, more than 300 million years old. Both sedimentary rocks such as sandstone and shale, and igneous rocks such as basalt lavas and tuffs and dolerite intrusions, are present there.

The Craigleith Sandstone and the Ravelston Sandstone, each 30 m or more thick, underlie large areas of the New Town and form outcrops in the Water of Leith. These hard, fine-grained sandstones provided excellent building stone which was much used in the New Town, being brought in from outlying quarries.

Interbedded mudstones, siltstones and shales with thin coals and limestones alternate with the sandstones. These can be seen on the banks of the Water of Leith gorge and form the tail of Calton Hill.

Calton Hill owes its prominence to the hard basalt lavas which form its craggy outcrops. Interbedded with the lavas are softer beds of volcanic ash (tuff). Intrusive sills and dykes of dolerite also occur but do not form features.

Several geological faults are known to cross the New Town. The Colinton Fault and the Leith Links Fault are large well-defined faults. Smaller faults, as proved in the Scotland Street Tunnel and other excavations, are likely to be only a proportion of the faults cutting the rocks and the exact position of even the best-known faults is mostly a matter of conjecture. Geological faults are associated with fractured strata, and rapid changes of rock across the faults.

The area is one of low seismic risk. There is no recorded evidence for geological faults moving in the area in recent times, although the earthquake of January 1889 has been attributed to the Colinton Fault. The Carlisle earthquake of December 1979 was felt in Edinburgh, but no significant damage was recorded.

Prospectus Civitatis EDINBURGENÆ a prædio DEAN dicto . The Prospect of EDINBURGH from ÿ DEAN.

2

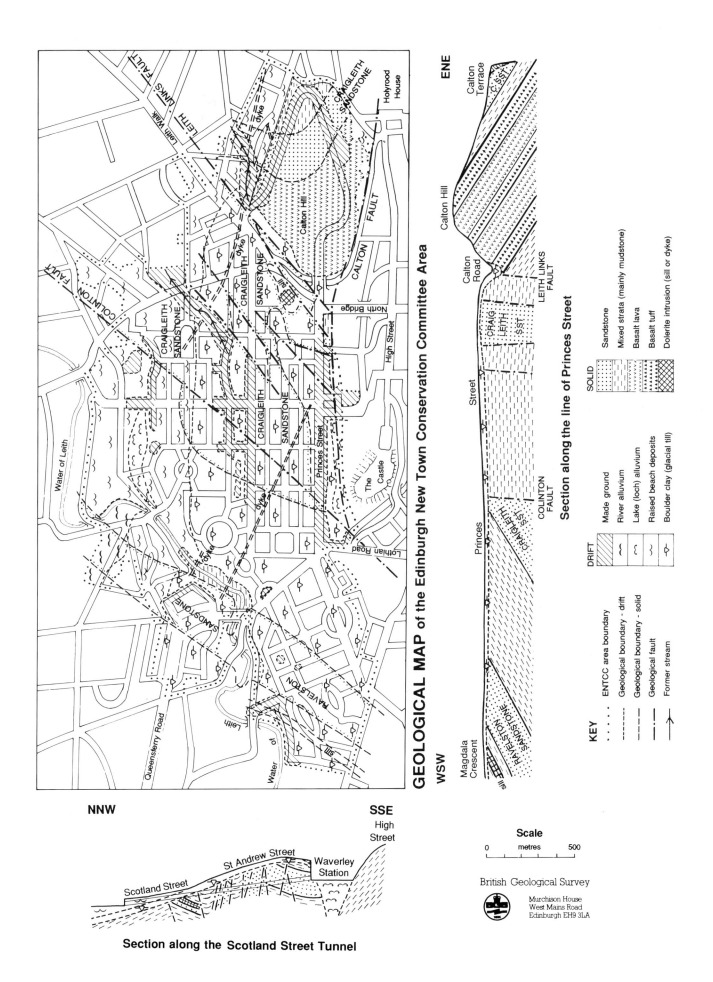

GEOLOGICAL MAP of the Edinburgh New Town Conservation Committee Area

Section along the line of Princes Street

Section along the Scotland Street Tunnel

KEY

⋯⋯ ENTCC area boundary

– – – Geological boundary - drift

– · – Geological boundary - solid

— · — Geological fault

→ Former stream

DRIFT

Made ground

River alluvium

Lake (loch) alluvium

Raised beach deposits

Boulder clay (glacial till)

SOLID

Sandstone

Mixed strata (mainly mudstone)

Basalt lava

Basalt tuff

Dolerite intrusion (sill or dyke)

Scale

0 ————— 500
metres

British Geological Survey

Murchison House
West Mains Road
Edinburgh EH9 3LA

47

Undermining for minerals is not a problem in the New Town, although it is possible that the thin Wardie coal was wrought in the gorge banks of the Water of Leith. The Scotland Street Tunnel underlies a section of the New Town; sewers and other tunnels must also exist and may exert their influence on the surface.

A report on the known cases of settlement, together with a geological map, may be studied at the offices of the ENTCC. These were prepared in 1973 with the help of the Department of Technical Services of the City of Edinburgh District Council and the British Geological Survey in Edinburgh, both of whom will advise on specific problems.

The majority of cases of recorded settlement occur within the northern area, although even here the incidence of settlement is remarkably low. Evidence of past movement does not necessarily imply a continuing structural problem, and a detailed report prepared by a structural engineer should allay the fears of most building society valuation surveyors.

FOUNDATIONS

GENERAL

The walls in the New Town are built on flat stones laid in shallow trenches sometimes less than 600 mm below the lowest floor level. These padstones, about 250 mm thick, are only slightly wider than the walls which they support and although not in accordance with modern practice, these traditional foundations have performed well.

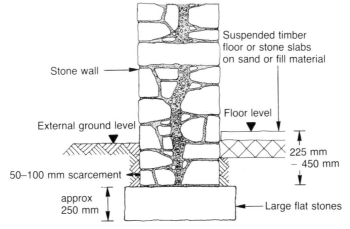

CROSS-SECTION THROUGH FOUNDATION

The foundations bear on the underlying clay and gravel which cover the bedrock in the New Town area but in some cases the foundations were laid on top of alluvial deposits, or on filled material used to level undulations or raise the finished ground level.

All subsoils, particularly filled material or alluvial deposits, tend to compress under load. When a building straddles a mixture of materials with varying bearing capacities, differential settlement will occur; this results in distortions of the superstructure.

The cracks seen in many buildings in the New Town are often the result of initial settlement which ceased when a state of equilibrium was reached. As such, they do not necessarily imply continuing structural instability.

Changes in the loadbearing capacity of the subsoils can result from saturation by underground streams, burst water mains and blocked drains, from the lowering of the water table due to pumping, tree roots, dry seasons, and from traffic vibration and excavation on adjoining sites. Any of these can cause foundation movement to occur or to be renewed.

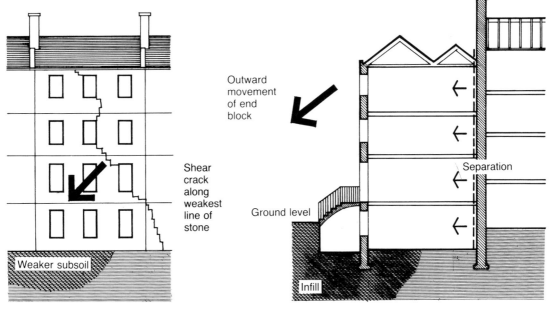

SETTLEMENT AT BROUGHTON PLACE   50

DIAGNOSIS

The interpretation of cracks and movements due to settlement of the foundations requires skill and experience. It is recommended that a structural engineer with knowledge of the locality and experience of old buildings should be employed. The engineer should be able to determine the likely causes of damage, to advise on whether the movement is progressive or cyclical, and to recommend any necessary remedial works. If movement has ceased, these remedial works should be confined to repair and reinstatement of individual stones to prevent further local decay.

Detailed inspections at intervals of several months may be necessary in order to establish whether structural movement is active or not. Trial pits may have to be dug to examine the construction of foundations, the condition of drains and so on, and samples may have to be taken from bore holes to establish the nature of the underlying strata. If defective foundations are suspected, these investigations are essential before any major repairs are carried out to the superstructure. Comparatively speaking, the costs of investigations are usually low in relation to any remedial works which may be required.

Building societies are naturally reluctant to lend on properties which are distorted or cracked, but they are sometimes prepared to do so if remedial work has been properly specified and supervised by a structural engineer, or if an engineer is willing to state that in his or her opinion the structure is stable.

The definition of structural stability should accept the possibility of further slight hair cracking (which the building may have suffered historically throughout the whole of its life), provided that this movement will not lead to collapse or severe structural damage. If such a definition is not acceptable, and if firm guarantees are required by building societies, the structural engineer may be driven into excessively elaborate solutions extending beyond the damage in question and absorbing a disproportionate amount of finance. The ENTCC requires investigations by experienced consultants before offering grants for major structural underpinning or piling.

REPAIR

The pattern of foundation movements can be complex; the nature of the underlying strata, the construction of the foundation and the pattern of settlement must all be taken into account before repairs are undertaken. Achieving overall stability and keeping the weather out is more important than trying to arrest deep-seated but minor movement. It is often wiser and more economical to accept a distorted appearance and to create flexible joints which can accommodate further movement. It is important to determine the relationship of the building to adjacent properties. It may be that halting the movement of one property in a terrace could lead to subsequent damage to adjacent properties as they continue to settle in relation to their newly stabilized neighbour. Cracking patterns often relate to differential settlement between properties. Underpinning or stabilizing work may not be possible on one part of a terrace alone. The creation of a hard spot would perpetuate differential effects, and the treatment of walls, such as mutual gables, may not be possible due to ownership problems. The solution can be a compromise to allow stability without rigidity and to accept small movement in the knowledge that this will allow preservation with minimum intervention.

The Building Control Division of the Property Services Department and a structural engineer should be consulted before undertaking any repairs. The repair can be carried out from the inside, outside or both sides, depending on specific requirements. The following are a few examples of possible repair techniques:

Underpinning

If there is sound loadbearing material a short distance below the foundations, any intermediate defective material such as decayed vegetation and rubbish can be dug out and replaced by mass concrete or brickwork. Excavation near foundations is expensive and can weaken the superstructure unless undertaken in short lengths of about 1.5 m at a time. If the defective area is extensive but shallow it may be possible to support the wall temporarily on steel beams, or needles, spanning between jacks off sound material on each side. All work must comply with the BS 8004: *Code of Practice for Foundations*. A building warrant is necessary and work should only be carried out under the supervision of a chartered structural engineer.

Piling

This method is expensive and rare in the New Town. Vibration from pile-drivers may cause movement in the superstructure; for this reason bored piles are preferred. There are a number of proprietary systems suitable for remedial work to traditional buildings which all require to be specified and supervised by technical experts.

Grouting

Movement of granular substrata can be arrested by pumping cement or PFA (pulverized fuel ash) grout under pressure through deep holes bored at frequent intervals around the defective foundations. When solidified, the mixture of grout and granular materials spreads the load from the building over a wider area, thus reducing the local pressure.

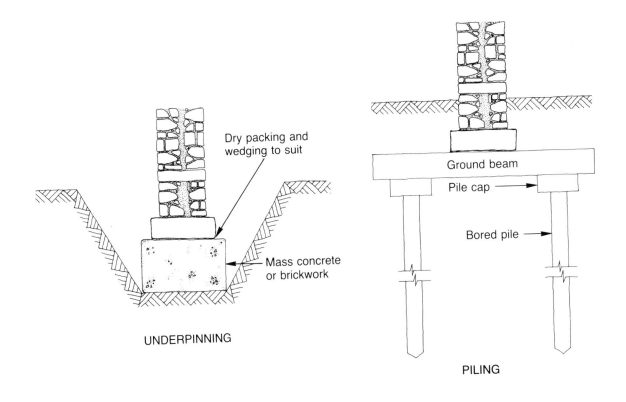

Dry packing and wedging to suit

Mass concrete or brickwork

UNDERPINNING

Ground beam

Pile cap

Bored pile

PILING

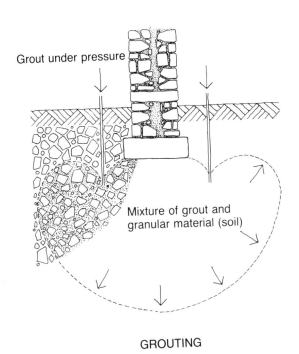

Grout under pressure

Mixture of grout and granular material (soil)

GROUTING

STONES OF THE NEW TOWN

The Edinburgh area was once well supplied with local quarries but most of them have now been infilled and built over. Areas of Upper Old Red Sandstone lie under the south of the Old Town, whereas to the north and west, and also in Fife and West Lothian, outcrops of lower carboniferous sandstone can be found. Much of the stone used for the Old Town came from quarries in the area now occupied by the University of Edinburgh. Sandstone for rubble masonry and paving was quarried on Salisbury Crags (until 1830), from the site now occupied by Waverley Station, and from quarries at Broughton and Easter Road and in the Dean Village area.

For dressed stone on prestigious buildings carboniferous sandstone was preferred. During the sixteenth century this was supplied by quarries at Cramond, Ravelston, South Queensferry and Culross, but by the early seventeenth century the hard and finely grained siliceous sandstone from Craigleith was available. Described as 'well nigh imperishable', Craigleith was for long the principal stone used for both private and public buildings in the city and was regarded as one of the most durable stones in Britain. The site of the old Craigleith Quarry is located about 2 miles north-west of the city centre. It was closed down at the beginning of this century and, although reopened for a short time between the wars, it could no longer supply materials fit for ashlar work. It was impossible to extend the quarry because by this time it was surrounded by housing. The quarry has now been infilled and is topped by a supermarket.

The following extract concerning the Craigleith quarry is taken from *Building Stones used in Edinburgh, 1892:*

Geological position	In the middle portion of the Lower Carboniferous series, below the base of the upper or oil-shale group.
Chemical analysis	It consists of quartz grains united by siliceous cement with small plates of mica.
	Silica – 98.3 per cent
	Carbonate of lime – 1.1 per cent
	Iron and alumina – 0.6 per cent
	Density – 2.22
	Specific gravity – 2.45
	Absorption of water – 3.61
	Crushing load – 861.9 tons/ft^2 (943 Kgf/cm^2)
	Weight – 138.6 lb/ft^3 (2220.37 Kg/m^3)
Tint	Grey
Texture	Very hard and close
Weathering	All the buildings of Craigleith are practically unaltered by the weather, and the polished faces retain their original gloss and clear appearance.

With the development of the New Town in the late eighteenth century there was an enormous increase in the demand for high quality stone and the famed Craigleith Quarry was worked with great intensity. Register House (1776-1826), St Andrew's Church in George Street (1785), the Old College of the University (1786-1834), St George's Church (1812), the National Monument on Calton Hill (1826), Dean Bridge (1832) and the whole of Charlotte Square (1791) are among the numerous buildings in Edinburgh constructed of Craigleith stone.

Craigleith was also used further afield for the Bank of England (1770), the British Museum (1828) and parts of Buckingham Palace (1825-35). It was one of the many stones considered by the Parliamentary Enquiry of 1839 into stones suitable for use in the new Houses of Parliament. The report stated the cost of Craigleith stone (including delivery by sea from Granton Harbour to London) as being one shilling and tenpence per cubic foot for bedrock, up to three shillings and sevenpence per cubic foot for liver-rock (i.e. between £3.30 and £6.30 per cubic metre). The best beds at Craigleith could provide blocks of huge length as is seen in the columns of the University Old College. After two centuries of exposure the quality of Craigleith sandstone can still be seen clearly in the sharp arrises and mouldings of many New Town buildings.

A number of other Edinburgh stones are described in *British and Foreign Building Stones* by John Watson (1911). These are all fit for structural purposes and were of the Barnton-quarried Craigleith-type stone (as opposed to Craigleith itself) and the blue, pink and white varieties of Hailes stone from Slateford. The so-called Craigleith stone from Barnton was used for building the Imperial Institute in London (1880), Blue Hailes for the Church of Scotland General Assembly Hall in Edinburgh (1846) and Pink Hailes for the Royal Infirmary in Edinburgh (1875).

Less durable stones – for instance from Culallo and Dalgety in Fife, from Ravelston, Murrayfield and Redhall in Edinburgh and from Binney in West Lothian – were often used for the sake of economy, until the middle of the nineteenth century, when rail transport encouraged the use of more easily quarried sandstones from Dumfriesshire and the north of England.

In conclusion, therefore, it can be said that the New Town is constructed of a wide variety of types of stone, of which Craigleith is predominant. It is important to identify the stone from individual buildings before carrying out repairs. There are several commercial laboratories and some university departments which undertake the identification and testing of samples.

STONES FOR REPAIR WORK

Today there are no quarries operating in the Edinburgh area. Stone for repairs in the New Town is now purchased from quarries in County Durham, Northumberland, Yorkshire, Derbyshire and Morayshire. These all have different characteristics from the original stones and it is important to get as close a match as possible for repairs to old buildings.

Stone tests

Sandstone for structural purposes is best selected on the basis of its previous reputation and appearance. However, the properties of stone from different beds in the same quarry may vary considerably. Where this data is not available certain scientific tests are usually carried out to try to assess the comparative durability of the stone:

1   *Acid immersion test:* By immersing stone in dilute sulphuric acid for ten days, it is possible to distinguish between limestone and sandstone and between calcareous and siliceous sandstones. Those sandstones which are not affected by the test can be regarded as resistant to acidic air pollution. Calcareous sandstones have a high proportion of calcium carbonate which is dissolved by sulphuric acid and therefore are less resistant to decay in polluted areas. Calcareous sandstones should not be built adjacent to siliceous sandstones because of the chemical reaction which accelerates decay in the latter.

2   *Sodium sulphate crystallization test:* The acid immersion test does not give any indication of the stone's resistance to salt crystallization, which is a major cause of stone decay. This test is the only method used for assessing the weather resistance of a stone. Samples are subjected to about fifteen cycles of immersion in salt solution, followed by drying in an oven; the ability to resist damage after immersion in a 14 per cent solution indicates a stone of exceptional durability. This test is particularly relevant to sandstone and should be applied whenever a stone is required to withstand severe conditions of exposure or is likely to become contaminated with soluble salts.

   The main disadvantage of the crystallization test is that it is time-consuming and rather expensive. However, because the resistance of stone to crystallization damage is closely related to its pore structure it is also possible to obtain an indication of durability by investigating the rate of water absorption, porosity and pore size distribution.

3   *Water absorption porosity and saturation coefficient:* Water absorption is defined as the volume of water absorbed in twenty-four hours when the sample is completely immersed. It is expressed as a percentage of the total volume of the sample. Porosity is defined as a volume of the stone's pore space expressed as a percentage of the stone's total volume; it is most conveniently measured by saturation with water in a vacuum. The ratio of water absorption to porosity is called the saturation coefficient, the value of which can range from about 0.4 to 0.95. If the value is low, the stone is generally very durable, containing adequate space for crystallization growth within its pores without disruption of the matrix. Unfortunately, within the middle range of values the saturation coefficient is an unreliable guide to durability and an assessment of the distribution of pore sizes in a stone is more useful.

4   *Pore size distribution:* If there is a high proportion of fine pores within a stone (i.e. high microporosity) its resistance to crystallization damage will be low. This is because salt enters the pore and on crystallization expands so that it is unable to be released. However, because most sandstones contain a high proportion of coarse pores (i.e. predominantly macroporous) their resistance to crystallization damage is generally very good. Various techniques are available for determining the distribution of pore sizes in stone. The water and air permeability test can be made on portable equipment which permits the measurement of a durability factor. The test is rapid, cheap and can be performed *in situ*.

Results of tests on Craigleith stone

Samples of Craigleith stone have been subjected to tests; the results listed below serve as criteria against which test results on new stones can be compared:

1 *Acid immersion test:* In general, surfaces showed no sign of corrosion or softening after ten days. Thicker carbon-rich layers of impurities exhibited clear signs of acid pitting, although loss in weight was negligible.

2 *Sodium sulphate crystallization test* (fifteen cycles; saturated solution; soaking temperature 20°C): This test was applied to five samples, of which three split due to flaws in the stone. All specimens registered a loss in weight, which varied from 20–26 per cent for non-split samples up to 32–35 per cent for split samples after the removal of flakes. On the whole, Craigleith responded well to the test, suffering slow, even and progressive surface disintegration with accelerated erosion around concentrations of impurities.

3 *Water absorption, porosity and saturation coefficient:*

	Test sample A %	Test sample B %
Water absorption	6.81	6.79
Porosity	13.84	13.18
Saturation coefficient	0.492	0.515

In both samples the low value of the saturation coefficient (below 0.55) indicates a stone of high durability.

New stones for repair work in the New Town

Each stone is different and quality and appearance may vary from year to year. It is therefore important to have up-to-date samples. Certain stones are better for certain locations and purposes and selection is all important. Below is a list of stones which are often used for repair work in the New Town of Edinburgh:

- *Blaxter:* Blaxter quarry, Elsdon, near Otterburn, Northumberland
  quarried by Dunhouse Quarry Ltd, Bishop Auckland, Co. Durham;
- *Cat Castle:* Cat Castle quarry, Deepdale, Lartingon
  quarried by Dunhouse Quarry Ltd, Bishop Auckland, Co. Durham;
- *Clashach:* Hopeman quarry, near Elgin, Morayshire
  quarried by Moray Stone Cutters, Elgin, Morayshire;
- *Crosland Moor:* Crosland Hill quarry, near Huddersfield, West Yorkshire
  quarried by Johnson's Wellfield Quarries Ltd, Huddersfield, West Yorkshire;
- *Darney:* Hexham quarry, Northumberland
  quarried by Natural Stone Products, Springwell, Newcastle-upon-Tyne;
- *Doddington:* Wooler quarry, Northumberland
  quarried by Natural Stone Products, Springwell, Newcastle-upon-Tyne;
- *Dunhouse:* Dunhouse quarry, Staindrop, Co. Durham
  quarried by Dunhouse Quarry Ltd, Bishop Auckland, Co. Durham;
- *Greenbrae:* Hopeman quarry, near Elgin, Morayshire
  quarried by Moray Stone Cutters, Elgin, Morayshire;
- *Springwell:* Springwell quarry, Gateshead, Sunderland
  quarried by Natural Stone Products, Springwell, Newcastle-upon-Tyne;
- *Stainton:* Stainton quarry, near Barnard Castle, Co. Durham
  quarried by Natural Stone Products, Springwell, Newcastle-upon-Tyne;
- *Stancliffe:* Stancliffe quarry, Darley Dale, near Matlock, Derbyshire
  quarried by Realstone Ltd, Darley Dale, near Matlock, Derbyshire;
- *Stanton Moor:* Stanton Moor quarry, near Matlock, Derbyshire
  quarried by Dimensional Stone Ltd, Matlock, Derbyshire and
  by Realstone Ltd, Darley Dale, near Matlock, Derbyshire;
- *Woodkirk, Brown York Stone:* Britannia quarry, Morley, Leeds
  quarried by Pawson Bros. Ltd, Morley, Leeds.

This list is not exhaustive and other stones may become available.

Second-hand stone

Stone is occasionally rescued from buildings in Edinburgh and can be useful in repairs. When used for repair work salvaged stone can match existing stone not only in colour and texture but also in durability. It may also display the same amount of weathering and will therefore blend in more readily. It is quite feasible to recut and redress a second-hand stone to suit its new home although the overall cost of using old stone can exceed that of new. The choice is between a good matching old stone with established and predictable weathering characteristics and a new stone which will not match and the durability of which can only be tested by time.

STONE DECAY AND DEFECTS

INTRODUCTION

Building stones, like all natural rocks, start to break down into their constituent parts as soon as they are exposed to the weather; the rate of degeneration will vary from stone to stone. It may be possible to delay decomposition if the processes of decay are understood and remedial action taken. This chapter, which deals only with sandstone, summarizes a large, complicated subject and further reading is recommended. Various studies are taking place at present which could change our understanding of this subject!

CAUSES OF DECAY

The decay of sandstone can usually be attributed to a combination of causes. The processes of decay – physical, chemical and organic – often work in conjunction and their destructive actions can be exacerbated or even precipitated by defective materials, careless workmanship or poor design.

Physical attack includes the continuous cycle of wetting and drying, expansion of water on freezing, thermal stresses and occasionally, but most aggressively, wind and wind-borne solids.

Building stone may be chemically contaminated by soluble salts as a result of poor construction or poor storage on site. Inadequate damp-proofing can allow salts from the soil to be carried in solution into the stone and this can be exacerbated by the salting of pavements in the winter. Other harmful chemicals may result from brick or concrete backings to masonry, unsuitable cleaning agents and dog urine.

Damage due to weeds, tree roots, lichens and other organic growths may also have chemical and physical effects on the stone.

Physical failure

The effects of wind erosion and thermal stresses on sandstone are relatively insignificant especially on well-detailed buildings and frost is rarely found to be the direct cause of deterioration even when stones have been wrongly bedded. Frost is only likely to cause damage where buildings are excessively wet before freezing. Even so, frost may give the final push to dislodge pieces of stone that have already been loosened by some other means. Great damage can be caused if water penetrates open masonry joints and expands under freezing conditions. It is also known that continuous cycles of wetting and drying have a harmful effect because soluble salts are deposited in the pores of the sandstone.

Stone is liable to suffer from inherent weaknesses such as cleavage planes, soft beds, vents, shakes and dries. Cleavage describes the tendency for a rock to split into parallel layers irrespective of its original geological structure. Bedding planes can be entirely destroyed by cleavage, although this is usually more of a problem in slate than in sandstone. Beds of varying structure and composition are sometimes contained within sedimentary rocks but these are often overlooked during quarrying. When exposed to the weather, the softer seams within a stone will erode more rapidly than the harder ones. In ashlar work, this differential weathering may be of little importance, provided that the stone is correctly bedded, but in projecting features, the decay of softer seams may have more serious consequences. Vents, shakes and dries are minute fissures in the stone, probably caused by movements in the earth. Where these fissures coincide with ornamental or moulded detail on a building they can be the source of considerable trouble. Iron pockets can cause discolouration and splitting in certain sandstones.

Apart from these natural defects, stones can also suffer from damage caused by careless quarrying or clumsy dressing. The cracks and bruises which result may affect the durability of stone as well as its appearance.

It is now becoming evident that certain sandstones contain volcanic clays. These were laid down as hard particles as the stone was formed but, on exposure, degenerate to liquid, expanding in volume at the same time. The result is a breakdown in the matrix of the stone in the area which receives the maximum wetting and drying, resulting in the formation of a contour scale (i.e. one which follows the contours of the stone) which becomes detached and falls off.

56

Chemical failure

Crystallization damage is closely associated with the size and distribution of pores which determine the capillary properties of the stone and the rate of evaporation from its surface. The resistance of stone to this damage decreases as the proportion of fine pores increases. A stone with high microporosity will attract more water through capillary action than a stone of low microporosity (in this respect the macroporous structure of Craigleith stone makes it one of the most durable sandstones, although it is by no means immune from crystallization damage). Crystallization of soluble salts occurs as the solvent (water) evaporates; salts are deposited in the surface pores where evaporation is rapid (due to sunlight, high ambient temperatures or wind movement). Crystallization produces changes in molecular volume; the accumulation of salts, coinciding with frequent crystallization cycles, can generate sufficient pressure within the pores of the stone to cause serious damage. Thus the ability to resist crystallization damage is one of the properties which should be checked when choosing a new stone for repair work.

Stone is affected by polluted atmospheres; harmful gases are emitted from power stations, factories and motor vehicles, but may also be produced within the building fabric, for instance by coal or gas fires which weaken the linings of the chimneys. According to information from the Environmental Health Department, the level of air pollution in Edinburgh has been reduced significantly in recent years, the mean concentration of sulphur dioxide dropping from more than 75 microgrammes per cubic metre in 1975/76 to 50 in 1983/84 and to a current level of about 40. This is set to be reduced again with the banning of non-smokeless coal-burning appliances throughout the city in 1995.

Soluble salts may result from the interaction of ammonia and sulphur dioxide in the atmosphere, but they are also derived from the interaction between sulphur dioxide, lime and other alkalis in the mortar. A significant cause of stone decay is the crystallization of these salts within the pores of the stone. Siliceous sandstone, in which the grains of sand are cemented together with silica, as used widely in the New Town, is not very susceptible to the direct attack of atmospheric pollutants – either gaseous or solid – and the thin film of insoluble dirt which forms on exposed surfaces is generally regarded as harmless. However, repeated wetting and drying gradually dissolves the silica, which is then carried in solution towards the outside of the stone. This moisture cycle can result in a hardening of the face and loss of cohesion behind it, so that the outer layer of stone falls off in a contour scale.

Calcareous sandstone, in which the grains of sand are cemented together by calcium carbonate, is susceptible to attack by sulphur dioxide and the matrix of this stone can be destroyed by sulphur dioxide in the atmosphere.

Limestones and sandstones must not be built into the same wall and nor should calcareous and siliceous sandstones be mixed.

Organic failure

Organic attack on masonry involves both physical and chemical processes and is due chiefly to visible vegetation growth. Weeds, small shrubs and ivy tendrils growing in the joints can extend their roots deep into the core of the wall, forcing the masonry apart and opening pathways for moisture. Some mosses, lichens and algaes secrete harmful acids which attack the stones. Like other forms of vegetation, they tend to encourage water retention and their presence usually indicates abnormally moist and potentially destructive conditions. Some microorganisms can form skins on the surface of the stone, causing soft pockets to develop behind.

Other factors

Sandstone, being laminar, must be correctly bedded. Incorrectly bedded stone, i.e. with the bedding planes laid parallel to the exposed surface, will always be prone to disintegration. Within a wrongly bedded stone, the natural movement of laminations – originally held under great pressure but unrestrained once removed from the quarry bed – will tend to cause the exposed layers to exfoliate. This action is accelerated by the processes of decay already described. It can be difficult to identify the natural bed of some stones after they have been dressed or sawn, although the presence of mica particles in a stone may be useful in identifying the bed, because they lie horizontally and shine when exposed. The quarry should always be instructed to mark the natural bed on all stones so that they can be identified by the mason and architect, to ensure correct building on site.

With hollow beds, the risk of spalling at the edges is increased as pressure is concentrated at the face of the stone; in serious cases it may be necessary to remove the whole stone and indent a new block. If the effects of spalling at joints are insignificant and the stones are basically sound, considered neglect is preferred to over-zealous repair. When such spalled stones are pointed, it is essential that the mortar is kept back from the spalled edge in order to avoid forming a wide joint.

Structural failures, settlement, inappropriate mortars, perished joints and badly designed flashings, drips, fixings and masonry details may predispose stonework to decay. In particular, the retention of moisture in open joints and the rusting of ferrous anchorages such as dowels and cramps can cause serious problems. (Iron security bars are frequently a cause of cracked and spalled cills.) Incorrect methods of repair and inappropriate methods of cleaning or waterproofing can permanently damage or deface masonry and accelerate decay at an alarming rate.

In most cases the decay of stone is due to the presence of moisture; persistent wetting can penetrate even a thick wall, either directly, through open joints or wide cracks (particularly on horizontal projecting stones), or by capillary action, through narrow hair cracks in mortar or stones. The moisture will often travel down through the wall and evaporate at low level causing greater decay at ground level.

INSPECTION

The majority of buildings in the First and Second New Towns were built to high standards, using very durable, locally quarried sandstone. However, as the New Town developed, these standards tended to decline and the use of softer stones from quarries outside Edinburgh became more common. Almost certainly, stone rejected as unfit for the most prestigious buildings would have been used elsewhere. Even on individual buildings, adjacent stones can vary appreciably in quality and durability.

A preliminary inspection from ground level with the use of binoculars locates problem areas but a close examination on a stone-by-stone basis is the only way to identify and diagnose defects properly. Inspection of the upper walls will usually require scaffolding or a mobile hoist. If scaffolding has been erected the opportunity can be taken to repair the upper parts of the building at the same time. Once the defects have been diagnosed a policy for repair should be formulated on a short, medium and long term basis (see SURVEY AND REPORT).

MASONRY WALL CONSTRUCTION

GENERAL CONSTRUCTION OF MASONRY WALLS

In the New Town the rubble and ashlar masonry used to construct external walls has two distinct skins of bonded stonework separated by a cavity filled with stone chippings and lime mortar. It was common practice to construct these compound walls with an inner backing of cheaper rubble. In some cases the cavity filling was omitted and the flanking skins were built in direct contact with each other.

The *internal skin* was constructed of random rubble finished with lath and plaster. The *external skin* of street or front elevations was normally of dressed ashlar while on rear elevations it was of random or squared rubble. Rubble, probably harled, was used on the front elevations of early New Town houses on the north side of St Andrew Square, but after 1780 all front elevations were finished in ashlar. *Bonding stones*, around door and window openings and between flues, provide lateral stability and also ensure the distribution of vertical loads between the outer and inner skin. The cutting of these stones is not recommended. *Through* stones extend the full thickness of the wall, whereas *headers* tie only part-way into the internal skin. The *cavity filling* normally consisted of stone chippings and lime mortar. *Joints* in ashlar were often worked hollow and filled with lime/sand mortar and pinnings.

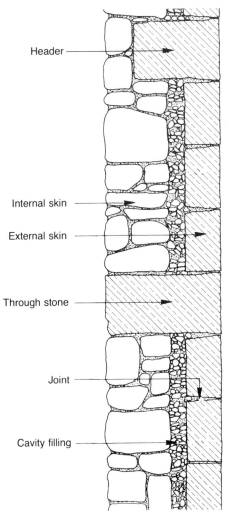

UNCOURSED RANDOM RUBBLE

COURSED RANDOM RUBBLE

Photos: The Royal Commission on the Ancient and Historical Monuments of Scotland

59

RUBBLE

Rubble work consists of undressed or roughly dressed blocks of stone with wide mortar joints containing many stone pinnings. Walls, or skins of walls, were built of coursed or uncoursed, random or squared rubble, often levelled up with small stones or pinnings, to give a horizontal bed for the stones above. Rubble masonry was most often used for basement, gable and rear elevation walls, as well as for internal loadbearing walls, party walls, garden walls, basement area retaining walls, and the inner skin of compound external walls.

Random rubble

The stones used for random rubblework were often laid in their quarried state, i.e. unsawn or undressed. Nowadays, with increasing mechanization, it is more difficult to get new rough stone; off-cuts of sawn stones are no substitute. Although smooth self-faced blocks require no dressing, other stones may be roughly bull-faced with a mash-hammer and splitter. Rubble stones are not of uniform size or shape, unlike those used for ashlar (see below), and great skill is needed to select stones of the correct size to ensure adequate transverse and longitudinal bonding. Throughs and headers should be used liberally and long continuous (risband) vertical joints avoided. The top bed of each stone should have a slight fall outwards to the face of the wall and stones should be level bedded to avoid creating a crazy paving effect.

Uncoursed random rubble is the roughest form of stone walling. Stones are built almost as they come to hand, inconvenient corners or bumps being knocked off as necessary. Larger blocks are level-bedded and wedged with small pieces of stone pinnings, or spalls, with no attempt to form accurate vertical or horizontal joints. Quoins and jambs are usually constructed with larger and roughly squared stones (for example, in stable buildings) or with broached ashlar.

Coursed random rubble is similar to the above, except that the work is roughly levelled up to form courses varying from 300 to 450 mm (12 inches to 18 inches) high. These courses usually coincide with the beds of dressed quoins and rybats.

Squared rubble

The blocks used for squared rubble were usually quarried from thin beds of stone, which was often heavily bedded (i.e. stone which displays marked lines at constant and regular intervals). The stones can be squared with relative ease and brought to a hammer-dressed or straight-cut finish. Squared rubble can be uncoursed, coursed or regular-coursed.

Uncoursed squared rubble, common in Victorian building, is otherwise known as square-snecked rubble. Walls are commonly built to a distinct pattern based on a unit of four stones, i.e. a large bonding, or through stone called a riser, two thinner stones or levellers, and a small stone termed a sneck. The vertical joint between each pair of levellers is more or less central over a riser. Adjacent risers are separated by a sneck, so that there are no long, continuous vertical joints.

Coursed squared rubble is levelled up to courses of varying height. Each course may consist of quoins, jamb stones and bonders of the same height, with smaller stones built up between them to complete the course.

Regular-coursed squared rubble is built in courses of varying height, but all the stones in any one course are the same height.

COURSED SQUARED RUBBLE   **REGULAR-COURSED SQUARED RUBBLE**

Photos: The Royal Commission on the Ancient and Historical Monuments of Scotland

60

ASHLAR

Ashlar masonry consists of blocks of stone finely dressed to given dimensions, laid in courses with thin joints (often as little as 3 mm). Ashlar was the most expensive and sophisticated grade of masonry and in the construction of the New Town it was reserved mainly for the walls on street or front elevations. The facing was secured by bonding stones, known as inband rybats, around openings.

Stones were hewn by hand, often on site, and the masons worked by stages from rough, self-faced blocks to broached, droved and polished finishes. The last, involving the use of carborundum, sand and water, was the most labour-intensive and therefore most expensive type of facing stone.

The beds of polished ashlar were often roughly undercut, or dressed off a hammer, to take stone wedges, or beeswing pinnings, inserted at the rear of the hollow beds to ensure accurate, plumb alignment of the stone and structural continuity. This technique allowed the mason to maintain a constantly narrow joint but it also placed a heavy load on the outer edge of the block, increasing the risk of the stone spalling at its edges.

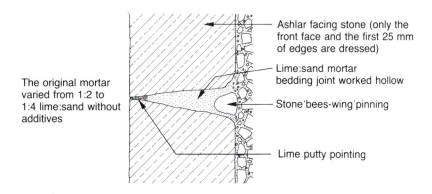

The original mortar varied from 1:2 to 1:4 lime:sand without additives

Ashlar facing stone (only the front face and the first 25 mm of edges are dressed)

Lime:sand mortar bedding joint worked hollow

Stone 'bees-wing' pinning

Lime putty pointing

TYPICAL SECTION THROUGH ORIGINAL ASHLAR JOINT

The jointing pattern of ashlar generally resembles that of regular-coursed squared rubble, although the joints are finer and more precise. Irregular coursing of ashlar can be found occasionally in the New Town. The pattern of ashlar joints was part of the overall design of elevations and new indents should always respect this by exactly matching the colour, tooling and size of the original stones. On less important elevations, perhaps with a broached finish, the ashlar was sometimes built in random lengths and great care must be exercised to ensure good bonding, particularly when replacing or repairing large areas of defective masonry.

The ground floor elevation of many New Town buildings is emphasized by the use of *rusticated V-jointed ashlar*. The margins of individual stones are chamfered to form V-shaped channels along the joints. A variation of rusticated ashlar consists of square-cut channels along the joints.

POLISHED ASHLAR

Photos: RCAHMS

RUSTICATED V-JOINTED ASHLAR

CHANNELLED V-JOINTED ASHLAR

Nowadays, stones are sawn by reciprocating and rotary diamond-toothed blades at the quarry or in the mason's yard; the sawn finish is polished to remove any saw marks using rubbing beds (stones which show the marks of the saw should be rejected). Polished ashlar produced by modern techniques is now the cheapest of the surface finishes, and tooling is an applied finish.

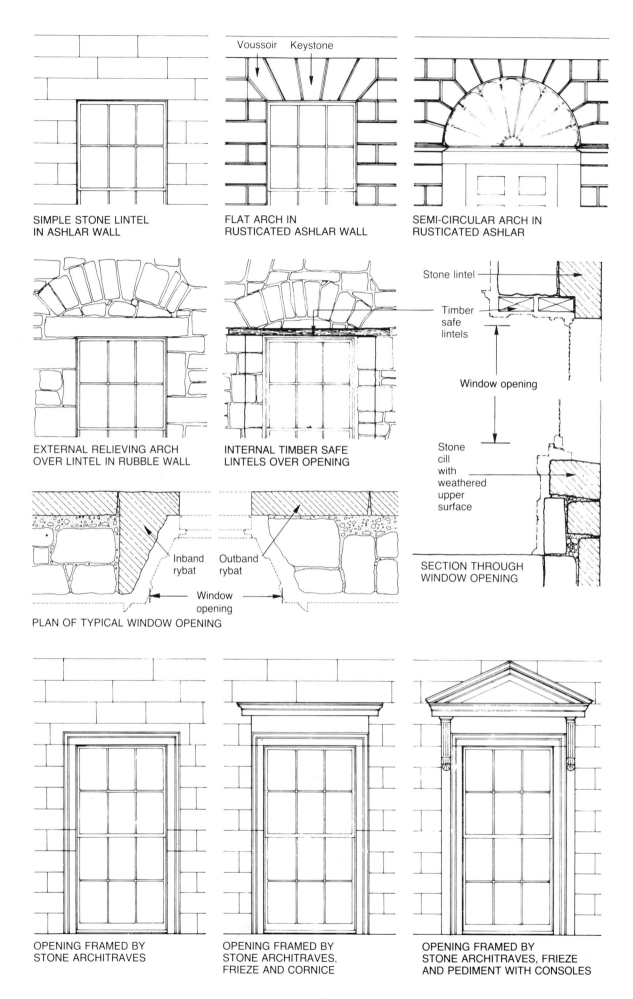

SIMPLE STONE LINTEL
IN ASHLAR WALL

FLAT ARCH IN
RUSTICATED ASHLAR WALL

Voussoir   Keystone

SEMI-CIRCULAR ARCH IN
RUSTICATED ASHLAR

EXTERNAL RELIEVING ARCH
OVER LINTEL IN RUBBLE WALL

INTERNAL TIMBER SAFE
LINTELS OVER OPENING

Stone lintel

Timber
safe
lintels

Window opening

Stone
cill
with
weathered
upper
surface

SECTION THROUGH
WINDOW OPENING

Inband
rybat

Outband
rybat

Window
opening

PLAN OF TYPICAL WINDOW OPENING

OPENING FRAMED BY
STONE ARCHITRAVES

OPENING FRAMED BY
STONE ARCHITRAVES,
FRIEZE AND CORNICE

OPENING FRAMED BY
STONE ARCHITRAVES, FRIEZE
AND PEDIMENT WITH CONSOLES

Tooled finishes include the following:

- *Broached finish:* a series of equally spaced furrows produced by a pointed chisel, usually between 80 and 260 per metre, the furrows stopping 15-30 mm short of the dressed edge. Individual stones treated in this way usually exhibit droved chisel-drafted margins, which were levelled through with sighting sticks to give a true or square surface. The finer the broaching, the narrower the draft. This finish is often found on platt arch stones (i.e. the downstand external face) and above first floor level on street elevations. Broaching was the first stage of refinement from the rough state of rubble to the sophistication of polished ashlar. Because broaching was applied to a roughly squared stone, the surface was slightly irregular; such an irregular surface cannot be reproduced when broaching is applied to a sawn block. Machine broaching is now possible, but the finish does not bear any resemblance to hand-tooling;
- *Droved finish:* produced by a hammer and a broad chisel to form a series of 35-50 mm high bands of more or less parallel tool marks, usually finer than broaching. The margin of broached or stugged ashlar is often droved, or chisel-drafted;
- *Stugged finish:* produced by forming random depressions on the rough stone surface with a mason's punch or point;
- *Rock-faced finish:* produced by a punch which gives a rugged, deeply profiled surface of semi-regular design which is difficult to imitate. The basements of street elevations are sometimes finished in this style, e.g. Charlotte Square, Heriot Row;
- *Vermiculated finish:* on ashlar gives the appearance of the surface having been eaten by worms with irregular shallow channels over the surface. This finish, derived from Italian Renaissance buildings, was most widely used in the Victorian period, but there are some earlier examples.

BROACHED FINISH   **DROVED FINISH**   **ROCK-FACED FINISH**

Photos: The Royal Commission on the Ancient and Historical Monuments of Scotland

SPECIAL FEATURES

Blocking courses and parapets

A blocking course is a plain course of stone surmounting the cornice in front of the wallhead gutter. A low wall above a cornice at roof level is usually referred to as a parapet.

Cornices

A cornice is a projecting moulded course, usually built near the top of the wall. As part of the classical entablature it is an important architectural feature which can vary considerably in detail. Its object is to shed water clear of the face of the building and the upper surface should be weathered so that water drains away from the wall face, with the joints saddled to prevent water penetration. A lead flashing over the cornice will also prevent water penetration, but traditionally these were not used in Edinburgh. Vertical joints may be joggled for stability. Cornice joints are a common source of moisture penetration which can cause rot in the wallplate and timber safe lintels and stone decay problems below the cornice.

Friezes

A frieze, the middle member of a classical entablature, is a stone course surmounted by a cornice. Originally decorated, emphasis is often given to a frieze in the New Town by projecting it slightly, especially if there is no string course or architrave immediately below it.

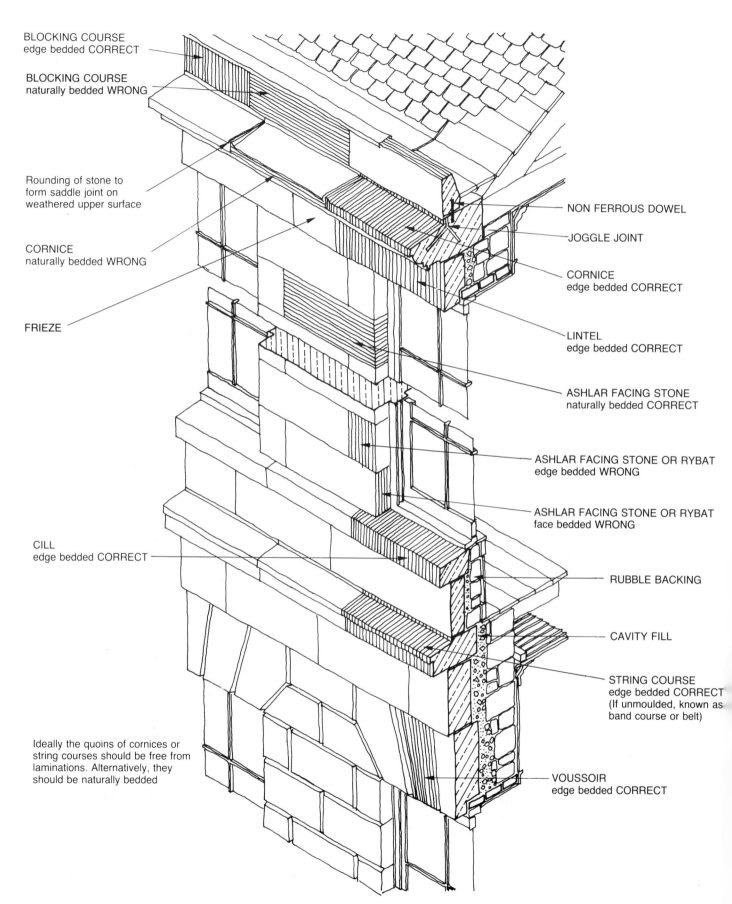

BLOCKING COURSE
edge bedded CORRECT

BLOCKING COURSE
naturally bedded WRONG

Rounding of stone to
form saddle joint on
weathered upper surface

CORNICE
naturally bedded WRONG

FRIEZE

CILL
edge bedded CORRECT

Ideally the quoins of cornices or
string courses should be free from
laminations. Alternatively, they
should be naturally bedded

NON FERROUS DOWEL

JOGGLE JOINT

CORNICE
edge bedded CORRECT

LINTEL
edge bedded CORRECT

ASHLAR FACING STONE
naturally bedded CORRECT

ASHLAR FACING STONE OR RYBAT
edge bedded WRONG

ASHLAR FACING STONE OR RYBAT
face bedded WRONG

RUBBLE BACKING

CAVITY FILL

STRING COURSE
edge bedded CORRECT
(If unmoulded, known as
band course or belt)

VOUSSOIR
edge bedded CORRECT

BEDDING OF STONEWORK

Architraves

An architrave is a horizontal projecting course of masonry which is the lowest member of the classical entabulature. (The term is also applied to the mouldings around openings.)

String and band courses

Usually constructed of bonding stones, these architectural features give horizontal emphasis to the elevation. Unmoulded string courses, known as band courses, are a common feature found at ground and first floor level throughout the New Town.

Pilasters

A pilaster is a shallow, pier-like column of classical proportions applied to the face of the wall. A pilaster strip has no base or capital. Pilasters extending through two storeys can be seen most commonly on the pavilion and centre blocks of terraces in the New Town, giving vertical emphasis to the elevation. Pilasters on ground floor elevations may also be built around door openings and on shop fronts, as at Circus Place, Howe Street and on tenements designed by Thomas Bonnar at India Street.

Wall openings

The head of an opening is generally finished externally with a simple stone lintel or, more elaborately, with a flat or semi-circular arch. Rusticated ashlar flat arches are a common feature over ground floor front elevation openings, the whole arch being formed of one large stone, with false V-joints conforming to the adjacent masonry, cut into its face to give the appearance of voussoirs. Semi-circular arches, on the other hand, are constructed of individual stones and are commonly seen over street entrance doorways.

In rubble walls, an external stone relieving arch is sometimes built above the main stone lintel. Internally, the rubble skin tends to support itself by natural arching over openings, although heavy timber safe lintels were invariably built into the wall for extra support, and to allow finishes to be fixed.

The masonry jambs of window openings are bonded by using alternative headers (inband rybats) and stretchers (outband rybats). The inband rybats are rebated to receive the window frame and chamfered internally to accommodate the window ingo, which is splayed to increase light transmission into the room. Another method of forming window reveals is to use starts or upstarts on the external masonry skin. These tall, thin reveal stones are all of equal width and usually project slightly from the face of the wall. A start generally refers to a stone which is not bonded into the jambs of the opening.

Stone architraves (often complete with surmounting entabulature or pediments with consoles) are used to provide a decorative framing around window openings and, in particular, to enhance the grander fenestration of the first floor or piano nobile. Usually constructed of moulded starts, they were a later fashion and are mostly found at the west end of the New Town.

In addition to semi-circular arches (described above), the importance of street entrance doorways is sometimes stressed by flanking the opening with columns or pilasters which carry a classical pediment or entabulature. Examples of these aediculated door openings are discussed further in PORTICOS.

Cills are single stones bedded hollow. They either project beyond or remain flush with the face of the wall. The upper surface of the cill is weathered (or washed) to discharge water clear of the wall. A throating should be cut into the underside of the cill.

BEDDING OF STONES

Sandstone, like all sedimentary rock, is a laminar material and it is important that the direction of the bedding planes is suited to the purpose and position of the stone. Stones can be naturally bedded, edge-bedded or face-bedded.

Many stones have pronounced bedding planes, the direction of which can be determined by examining the mica particles within the stone. The plane of these small, flat, glittering discs coincides with that of the laminations and therefore the bed of the stone.

Naturally bedded stones, laid with their bedding planes horizontal as in their natural state, should always be used for plain ashlar walling. Because the bedding planes are laid at right angles to the downward pressures in the wall and are protected from the weather, the risk of surface disintegration is minimized.

Edge-bedded (or *joint-bedded*) stones are laid with their bedding vertical and perpendicular to any exposed surface. Edge-bedding is desirable for string courses, cornices (except on an external corner where it is not acceptable) cills and lintels, where natural bedding would tend to encourage erosion of the mouldings, throatings and projecting horizontal surfaces. Arched stones should be carefully arranged so that the bedding planes are at right angles to the thrust (that is, parallel to the direction of the joints). The natural bed height of stone within any particular quarry will determine the maximum possible length of an edge-bedded stone.

Face-bedded stones are laid with the bedding planes vertical and parallel to the face of the wall. This is often referred to as building on cant. The surface layers of face-bedded stones will always tend to disintegrate, especially when exposed to the weather, and this can seriously affect the durability of the wall.

STONE CLEANING

INTRODUCTION

All external stone in the New Town is covered with dirt and soot to some degree although much of the blackness can be attributed to the fruit crop bodies of minute crustose lichens or fungi, or chemical oxidization. There has over the last thirty years been a movement towards the cleaning of the facades of buildings with the aim of upgrading the appearance of individual buildings and streetscapes. This has to some extent been successful although in other areas it has led to individual buildings in street blocks being cleaned separately using different methods, with the result that the facades are out of step and terraces are no longer homogeneous in appearance.

Dirt is attached to sandstone by a chemical bond which cannot be broken by water washing, unlike dirt on limestone. Aggressive methods are therefore required to break this bond if the dirt has to be removed.

It has become apparent in more recent years that there are positive disadvantages to cleaning, and that no cleaning method can be used without some damage to the sandstone. The degree of damage depends on the method used and the skill of the operative carrying out the works.

The erosion of even small particles of stone from important listed buildings is neither conservation nor restoration. Historic Scotland are at present preparing a memorandum on the subject of stone cleaning which will explain their philosophy in detail. The general principle, however, is that those owners wishing to stone clean their buildings will need to prove that it is beneficial and that no damage will occur.

No study has been undertaken to assess the extent of stonework repairs occasioned by damage caused by earlier cleaning exercises. However, it is apparent that in other cities some buildings, cleaned within the last twenty years, are undergoing or are requiring stonework repairs, often on a large scale.

Cleaning cannot be undertaken without Listed Building Consent. It is the policy of the Edinburgh New Town Conservation Committee that cleaning be positively discouraged and grants will not be available even where permission is granted for cleaning to take place. It must also be pointed out that at the time of writing, Edinburgh District Council has declared a moratorium on stone cleaning within the city, along with other authorities in Scotland.

Recent research into the public's perception of buildings indicates that people generally do not wish to see totally clean buildings and that some degree of soiling, softening the architecture, is more appealing.

New methods of carrying out cleaning are constantly being put forward by different commercial organizations and it may well be that in the future a method will become available which could remove dirt deposits without damaging the stone underneath.

Uncleaned string course in Rutland Square with original polish and tooling

Cleaned string course in Rutland Street with polish and tooling much reduced. The surface now has a series of ledges which will retain moisture. Note that the geological bedding of the stone (from Binny quarry) has been opened up

AFTER ACID CLEANING

AFTER ABRASIVE CLEANING 68 Photo: Peter Davey, *The Architect's Journal*

CONTROL OF VEGETATION

INTRODUCTION

Plants, mosses and algae are not a serious problem on buildings in the New Town. However, when organic growth does occur, the result can be unsightly and, in some cases, detrimental.

Although only a few plants are directly harmful to masonry, vegetation may cause decay indirectly by retaining moisture or dirt in which other plants can root and its presence indicates excessively damp conditions. Some of the lower forms of plant life, such as lichens, produce weak organic acids which may attack metal roof coverings and flashings, while creepers make it difficult to inspect the fabric of a building.

This chapter outlines the types of vegetation that commonly thrive on buildings, describes their effects on masonry and roof coverings and gives recommendations for removal and prevention of growth.

TYPES OF VEGETATION

Algae (see also notes on lichens, below)

Algae are the most common forms of vegetation on buildings in Edinburgh, and they will develop on porous stone whenever a suitable combination of dampness, warmth and light occurs. Their presence varies according to the geology of certain stones. They appear as green, red or brown powders or filaments, and may be slimy under moist conditions. The presence of algae is usually more noticeable and prolific on buildings which have been chemically cleaned.

- *Effects on masonry:* Although algal growth may be aesthetically undesirable, there is little indication that it is directly harmful to stone. Research is currently being undertaken by Robert Gordon's University; when this is complete our understanding of the mechanisms should be greater. However, in common with other forms of vegetation, algae collect dirt and water, and they tend to increase the transpiration rate of stone, thereby contributing to its ultimate destruction. Although Craigleith stone, being macroporous, is not unduly susceptible to this type of decay, the presence of algae on masonry is generally considered to be undesirable. Algal growth on wet and slippery stone steps can be extremely hazardous.
- *Removal:* Algae generally occur with lichens, but it may be necessary to control algae alone, usually on horizontal surfaces, such as steps, string courses, cills, cornices and the tops of monuments (which become covered with bird droppings, dead leaves and dirt) or on vertical surfaces subject to water runs. Most algicidal treatments are unsuitable for use on stonework for various reasons: ordinary household bleach is effective but growth will usually recur within a matter of weeks; other chemicals, like sodium pentachlorophenate, can damage stonework through salt crystallization. Biocides such those based on quaternary ammonium compounds are generally considered to be the most effective – they are relatively inexpensive, easy to apply and safe to use – in that they have no ill effects on stonework. The ENTCC will supply the names of suitable proprietary products on request. The liquid is usually supplied as a 50 per cent solution which should be diluted according to the manufacturer's instructions. Applications by generous flood spray, giving a coverage of about 1.5 square metres per litre, will normally prevent further growth for about two years. Sparing application will give poorer control for a shorter duration. High pressure sprays and application in windy weather should be avoided, because drifting spray can scorch surrounding plants.

Lichens

Lichens, a symbiotic co-existence of fungi and algae, are better able to withstand dry conditions than either of the components alone and are capable of deriving nutrients from a wider range of materials. Common 'dirt' on masonry is often made up of deposits of dead algae and the surface fruiting bodies of minute lichens within the stone pores. Many common lichens cannot tolerate a polluted atmosphere, but there are others which can survive.

69

ANN STREET
SIR HENRY RAEBURN'S ESTATE
1815-25

Photo: John Johnson

- *Effects on masonry:* Lichens are far more harmful than algae alone and tend to absorb elements from mortar and stone by the secretion of acids. In this way the stone is weakened and, with some crustose lichens, the densification of the surface which occurs may be followed by spalling. For this reason, lichens should not be tolerated on finely carved, lettered or otherwise valuable masonry.
- *Effects on roof coverings:* It has been suggested that rainwater running off a lichen-covered slated roof picks up the organic acids secreted by the lichens, and that even in heavily diluted form these acids will slowly corrode most non-ferrous metals. The extent of attack is uncertain, however, due to the lack of research into the subject of acid rain. On the typical 'M' shaped roofs of the New Town, this problem is manifest at wallhead and valley gutters as narrow, clean-cut grooves or round spots where water drips off the slates onto the lead or zinc. The acidic water can damage the normal protective patina of the lead and the repetitive dissolving and reforming of the patina eventually results in the failure of the lead sheet. With leadwork it may be many years before actual failure occurs; with thinner metals, such as copper and zinc, failure from dilute acid attack can be surprisingly rapid.
- *Removal:* There are a number of toxicants currently available, but not all are recommended for use on stone or slate. Details of proprietary products are available from the ENTCC. Robert Gordon's University research will evaluate the available chemicals for efficacy and friendliness to both stone and the environment. As with the algicides, these products are diluted (usually one litre diluted to twenty litres with tap water) and applied by generous flood spray. To eradicate heavy encrustations of lichen growth, a second inhibitory application (at least twenty-eight days later) may be needed. Just prior to this, the affected area should be dry-brushed with bristle brushes to remove the dead lichens. Where lichen and algae occur together, the algal growth may soon reappear on its own, but can be controlled if necessary using the less expensive algicides.
- *Prevention:* Organic acid attack on gutter linings can be such a problem that permanent measures to prevent future lichen growth on slates are required. This can be achieved by inserting copper strips at regular intervals between the slates; the run-off from the copper contains salts which prevent the incipient growth of lichens and mosses. Where copper strips are undesirable for aesthetic reasons, spraying with a recommended toxicant may be necessary as part of the annual roof maintenance programme. If all else fails, it may be necessary to insert sacrificial lead flashings which corrode in preference to the vulnerable gutter linings themselves. For further details on sacrificial flashings and the insertion of copper strips, refer to FLASHINGS.

Mosses

- *Effects on masonry:* The growth of mosses on stone or in the joints of stonework usually indicates abnormally wet conditions and sometimes serious water penetration. Their ability to retain moisture provides an ideal medium for transmitting harmful chemical agents and may mean that frost action will break up the surface of stone. Mosses, like lichens, secrete acids which may cause deterioration of carbonaceous stones.
- *Effects on roof coverings:* Moss will frequently take hold in the joints between slates and cause blockages in gutters and downpipes.
- *Removal:* Treatment should be as for lichens.

Ivy (*Hedera helix*)

- *Effects on masonry:* By rooting in the joints in stonework and deriving moisture from the mortar, ivy can dislodge stones and facilitate the passage of water into the heart of the wall. Sometimes fine rootlets even penetrate minute crevices in the blocks and exert enough force to split hard stones.
- *Removal:* The thick roots at ground level should be cut and the protruding stumps 'frill girdled' in order to expose a greater area to the toxic material. An ammonium sulphamate paste should be painted onto the root, which will eventually wither and die. If the roots of the ivy have penetrated the building, they should be cut at their points of entry and treated in a similar fashion. Ammonium sulphamate should be applied with care. Contact with stone should be avoided because the chemical decomposes to form ammonium sulphate, which encourages the growth of algae and other vegetation. Some bacteria can convert this ammonium sulphate to nitric and sulphuric acids, which will damage carbonaceous stones. In addition, the crystallization of ammonium sulphate will affect all but the most durable of stones.

  Sodium chlorate should not be used as a herbicide. Roots treated with this chemical can ignite spontaneously and may lead to fire damage if they penetrate into the building. Accidental spillage on clothing introduces a severe fire risk.

  The main plant should not be removed from the building until it has died back completely. Left untreated, a sizeable plant may take up to two years before it relinquishes its tenacious hold, and earlier removal may result in damage to the stonework. Treatment of the upper leafy growths with 2,4,5,T formulation is no longer recommended, because the health risks associated with the chemical are unacceptable. No safer chemical substitute has been developed.

  Particular care should be taken when removing ivy from stonework in poor condition, because the wall may be relying on the structure of the plant itself for stability.

Virginia Creeper (*Parthenocissus quinquefolia*) and Boston Ivy (*Parthenocissus tricuspidata*)

- *Effects on masonry:* These creepers are relatively harmless, attaching themselves to the wall by suckers rather than by roots. The only dangers result from the effects of roots at foundation level and growth under slating and into gutters and hopperheads at roof level. Growth prevents inspection of the wall, and may result in neglect.
- *Removal:* In general, the creepers can be safely retained, although careful trimming is essential to prevent the spread of growth. To remove creeper, the same treatment as for ivy can be used.

Other plants

Plants whose roots can penetrate mortar and may dislodge masonry, such as ferns, nettles, young seedlings and wallflowers, can all be controlled using glyphosate during the growing season. Elder, brushwood, brambles and so on should be cut down and the stumps treated with ammonium sulphamate.

Tree roots can be damaging but their removal is not recommended without specialist arboricultural advice; root removal may cause further root growth and the removal of the wrong roots can make a tree unstable and dangerous. Within Conservation Areas and where Tree Preservation Orders apply, trees are protected and the Planning Department should be consulted if there is any intention to uproot, fell, lop or top a tree or to carry out any action which might damage it.

RECOMMENDATIONS

In general, the growth of all types of vegetation on buildings should be discouraged and existing growths removed as quickly as possible, if feasible without damage to the fabric. To reduce the occurrence of vegetation, it is important to identify sources of dampness, and to carry out remedial work.

Proprietary chemicals should always be applied in accordance with the manufacturer's instructions; fungicides and weed-killers are poisons and should always be treated with care. Prior to use, it is wise to check with the manufacturer of any chemical that it will not have any adverse effect on stonework. Treatments containing alkali or metal salts of any kind (e.g. copper, mercury and arsenic) should always be avoided.

Since vegetation thrives in damp conditions, it is important to prevent water penetrating the fabric of buildings or accumulating on the surface of stone. The maintenance of gutters and downpipes should help keep excessive water off the stonework and therefore lessen the presence of vegetation. The use of water-repellent treatments such as silicones and silanes should be actively discouraged because they prevent moisture evaporating and trap it inside the stone.

If creepers are desired for visual effect, a harmless variety from the Parthenocissus family should be chosen and trained on a frame which can be unfixed and bent forward with the plant intact, enabling routine repairs and inspections to be carried out. For further information on the effect of fungal growths on timber, refer to TIMBER DECAY.

REMOVAL OF PAINT FROM STONEWORK

INTRODUCTION

Painted masonry is alien to the architectural character of the New Town; Listed Building Consent is required and is unlikely to be given for painting any masonry except for some original shop fronts and area retaining walls which reflect light into basement rooms. Removal of paint has its hazards too; for instance, if it is taken from part of an otherwise dark stone facade, the cleaned section can be so conspicuous that the architectural composition of the whole building is spoiled. Cleaning may also reveal blemishes that the paint was meant to hide, and the cleaned surface may encourage graffiti.

Soft sandstone, particularly if damp, will not accept paint for long. Bleaching, discolouration, loss of adhesion, blistering, peeling, flaking, cracking and pitting of paint are all signs of the breakdown caused by dampness, soluble salts, alkali in lime mortar, powdery stone, poor preparation and paint solvents. Any barrier to the natural breathing of the building will be thrown off by the build-up of moisture on the inner face of the applied material, resulting in the need to repaint the walls frequently, often at only two or three year intervals.

REMOVAL OF PAINT

It is difficult to remove paint from sound dry stonework, and the exact type of paint must be identified before deciding how to remove it.

Bituminous paints

These will usually have been applied to stone as an undercoat for a sand paint. Any excess should be scraped off and the stone scrubbed with water containing liquid detergent. Powdered detergents should not be used because they contain large quantities of soluble salts which will be harmful to masonry. The surface should be left to dry and, if necessary, poulticed with white spirit.

Cementitious paints

These will respond to dilute hydrochloric acid which must be thoroughly washed off after application to prevent damage to the substrate. Low pressure wet abrasive cleaning methods, however, may be less hazardous to operatives and to masonry. Investigation of both techniques on part of the affected area is recommended.

Emulsion paints

The removal of emulsion paint is both laborious and, if carried out by a contractor, expensive; for this reason emulsion is often painted over. Where removal is required, the paint has to be softened with methylated spirits and steam before being carefully scraped away. Steam wallpaper strippers, which can be hired, may be useful. The surface of the emulsion should be well scored or scratched to assist penetration of the steam. If layers are heavily built up, proprietary paint strippers may be used, but only after the advice of the manufacturer or specialist supplier has been sought.

Limewash

This is a traditional finish which, in common with all lime-based applications, develops a surface hardness related to age and exposure. Old sulphated limewash, applied in several coats, is often impossible to scrub off and may need to be softened by a wet poultice over several days. If the limewash is on sound rubble masonry, rather than ashlar, it may be removed by light abrasive blasting. When limewash has been overlaid or interspersed with coats of oil or other paints, removal is difficult, because each layer may require different solvents.

Sand paints

These will usually respond to proprietary paint strippers or in some cases to naphtha.

Oil-based paints

Old oil-based paints are difficult to remove from porous masonry and the Historic Scotland Conservation Centre can advise on techniques. Any loose paint can be removed by careful wire brushing by an experienced contractor except on details of mouldings and so on. Many proprietary paint removers are available, most of which may be thinned with a little methylated spirits. The surfaces should first be washed down with lime water or sugar soap solutions, well rinsed and allowed to dry thoroughly. Blow-lamps and hot-air strippers should not be used, because there is a possibility that heat may drive the oil into the pores of the stone. For stubborn oil marks, some lime putty or liquid ammonia may be added to the water. Alternatively, in ventilated conditions, carbon tetrachloride may be applied in poultices of whiting or natural clay.

SPECIAL CLEANING PROBLEMS

Cement slurry

Splashes of cement slurry or mortar should always be dealt with immediately they occur by using copious amounts of water and a soft brush. Once the cement has set, there is no easy way to remove slurry stains, though it may be possible to remove some of the deposit with a wooden spatula or a brush. The remainder can be treated with diluted hydrochloric acid, applied with a swab, the area being thoroughly wetted first to reduce acid absorption. The acid should not be allowed to remain on the surface for more than one hour. The masonry should then be thoroughly hosed down and, once it has dried, the process repeated if necessary. Great care is needed when using acid; protective clothing, rubber gloves and eye guards should be used and a first aid kit, including an eye-bath solution, should be available nearby.

Graffiti and aerosol paints

This topic is the subject of a current BRE research project (1993). There are many proprietary cleaners which remove nitrocellulose aerosol paint. These are applied to the paint by brush, low pressure spray or a soft cloth and allowed to stand before being thoroughly washed off with water. When graffiti is removed, the ghost of the slogan may still be visible, and artificial dirtying of the cleaned stone may have to be considered, though it is difficult to match the original appearance.

If small areas of masonry are repeatedly vandalized by graffiti they can be treated, using a water-sensitive acrylic polymer solution and a curing agent, so that any further graffiti can be removed using normal paint strippers. This treatment should not be used over large areas, because it gives a shiny finish to masonry.

Before trying the methods of paint removal described above, it is well worth attempting to discover the brand name of the paint used. Finding out what aerosols are sold locally or even searching dustbins or side streets for a discarded paint-spray can be helpful! If the paint is identified, the manufacturer's chemist may be able to provide information on the exact type and sequence of solvents required to remove the paint, thus saving a great deal of time.

Metal stains

Long-standing rust stains are extremely difficult to remove, though they can be reduced by the application of some acids, which must be thoroughly washed off afterwards. Repeated applications of an ammonia-solution poultice can help to remove copper stains. Specialist firms undertake this work and should always be consulted.

Oil and grease stains

Greasy stains from the residue of oil-based paints are often noticed on stone after cleaning. Removal may be attempted using liquid detergents (see Oil-based paints above) or, in more stubborn cases and where ventilation is good, carbon tetrachloride. A poultice may be required if stains are deep.

Poulticing

A poultice may be used to remove stubborn areas of paintwork where, for example, paint has been retained in the grooves and pits of droved or broached ashlar. In some cases it may be wise to use a poultice from the start for tooled stonework.

Poulticing is carried out with a mud-pack made of whiting or sawdust, or one of two naturally occurring magnesium silicates (sepiolite or altapulgite), containing a solvent appropriate to the paint type, e.g. a proprietary solvent for an oil-based paint. The mixture is trowelled onto the wall and covered with light polythene, such as cling film or a similar material, to delay drying out. Poulticing has generally been found to be the most successful method of paint removal in the New Town.

Chewing gum

Small areas of chewing gum can be frozen with dry ice and picked off. For larger areas, or where the gum has been trodden in, pressure steam cleaning should be used.

STRUCTURAL DEFECTS

INTRODUCTION

People are often alarmed by the discovery of cracks, which seem to signify structural movement. Cracks require skilled interpretation and, if active movement is suspected, a structural engineer should be consulted. Many cracks and other signs of movement turn out to be insignificant in terms of stability but nevertheless should be checked. This chapter describes some common structural defects, without claiming to cover all possible faults or to be a substitute for professional advice on specific problems.

Movement in a building can be caused by overloading the floors, by mechanical vibration, by the failure of the structure, by wind, water, changes of temperature and so on. The most common cause in the New Town is settlement; if foundations rest on strata of poor bearing quality, or strata which have been disturbed by underground streams or tunnels, they may have to be strengthened by one of the methods described in FOUNDATIONS. Injudicious alterations involving the removal of bearing walls, for instance, are another source of cracks.

Cracks on horizontal and projecting external surfaces, like cornices and cills, enable damp and frost to penetrate the building and accelerate decay.

COMMON DEFECTS AND TREATMENTS

Foundations

All buildings settle during and after their construction as the load compresses the ground. If settlement is uniform it will not damage the building, but differential settlement may produce cracks in the masonry. This sort of settlement took place in several New Town houses many years ago.

The effects of differential settlement can sometimes be seen in wide risband joints between terrace houses built at different times. This relative movement need not by itself cause concern unless it is active, but open joints should be pointed in weak lime mortar to keep the weather out.

Once initial settlement has occurred, there should be no further cracking or distortion unless something happens to weaken the ground and cause further settlement. Common causes are:

- Excavation for drains or new foundations near the building;
- New building on an adjoining site;
- Saturation of the ground due to choked and leaking pipes and drains;
- Shrinkage of the ground dried out by the roots of trees planted too close to the building, or following periods of drought;
- Change in the water table due to piling nearby.

The extent to which traffic vibration has a compacting effect on material beneath the foundations of buildings is not clearly understood. Studies have shown that vibration caused by modern traffic may cause road settlement but is unlikely to damage buildings, although there is still much controversy about the possible effects of heavy traffic.

Openings

Movement in masonry is normally accommodated by the mortar joints. However, long stones, like cills and lintels, are more likely to crack than several smaller stones subjected to the same stresses. Lintels do more than just span across windows and doors. Together with the rybats which support them, they tie the whole facade together and so the structure of the whole wall should be considered before repairing a single lintel.

Radial cracks in lintels can often be treated simply by pointing up in lime mortar. Cracks immediately adjacent to the supporting points can be pinned by drilling in stainless steel dowels set in epoxy resin or by inserting a stainless steel corbel plate; these methods are considerably cheaper than renewing the whole lintel. Cracks not adjacent to the supports can be supported on a stainless steel flat bar inserted below the lintel (the bar should return round the ends of the lintel or be drill-fixed to the underside) but this can be difficult to achieve in ashlar.

Cracks in stone cills (traditionally not bedded on a damp-proof course) are likely to allow troublesome water penetration. Fine cracks can be pointed up with lime mortar but with wider cracks it is usually prudent to renew the cill. Cills should be hollow-bedded.

A decayed timber safe lintel will reduce the strength of the inner rubble skin of an external wall, particularly if there is no stone relieving-arch above the opening. If the load normally carried by the inner skin is transferred to the outer skin, it may result in spalling and cracking of the ashlar rybats, especially if the beds are worked hollow. After the source of the damp has been removed, replacement of decayed timber safe lintels should be carried out on a like-for-like basis; alternatively, precast and prestressed concrete lintels might be considered.

Continuous bonding or levelling timbers have been found in external walls at Forres Street, Mansfield Place and Moray Place and they may be widespread throughout the New Town. These timbers, up to 450 mm × 150 mm, were built into the inner rubble of external walls level with safe lintels or below floor joists. Damp penetrating the external walls can cause wet or dry rot in bonding timbers with serious consequences for the walls which they support. All decayed timbers must be removed in short sections and replaced by concrete (see TIMBER DECAY).

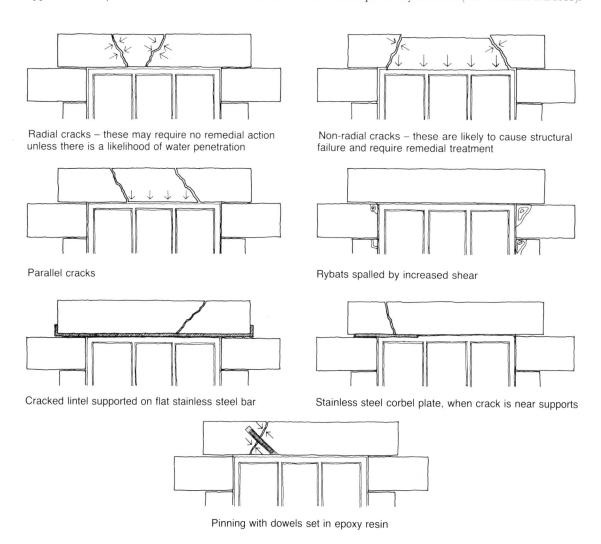

Radial cracks – these may require no remedial action unless there is a likelihood of water penetration

Non-radial cracks – these are likely to cause structural failure and require remedial treatment

Parallel cracks

Rybats spalled by increased shear

Cracked lintel supported on flat stainless steel bar

Stainless steel corbel plate, when crack is near supports

Pinning with dowels set in epoxy resin

Structural timbers

The decay of the ends of timber joists will obviously reduce the strength of a floor and the loss of such horizontal ties between external walls can weaken the entire structure. Outward movement in the walls will reduce the bearing for floor joists, and failure of a timber safe lintel can reduce the effective support of the floor joists above it.

Rafters are prevented from spreading outwards by horizontal collars or ties. If these ties lose their strength through decay, accidental damage or injudicious alterations, the outward thrust of the roof can distort and overturn the upper part of the wall. Solutions to some of these problems are covered in ROOF STRUCTURE and FLOORS.

Basement area walls, platts and arches

Movement of the area retaining wall or water penetration through the platt may cause opening or erosion of voussoir joints in the supporting arches. If the arch has flattened to a dangerous extent it should be taken down and rebuilt. The weight of heavy vehicles on the footpath can distort the coping and upper masonry of area walls, resulting in the walls bowing into the area and possibly collapsing. Cellar arches are seldom bonded to area walls, so a system of ties should be devised to rectify this problem.

External walls

The strength of existing masonry walls may be calculated by following BS 5628, which treats masonry as block-work. Where the wall is constructed from large, carefully shaped pieces with relatively thin joints, i.e. ashlar, its loadbearing capacity is more closely related to the intrinsic strength of the stone. This is not the case where small structural units are used. Design stresses in excess of those obtained from this code may be allowed in massive stonework, provided that the designer is satisfied that the properties of the stone warrant an increase.

The inner and outer skins of traditionally constructed walls can separate if the lime mortar and small stones used to fill the central core are disrupted through settlement or water penetration. Delamination and internal erosion of the masonry may not be immediately obvious because the external skin of the wall can still appear to be sound. Walls should be tested with a hammer; the strike note and bounce-back are good indicators of the condition of the masonry.

Examination of several walls in the New Town has shown that the central core is usually sound and that the inner and outer skins are tied to each other at openings by bonding stones. In 1975, tests on sample piers removed from buildings in the New Town were undertaken for the ENTCC by Edinburgh University's Department of Civil Engineering. Mortars of various strengths and specifications were tried, and even when the stones were bedded in sand alone, the tests proved that the traditional masonry construction was more than adequate to support normal domestic loads.

External walls lacking bonding stones or with disintegrating mortar can sometimes be strengthened by grouting. This specialized technique can burst a wall if used wrongly; it should only be used if traditional masonry construction proves impractical and then only by skilled tradesmen under professional supervision. The report by the Board of Agrément commissioned by the ENTCC in 1975 on the structural condition of 27 Clarence Street/97 St Stephen's Street recommended the use of cement/PFA (pulverized fuel ash) grout injections into areas of the external walls which were thought to contain voids in the central core. However, due to the density of the hearting, the experimental grouting failed to penetrate the core of the worst sections and the building was subsequently repaired by traditional methods.

Outward bulging of walls can be restrained by inserting tie rods through the building, or by using the floor construction as a rigid horizontal plate to which the walls can be tied. A traditional method of tying walls together uses face plates and consideration may be given to their use in certain situations but there will be both aesthetic and planning implications. Face plates can be recessed and concealed by stone indents not smaller or narrower than the course height.

Similar techniques can be used for stitching cracks or for tying across risband or straight butt joints. Normally about three straps per storey will be required across any risband joint or serious crack. Another method of stitching major cracks in rubble or internal walls is to use prestressed concrete lintels built into the masonry at regular intervals. Minor cracks in masonry can be stabilized by inserting stainless steel wall ties into matching joints across cracks. Many cracks in walls are of no serious significance and can be dealt with simply by pointing up with a weak lime mortar which will accommodate further minor movement.

Differential movement between stone external walls and brick internal walls is often apparent in common entrance passages. These cracks may require remedial action to restore the integrity of the junction.

Internal walls

It is not uncommon to find internal walls only one brick (9 inches) or even half a brick (4 inches) thick, extending the full height of a four-storey tenement and supporting the intermediate floors. Such walls may be liable to buckle at basement and ground floor levels where the load is greatest, and this may transmit extra load to adjacent walls. Reinforcement was often provided by the use of brickfilled studs and there is a danger of loosening the bricks in such a wall if plaster is hacked off or a new opening made.

Thermal movement

The long terraces of the New Town were built without any expansion joints, and the scarcity of distortion due to thermal movement illustrates the flexibility of traditional lime mortar. A rare example of the effects of thermal movement can be seen on the gable wall at 24 Fettes Row. This random rubble gable was intended to be an internal party wall but, because the terrace was never finished, it has been exposed to sun and rain from the south-west for at least 150 years. Because the gable is inadequately bonded to the front, back and internal walls, external changes in temperature and humidity have caused cracks at the junctions with the stair walls. The recommended remedy, though never carried out, was to protect the gable from the weather. This could have been achieved by applying a render or, ideally, an ashlar facing.

Chimneys

Cracks on the lines of internal flues may suggest weaknesses due to breakdown of the internal bridges between adjacent flues. If these are severe, or are accompanied by bulges on one or both sides of the chimney, then it is certain that structural failure has occurred.

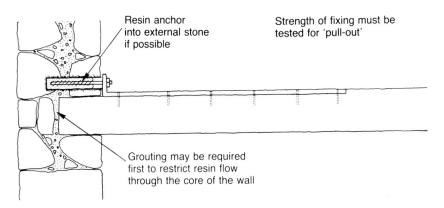

Resin anchor into external stone if possible

Strength of fixing must be tested for 'pull-out'

Grouting may be required first to restrict resin flow through the core of the wall

SECTION ALONG JOIST

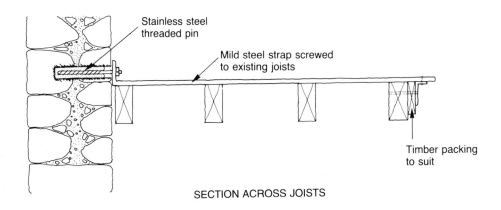

Stainless steel threaded pin

Mild steel strap screwed to existing joists

Timber packing to suit

SECTION ACROSS JOISTS

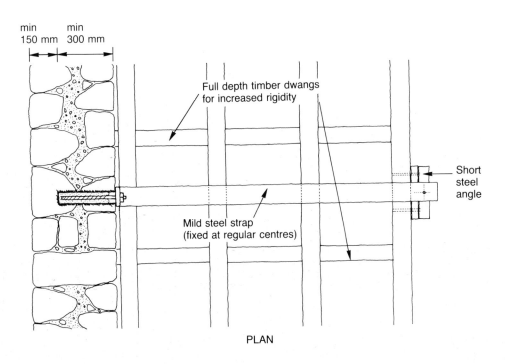

min 150 mm   min 300 mm

Full depth timber dwangs for increased rigidity

Short steel angle

Mild steel strap (fixed at regular centres)

PLAN

DETAIL SHOWING A COMMON METHOD OF ANCHORING FLOOR TO PERIMETER WALL

BEFORE RESTORATION

Photo: The Royal Commission on the Ancient
and Historical Monuments of Scotland

AFTER RESTORATION

79

Photo: Stewart Guthrie

1-8 HADDINGTON PLACE/1-9 ANNANDALE STREET

In certain cases these cracks and bulges may extend down a gable or split-level party wall and combine with external erosion to produce a very weak structure. Expert analysis is essential as a bulge may indicate a danger of collapse of the outer skin of the stone wall. With the flues possibly full of rubble and debris, the outer skin may be acting as a retaining wall. Where there is moderate cracking and bulging in the gable flue area, it may be possible to tie back the outer skin of stone, but in many cases rebuilding is the only answer.

Failure of mortar and render is a common occurrence on exposed chimney-heads which, together with erosion of the stone on both faces, can lead to a significant reduction in strength and ability to resist wind forces. Inward-leaning chimneys require careful assessment. Close-up inspection is essential; chimneys can appear sound from ground level but subsequent detailed inspection can often reveal a completely different picture.

Stairs

Diagonal cracks on the internal walls flanking common stairs are seldom active. Any evidence of active movement should be referred to a structural engineer, because it may indicate slight movement in an outer wall.

Movement in pen-checked stairs often indicates stress in the internal walls of a building. Because of the interdependent nature of these steps, it is critical that movement is minimal because individual steps may crack and could result in the collapse of the stair.

ANALYSIS OF MOVEMENT

It is essential to discover whether the suspected movement is active. Fresh cracks in the stone will be bright and clean, whereas old cracks will have dirt on the broken edge. Torn paper or cracks on walls which have recently been repapered are a sign of active movement. If this is suspected, the cracks should be examined regularly to determine the rate of movement. Pencil marks at the limit of each crack, with the date written alongside, are a simple way of recording movement. Another simple method is to fill a short length of an internal crack with plaster and date it; if the crack widens, another short length can be filled and dated. (Note: there will be initial cracking as the plaster sets.)

If more precise measurement is required, there are a number of proprietary telltales which can be used. One type, for example, gives an accurate reading of the relative movements between small metal studs, fixed in a triangular pattern around a crack, while another consists of two pieces of clear acrylic, calibrated so that any movement between the two can be monitored. For accurate and repeated monitoring, both these types of telltale have to be accessible over a period of time, which can be difficult if they are high on the external wall and scaffolding has been removed. The traditional glass-strip telltales fixed across a crack are unreliable; they are likely to be broken accidentally or by normal thermal movements. A special theodolite, fixed at an external datum, can be used to take readings on permanent plumb points and is an extremely accurate method of monitoring.

Bulges, although rare in the New Town, can be dangerous and may indicate loss of integrity between internal and external walls, or occasionally the loss of bond between the core and the external leaf of the wall. Large areas of plain gable are prone to bulging due to lack of through stones.

MORTAR JOINTS

INTRODUCTION

The original mortar used for the sandstone walling of the New Town was made of slaked lime putty and sand. The lime may have come from the Burdiehouse quarry near Loanhead and the sand, a beach gravel, from Gullane in East Lothian. These were mixed in proportions varying from 1:2 to 1:4 lime:sand, without any known additives. The result was a mortar which did not shrink when it set, and yet was elastic enough to accommodate slight local settlement.

All stones, both ashlar and rubble, were bedded in this mortar, and in pure lime putty along the front edge of squared ashlar stones. Rubblework was not pointed as a separate exercise but the building mortar was exposed. This means that any repointing may result in the removal of original mortar from the wall.

COMMON DEFECTS

The gradual loss of mortar from the face of a wall, particularly in vertical joints, is to be expected with normal weathering processes. Considerable damage is caused by unsuitable pointing mixes and poor pointing techniques, which damage not only the joints but often the stones themselves. Perished or faulty joints can permit water to penetrate the core of the wall, loosening the stones and allowing frost to damage the wall.

These dangers can be avoided by ensuring that joints are well filled with mortar and free from cracks. A weather eye should be kept open for any signs of decay or open joints which could allow water to penetrate the joints in cills, cornices, string courses and copings. Occasionally, it may be advisable to protect these projecting features with a lead flashing, although this was not the traditional practice in the New Town.

MORTARS

The purpose of mortar is to keep the stones apart, not to stick them together. A mortar must be durable, giving protection against the weather and be able to support superimposed loads; it should also possess good working qualities for bedding and pointing.

Mortar should be sufficiently resilient to accommodate minor movements in the masonry, but it should never be stronger or denser than the adjoining stones. Over-strong mortar will not only craze but, if it is more impervious than the stone, it will prevent drying-out through the joints and cause moisture to evaporate through the stones themselves thereby accelerating decay.

Mortar is composed of sand and a binding agent. Originally the binder was lime putty with or without additives, and the majority of buildings over 150 years old are still performing well with these mortars. After the 1830s the use of Portland cements in conjunction with or instead of lime became increasingly popular and the understanding of traditional mortars declined.

There is now a reawakening to the logic of using mortar similar to the original when repairing an historic building, and this approach is to be encouraged. However, as the technology, science and physical properties of pure lime mortars vary considerably from cement-gauged mortars the logic must be tempered with the practicalities. Insufficient knowledge will result in failures and failures will result in mistrust of the materials and disillusionment of the workers on site.

The only way to fully appreciate the methods of using lime is by hands-on experience beside an experienced craftsman.

MATERIALS

Sand

Sand for mortars should be sharp, durable, clean and free from coal, iron pyrites, soluble salts, clay and other impurities, all in accordance with BS 1200 and with a good spread of gradings. Thorough washing is essential to remove silt, sea salt or other organic matter which may weaken the mortar, affect porous masonry or cause efflorescence, especially in gauged mixes (lime mixes are more tolerant of impurities and are sometimes enhanced by them). Muddy sand which stains the fingers or balls up when rubbed should be rejected.

Sharp sand, i.e. with angular grains, produces a mortar of greater strength and tougher workability than a soft sand with rounded grains. The sand should be graded to produce a mixture of fine and coarse grains which will prevent voids in the aggregate. For pointing ashlar, a sieved fine sand should be used, because in narrow beds even a small pebble can crack a stone. For wider joints such as in rubblework the graded aggregate should include about 25 per cent of 6 mm diameter grit or river gravel.

Sand, particularly fine-grained sand, is affected by a phenomenon called bulking whereby damp sand has a greater volume than when it is either dry or very wet. However, this problem can be avoided by the use of ready-mixed coarse stuff.

The colour of the fine particles in the sand determines the colour of the mortar, and aggregates should therefore be chosen with care. It is important to ask for samples of sand before issuing the specification.

Colour and pigments

The colour and texture of existing mortar can be matched with carefully chosen sand and lime. Artificial pigments will fade after a while, especially if they contain fugitive dyes, and may produce dead colours which are unsuitable for historic buildings. Mortar should never be darker than the colour of the stone.

Limes

Lime is available commercially in three basic forms: as bagged dry hydrated lime, bagged dry hydraulic lime (French lime) or lime putty usually in plastic buckets. Each form has its own distinct qualities and method of use:

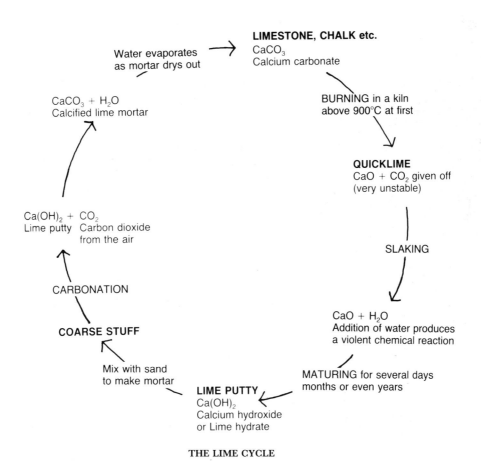

THE LIME CYCLE

1   *Lime putty:* Traditionally, lime putty for building work was slaked on site, either by adding quicklime (burnt lime) to water or by mixing crushed quicklime and sand together in a pit and then adding water to slake. It is now more usual to obtain matured lime putty from a supplier and to add sand to form coarse stuff in a variety of specifications. Some suppliers have ready-mixed coarse stuff available.

The essence of a successful mix is the age of the putty at time of mixing and that of the coarse stuff when used. The former should be at least three months and the latter two months. It must be noted that these are minimum times because the longer both are left to mature, the better the final result. It is therefore essential that the suppliers declare the true age of their putty/coarse stuff at time of buying.

The coarse stuff will require to be knocked up, i.e. rammed and beaten before use, to make it workable and to drive off excessive water; water should never be added. Mechanical cement-mixers are unsuitable for this task.

Lime putty for use in the New Town usually originates from Derbyshire, which is very white and pure, or Cumbria, which is greyer because of (desirable) impurities.

Coarse stuff can be used by itself and will set (carbonate) without additives. However, although this is the preferred method, it is very sensitive to changes in humidity and temperature, and in order to obtain a quicker set very small proportions of cement, brick dust, or HTI powder can be added. The density and hardness of the final result will be altered by these additives and experimentation is therefore required before specifying the proportions of the mix to be used.

2 *Dry hydrated lime:* Lime mortars can be made using a dry powder of calcium hydroxide (also known as builders' lime or hydrated lime). This can be used by mixing the powder with a small amount of water to form a thick paste which should be stored for as long as possible before use. However, this does not produce a self-setting mortar and an additive, usually Ordinary Portland or White Portland cement, is essential for a set to take place.

3 *Dry hydraulic lime:* Hydraulic limes are made from limestone containing a proportion of clay, calcium and magnesium carbonates, with impurities such as iron and sulphur. Because of the clay content, the chemical changes which take place during the kilning process are more complex than with non-hydraulic lime, producing a material which will set under water (hence hydraulic).

The raw material occurs naturally in Britain but is, at the time of writing, no longer commercially available. An eminently hydraulic lime is currently imported from France and the ENTCC can give further details of suppliers. This is light buff in colour and available as a powdered hydrate which is delivered in sealed plastic sacks and must be kept dry.

In his book *Mortars, Plasters and Renders in Conservation,* John Ashurst gives recommendations for mixing thus: 'The lime must be mixed very thoroughly with the selected aggregates and with the minimum amount of water to make the coarse stuff workable, the mixed material being able to take a 'polish' from the back of a shovel. Mixing should take place on a clean, boarded platform before any water is added and then again after watering. This coarse stuff must be used within four hours and must not be knocked up after stiffening has taken place. Correct judgement on the quantity required for each working phase is therefore important.'

Mortar made with hydraulic lime should not be gauged with cement.

Water

Water gives mortar its plasticity but plays no chemical part in the mix. It should be clean, potable and comply with BS 3148. Only the minimum amount should be added for workability. Too much causes excessive shrinking and cracking as it dries out or inhibits the set.

Cement

If used at all, cement will be Ordinary Portland Cement, conforming to BS 12. White Portland Cement can be useful for gauging where colour matching is required; it is weaker than Ordinary, which is probably an advantage.

Sulphate-resisting cement is recommended for use in flues and chimneys, and below external ground level in places where stonework is liable to sulphate attack.

MORTAR MIXES

For building and pointing work in the New Town the following mortar mixes can be recommended:

1 *Lime putty:sand*
   (a) 1:3 coarse stuff for pointing rubble and some wider joints in ashlar (dependent on aggregate size). Very well matured pure lime putty, with no sand, can be used for fine joints.
2 *Cement:non-hydraulic lime:sand*
   (a) 1:2:9 for building and pointing as outlined under 3(a).
   (b) 1:3:12 an alternative, weaker mix for pointing work, as at 3(b).
   (c) 1:1:6 a stronger mix for mortar fillets, building and pointing chimneys and other exposed roof work such as open verges.

These mortars are best made up by gauging ready-mixed coarse stuff with cement. Site gauging should be carried out accurately by volume using gauge boxes: e.g. a 1:2:9 mortar is produced by mixing one volume of cement with nine volumes of 2:9 lime/sand. (The lime fills the voids between the sand grains and therefore does not increase the volume.)

3   *Hydraulic lime:sand*
  (a)   2:5 for building rubble and ashlar work (including indents) and for general pointing of rubble masonry and wide-jointed ashlar.
  (b)   1:3 an alternative, weaker mix for pointing work; useful if stonework is soft or badly weathered.

JOINTING AND POINTING

All joints on an elevation should be pointed in one continuous operation after other masonry work is complete.

Ashlar

New stones should be cut exactly to the required shape with beds machine-sawn square to the face. Stone should not be hollow-bedded, and joints should not be more than 5 mm or less than 3 mm wide. Repointing of very fine joints is not recommended; the beds will have been protected against the weather even though the outer 10 mm or so may have perished, and it is difficult to avoid damage to the arrises when raking out the joints. Wider joints can be repointed as follows:

Rake out beds and joints for a depth of at least 25 mm or the width of the joint (whichever is the greater) until all soft mortar has been removed. A plasterer's joint tool, a thin steel plate, or a hacksaw blade marked with the correct depth should be used, taking care not to lip the stone. Hard mortar should not be disturbed and on no account should the stone be cut to widen the joint. Flush out the joints with clean water (using a hypodermic syringe for fine joints) and thoroughly hose down the wall below the joint to avoid mortar staining. A soft hairbrush should then be used to clean loose mortar and sand out of the joints. While the stone is damp apply masking tape or carpet tape to either side of joint. Form lime putty or lime putty:sand into a roll and press well into the joint using the ball of the thumb (lightweight rubber gloves should be worn); alternatively, a very flexible putty or palette knife can be used. The tape is removed when the mortar surface has firmed up sufficiently.

In dry weather a fine spray may be required to inhibit rapid set. Further working may also be required while the mix is green to prevent shrinkage cracking. Protection of the work with damp sacking can also be beneficial.

The pointing should be neatly finished, flush with the face or slightly recessed. When filling a deep joint, tamp with a 1:3 lime:sand mixture to within 38 mm of the surface.

If an arris is irregular due to spalling or mechanical damage and the stone is not to be replaced, the joint should be contour-pointed by recessing the mortar to conform with the chipped edge of the stone. Wider joints may be pointed with a 2:5 or 1:2:9 mortar as already discussed.

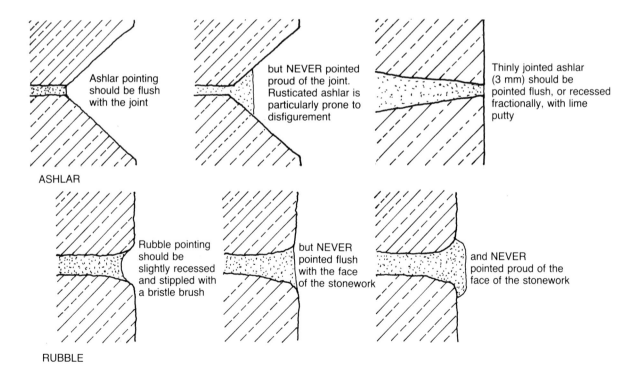

ASHLAR

Ashlar pointing should be flush with the joint

but NEVER pointed proud of the joint. Rusticated ashlar is particularly prone to disfigurement

Thinly jointed ashlar (3 mm) should be pointed flush, or recessed fractionally, with lime putty

RUBBLE

Rubble pointing should be slightly recessed and stippled with a bristle brush

but NEVER pointed flush with the face of the stonework

and NEVER pointed proud of the face of the stonework

Rubble masonry

The original mortar used in the building should be analysed for its aggregate grading and lime proportion. It is likely to contain a locally quarried river or beach sand, well graded from 6 mm down. It is best to match the original mix using sands which are available commercially from the Gifford and Eddlestone areas. The method of pointing will depend on the type of lime/gauged mix specified.

If pure lime mortars are being used then it should be possible to repoint defective areas without raking out, unless soft pockets are encountered. This allows the maximum of original fabric to be retained.

For hydraulic lime and gauged mortars, raking out to a depth to match the width of the joint or deeper is essential. The tool used for raking should be a fine chisel, no wider than the joint. Mechanical tools should never be used. Sometimes it may be necessary to remove small stone pinnings when raking out, and these should be replaced during tamping or pointing. The raked-out joints should be thoroughly washed with a hose or spray. Treatment with a herbicide at this stage may sometimes be useful if weeds or mosses have started to grow (see CONTROL OF VEGETATION). All deep holes should then be tamped with mortar brought forward in stages of 10 mm, adding small stone gallets to wider joints or holes. The final 40-50 mm should be left sufficiently rough as a key for the pointing mortar. The joints should then be thoroughly filled with mortar and left recessed to within 3 mm of the face. The pointing should not spread over the rounded edges of the stone. If a stone is 'buttered over' in this way, a thin skin or feather edge is formed which will eventually fall off, leaving a pocket to hold moisture and accelerate decay.

New pointing should harmonize in colour and texture with the old. A smooth joint struck with a steel trowel should be avoided and green partially-set mortar should be compacted with a bristle brush or pot scourer to expose the aggregate. Alternatively, after the initial set, the mortar can be sprayed with a fine jet of water (not too strong or it can scour the mortar) to remove the whiteness of the lime and to bring out the texture and colour of the sand.

Fresh pointing must be protected from frost, rain, strong heat and local draughts. During hot weather, periodic wetting is necessary to retard the drying-out process, which if too rapid will result in weak mortar.

29-31A SCOTLAND STREET
AFTER COMPREHENSIVE REPAIRS

87

Photo: John Johnson

STONE REPAIR

INTRODUCTION

Before contemplating stonework repairs it is essential to be sure that intervention is necessary and then to form a clear philosophy of repair. Excessive stone replacement or repair should be resisted. Every new stone represents the loss of part of the original historic fabric.

The three requirements for stonework are structural, functional and aesthetic. If the stone is structurally undermined then this can have a detrimental effect on the surrounding stones and on the building as a whole. The function of an external stone is to keep out the weather. Weathering and decay can upset this, and lead to further damage to stones and also to the general fabric. The aesthetic requirement of stonework is to display the architecture of the building, and is usually secondary to the first two, although in some cases it can be the most important requirement when the building is viewed as a work of art.

Damaged or decayed stonework can be repaired by indenting (piecing in), which involves cutting out the defective stone and building in a new stone. Patch repairs using a gauged mortar are seldom successful and often cause more problems than they solve because the presence of cement in the mix will accelerate decay in the neighbouring stones.

INSPECTION AND DEFECTS

High-level masonry can be inspected from scaffolding or mobile hoists. Once the defective stones have been pinpointed they should be indelibly marked and scheduled on detailed elevational drawings or enlarged photographs. Structural cracking and the disintegration of stone surfaces are the most common and easily perceived defects; face-bedded stones are often the first to reveal this kind of damage. The loss of an outer layer of stone is almost certainly due to an accumulation of soluble salts, or degenerating clays, giving rise to contour scaling and delamination. Spalling may occur near to mortar joints and along arrises, due to salt or frost action or to eccentric loadings. Cornices, string courses and other projections usually erode at a greater rate than the surrounding stones which they protect.

It may not be necessary to replace a stone just because it has a defect. It is tempting to over-repair and it is important to distinguish between superficial and structural defects. The residual structural life of each stone should be considered and if it can reasonably be expected to survive for another thirty years, it should be left, regardless of its appearance. A defective but viable original stone is often preferable to a new one.

CHOICE OF METHOD

Opinions vary as to whether indenting, which involves the loss of original stone, is as true to the spirit of conservation as mortar repair. Mortar repairs are drab in comparison with the quality and texture of natural stone. It is also usually necessary to cut back original stone before building back with mortar. Because mortar is a semi-porous material with properties significantly different from stone and because of the difficulty of ensuring a permanent bond between the two materials a mortar repair will have a considerably shorter life than a stone indent. The ENTCC actively discourages mortar repairs unless the mortar is used almost as an extension of pointing to fill in small areas of decay and thereby extend the life of a stone which would otherwise have to be replaced.

The use of silane and silicone treatments to weather-proof stone is not recommended because serious damage can occur if interstitial condensation builds up within a stone. In traditional construction, the free movement of water vapour through the fabric of a building in both directions is essential. Silanes can be used to consolidate stone but are really only justifiable on irreplaceable stone ornaments. Knowledge of long-term effects of these consolidates is small, and as they are non-reversible, their use should be resisted.

INDENTING

Stone for replacement

It is best to reuse stone from the same building or to use new stone from the same quarry; but if stone has to be brought from another quarry, samples should first be compared with and, if possible, built adjacent to the existing masonry before a final choice is made. Stone should compare as closely as possible with the original in terms of both chemical and statisgraphical composition and colour. They can be more porous but should never be less. Stones used for indents in ashlar facing or for any replacement work must be correctly bedded.

Cutting out

Before repairing, it is imperative to remedy the cause of the decay by eliminating sources of soluble salts, preventing the passage of moisture, rectifying active structural faults and so on. When repairing badly decayed ashlar it is normal practice to cut away the defective stone to its full depth and bed. This should be done by a skilled mason using a hammer and chisel. Power saws should only be used to take away stone from the heart of a block because of the likelihood of causing damage to surrounding stones. Power hammers should not be used although small air tools such as pneumatic chisels can be effective in the correct hands.

If decay does not necessitate the removal of the whole depth of a stone or if the stone is bonding the outer and inner skins of masonry (throughs, headers, inbands and so on) then the surface can be cut back, usually by 100 mm on the face but never by less than 75 mm. It must be noted that this can have a serious effect on the masonry structure, weakening the loadbearing capacity of the wall. The through stones around openings are, in a well-fenestrated wall, the only means of tying outer and inner skins of masonry together. Any serious weakening of the tie between the two faces can result in the parting of the skins of the wall, and the resulting pattern of cracking and deterioration may be worse than the original defect which the indent was intended to overcome. Worst of all is the effect of relieving the well-wrought outer skin of ashlar or rubble masonry of its share of the load, throwing the whole of the floor load onto the inner skin, which invariably is of a much poorer quality of random rubble masonry, with wide, soft joints. This results in compression of the inner skin, which in turn can cause bowing of the wall, separation of the outer skin and extensive internal cracking.

It is therefore essential that when the new stone is built into the wall it becomes a structural part of the fabric. Correct mechanical bonding between the new and old will be necessary to achieve this using stainless steel cramps, dowels, restraints and other anchorages. It will also be necessary to ensure that the old stone which is to receive the mechanical fixing is of adequate strength for the job it is required to do. It is recommended that a structural engineer be consulted before carrying out indent repairs to ensure that the correct fixings are used.

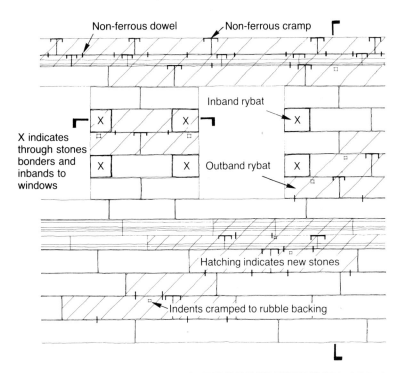

UPPER PART OF TYPICAL FRONT ELEVATION EXTENSIVELY INDENTED

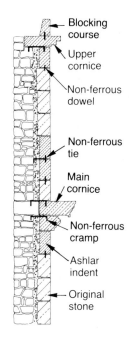

SECTION

PLAN OF PIER

89

One facing stone can be removed safely without propping, because the weight of the wall is supported by the adjoining facing stones and the inner rubble wall. If several adjoining stones have to be removed simultaneously, the adjoining masonry must be securely propped and wedged. A structural engineer should always be consulted if large areas of masonry have to be replaced. There are several examples in the New Town of total replacement while retaining the inner rubble skin of the front wall.

Indents should always be a full course in height, at least square in elevation and normally the same length as those stones which they replace. If the vertical joints (perpends) in a masonry wall are not to a regular pattern, and if only part of a long stone is defective, it may be possible to cut out and indent the defective part, although small-scale piecing in should generally be avoided. In such cases the vertical joints between new and existing stone should break bond by at least 75 mm but preferably by the course height.

Settings and anchorages

When bedding and backing stone indents with lime or cement/lime mortar, reference should be made to MORTAR JOINTS. Mortar alone can rarely be relied upon to give sufficient stability to new stone indents, especially where fairly large areas of repair work are undertaken. Dowels, cramps, joggle joints or other fixings will normally be required to tie the indent to adjacent stones. Metal anchorages should not rust and should be austenitic stainless steel or a non-ferrous metal such as phosphor-bronze; copper and aluminium fixings should be avoided. Galvanizing or coating with bitumen affords only temporary protection to iron or mild steel.

Cramps are used to secure cornices, blocking courses, balustrades and decorative masonry. They are hooked across the joint into slots in adjoining stones, and are usually recessed into the top of the stones. Anchor-bolts set in epoxy resin can tie indents to the rubble backing or to adjoining bonding stones. Joggle joints cut male and female fashion into the ends of adjoining stones, or formed in grout poured into a 'V' joint between stones, will prevent local movement. Non-ferrous dowels will positively locate mullions, columns and balusters.

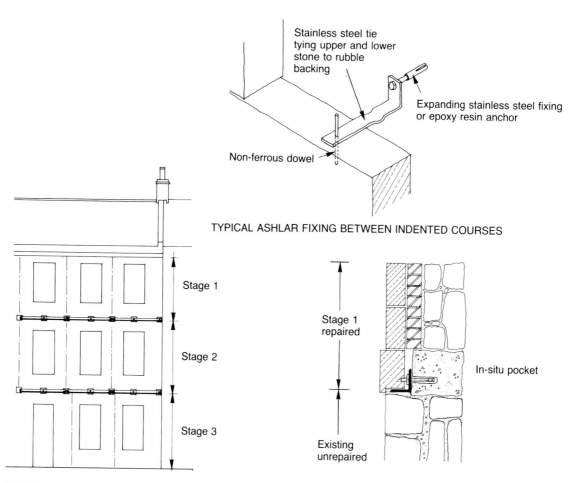

TYPICAL ELEVATION TO BE REFACED ON CONTINUOUS
STEEL ANGLES BOLTED TO IN-SITU CONCRETE POURED
INTO POCKETS IN RUBBLE BACKING

Finished appearance

For the appearance of mortar joints and pointing see MORTAR JOINTS. Surface finishes to ashlar blocks and so on are discussed in MASONRY WALL CONSTRUCTION. Where possible, hand tooling of finishes is preferable to machine tooling.

If soiled stonework is indented, a marked contrast between old and new work will result. To purists the honest appearance of a repaired facade can be readily accepted as part of the evolution of the building as a whole but others dislike the patched effect of newly indented masonry. The ENTCC advises against any cosmetic treatment of indented stone, either cleaning the old stone or distressing the new. This is because within a few years of repair, the effects of natural weathering will have gone a long way to regularize its appearance. Any contrast can be reduced by washing the facade after repair, using only soft bristle brushes and clean water with no additives.

PORTICOS

INTRODUCTION

Most houses and main-door flats in the New Town have plain door openings, but some are adorned with aedicules and a few have porticos. An aedicule has columns or pilasters supporting an entablature and pediment, directly applied to the wall face, while the entablature of a portico projects from a building, creating a roof supported by free-standing columns.

Aedicules are found at Northumberland Street (West), Wemyss Place, Queen Street, Dublin Street, Gayfield Square, Hillside Crescent, Windsor Street, Annandale Street and Regent Terrace. There are original porticos by W. H. Playfair at Brunswick Street (1826) and by William Burn at Henderson Row (1825). Porticos were part of David Bryce's designs for the Royal Society of Edinburgh (1847) at 22-24 George Street and for the Union Bank at 62-66. A most elaborate two-level design was produced in 1845 by Thomas Hamilton for the Royal College of Physicians at 9 Queen Street and porticos were added later to the houses on the north side of St Andrew Square and at the end of Queen Street.

All these ornamental porches are in the classical style, with Doric or Ionic orders on the earlier domestic buildings and the Corinthian order on later and more elaborate examples.

INSPECTION

Porticos and aedicules should be included in annual and quinquennial inspections, paying particular attention to rainwater conduction and damp penetration which can cause staining and spalling of masonry and decay of the portico ceiling.

DEFECTS AND REMEDIES

Where porticos and aedicules are part of the original design and bonded to the front wall of the house, defects will probably be limited to open joints which enable water to enter and damage the masonry. Rainwater must be adequately shed from the roof which is normally covered with lead. Rainwater outlets and pipes must be kept clear and overflows should be provided. Flowerpots and window boxes set on the roof of porticos add colour to the street scene but, unless they are kept tidy, the soil and foliage can block rainwater outlets with disastrous results.

Where a portico is a subsequent addition there may be an open risband joint at the face of the wall; this joint can be tied by stainless steel cramps if major movement is evident. Open joints should be filled with lime mortar. Lead flashings have been used to protect the joint between the wall face and parapets, but their insertion should be discouraged.

Open vertical joints, such as exist on the upper surfaces of the entablatures, are the most common defect, allowing water to penetrate and damage the structure. Such joints should be carefully raked out and pointed with lime mortar.

Stone repairs should be like for like; the use of precast concrete (artificial stone) is not recommended. Repairs such as these have not proved satisfactory because, over a period of time, their different weathering properties become conspicuous, and the use of materials harder than the stone itself can encourage further erosion of the original fabric.

Rain dripping on the platt splashes against columns whose bases, being constantly damp, tend to erode, especially as this is the drain point for any rainwater which has been taken into the stone at higher levels. Tapered and fluted columns, capitals and consoles can only be carved by hand; replacement in stone is essential even although it may be expensive. There are few carvers capable of doing this sort of work to the required standard, and so proper maintenance is very important.

Single replacement stones for large monolithic columns may not be available from the quarry, but if only part of the column is eroded, indenting is possible. If a new stone is to be indented around the whole of the column shaft it is wise to ask a structural engineer to check the strength of the remaining sound stone. Very careful mortar repairs may be appropriate in certain cases because they allow the maximum original fabric to be retained, although with a risk of frost and salt damage they may need regular attention.

Structural damage to porticos is most likely to be due to movement in the arch supporting the platt (see STEPS, PLATTS AND ARCHES). Strengthening may be required, but if the platt has to be rebuilt it may also be necessary to dismantle and rebuild the portico.

Some of the lath and plaster ceilings have elaborate mouldings. If timber lathing has rotted due to exposure to damp conditions, keys may be damaged; for remedial action see PLASTERWORK.

Masonry should not be painted or rendered and existing paintwork should be removed if this is feasible without damage (see REMOVAL OF PAINT FROM STONEWORK); if it is not, it may be safer to repaint in a stone-coloured oil paint. In an area covered by an Article 4 Direction, permission for painting or repainting will be required from the Planning Department (see LISTED BUILDINGS AND CONSERVATION AREAS). Plaques, nameplates and signs should not be fixed to pilasters or columns.

When repair work is being carried out on a building, all mouldings and carved stone should be protected with timber against accidental damage by scaffolding.

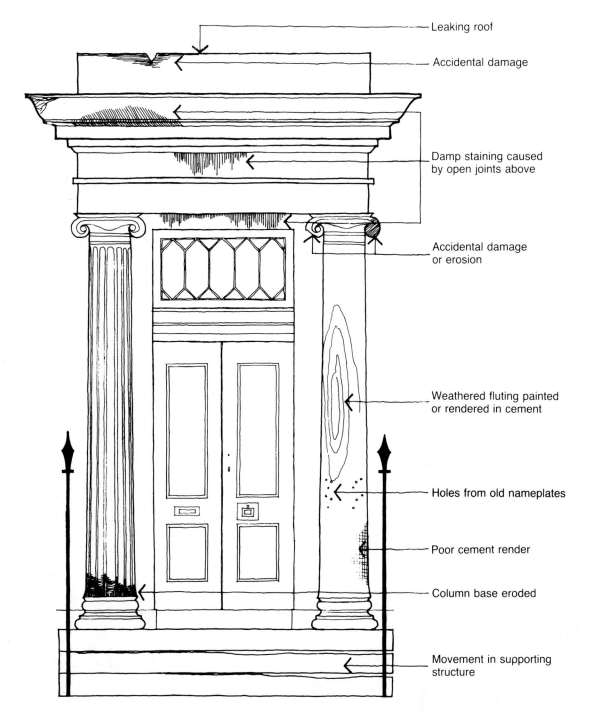

POSSIBLE DEFECTS

NEW PORTICO AT HENDERSON ROW

Photos: André Sypestyn

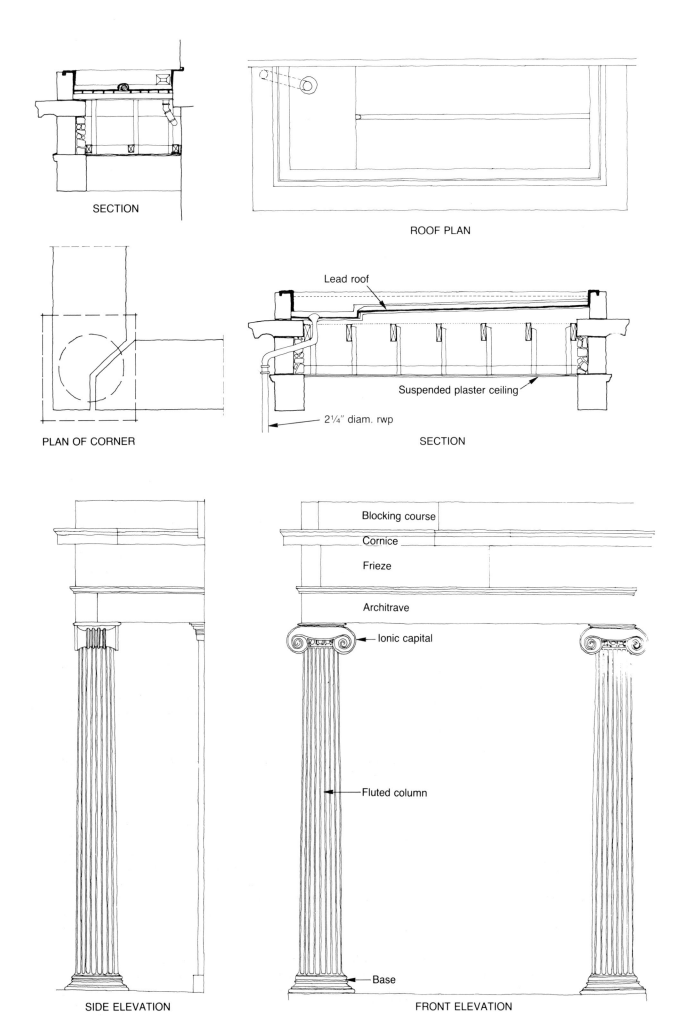

SECTION

ROOF PLAN

PLAN OF CORNER

Lead roof

Suspended plaster ceiling

2¼" diam. rwp

SECTION

Blocking course

Cornice

Frieze

Architrave

Ionic capital

Fluted column

Base

SIDE ELEVATION

FRONT ELEVATION

NEW PORTICOS AT 16-26 HENDERSON ROW   96

DOORS

INTRODUCTION

Doors, with their distinctive ironmongery, are an important feature of New Town buildings. The majority of doors are simple four or six panel type and great variety is achieved by the different configurations of the panels and mouldings, but curved doors, jib (concealed) doors and dummy doors are also occasionally found. Many doors are in excellent order and careful maintenance will keep them so; others can be improved and renovated for a very small cost. Good maintenance will avoid the need for authentic replacement, which can be expensive.

CONSTRUCTION

Generally, external doors were made of Baltic pine and internal doors of yellow pine; hardwood, usually oak or mahogany, is only found occasionally in the New Town. All doors were painted with lead-based oil paint and in principal rooms they were often grained to resemble mahogany or other hardwoods. Doors were framed with mortice and tenon joints secured by small wedges but not dowelled. Panels varied in size and section but were limited in width to available timber sizes. They were fitted into grooves or secured by loose mouldings. Simple flush beads are usually found on basement doors and are frequently seen on common stair doors. Bolection mouldings are more common on doors at ground level, and fielded panels are found on doors to the more important rooms.

When repairing or replacing doors the same construction should be used and moulding profiles must be copied accurately. Good quality redwood is recommended for painted doors but if this is not available other timbers, preferably softwood, may be considered. Timber should be fine-grained, relatively free of knots and shakes, and quarter sawn to prevent rails and stiles warping. It is always wise to check the available types and quality of timber stocked by local suppliers before specifying. Panels should be made in solid timber and not plywood for the best appearance.

DEFECTS AND REMEDIES

Moisture penetration may lead to a number of problems. Doors and frames can warp and twist, or timber may swell, causing doors to bind. Rot, particularly wet rot, is usually caused by contact with wet stonework (sometimes found at the bottom of the door standards where they touch the step or at a flagged basement floor), at a faulty joint between a lining and a wall, or where the paintwork has not been properly maintained, resulting in the breakdown of the protective finish and allowing water into the joints. While redecoration is important there is also a danger that successive layers of paint will build up over mouldings and blur the detail.

Structural and thermal movement can also cause a door to warp or distort. Thermal movement is often the cause of cracked panels, but it may also be responsible for the weakening of a door because joints have opened. Joints, rails and standards can be weakened by alterations, for example to fit new locks or door furniture such as large letter plates, or by accidental damage through strain or impact.

Ill-fitting doors can be remedied by planing or building up the door to suit the opening, or by similarly adjusting the frame. In some cases a warped door can be twisted back and clamped (and occasionally wetted) to bring it back into true. Alternatively, the offending stile or rail can sometimes be replaced with new quarter sawn timber. Where part of a door has been damaged the affected areas can generally be cut out and a new section of timber patched in. In some extreme cases it may be necessary to dismantle and reassemble the door, for example where panels which are fitted into grooves in the door frame are cracked and need to be replaced. If timber replacement is required because of rot it is pointless to repair the defect without also remedying the cause. For example a new patch at the bottom of a door standard should be kept away from the stone step or isolated from wet masonry by a strip of damp-proof course.

External doors should be repainted at least every five years (see MAINTENANCE). The excessive build-up of paint can be removed by caustic dipping but while this method protects the detailed mouldings from damage, caustic stripper may soften the wood and weaken the joints. For further information on stripping, preparing and repainting doors see EXTERNAL PAINTWORK and DOOR FURNITURE.

**INTERNAL DOOR ON FIRST FLOOR
NO. 13 HERIOT ROW, GRAINED TO
RESEMBLE SATINWOOD**

98

Photo: Alan Forbes
by kind permission of the
National Galleries of Scotland

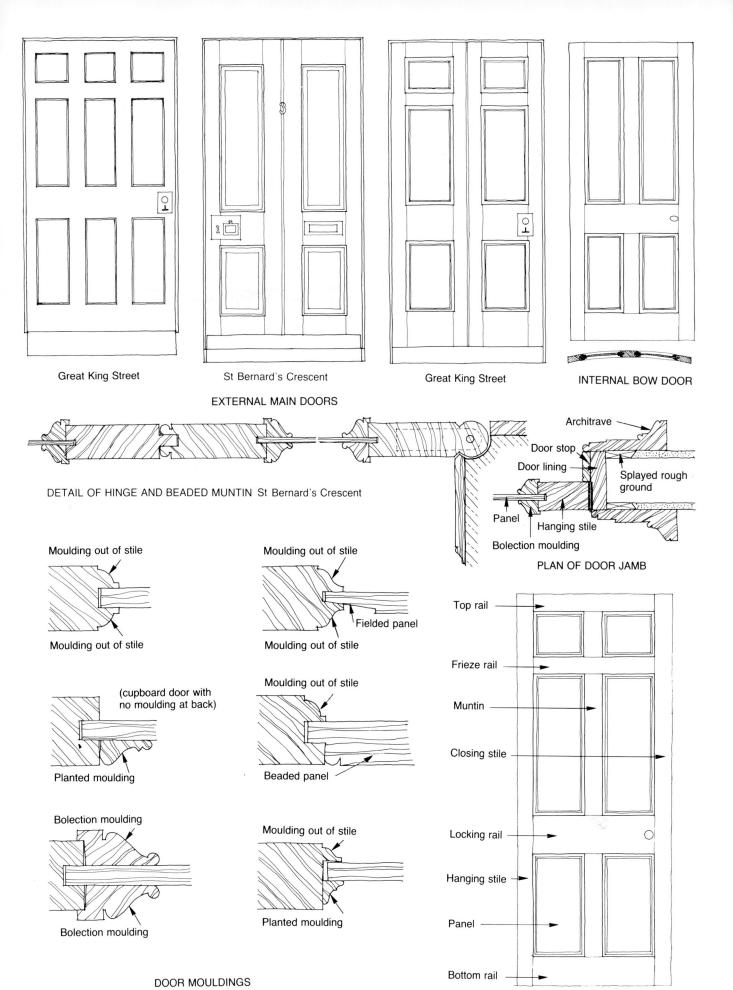

Great King Street

St Bernard's Crescent

Great King Street

INTERNAL BOW DOOR

EXTERNAL MAIN DOORS

DETAIL OF HINGE AND BEADED MUNTIN St Bernard's Crescent

Architrave

Door stop

Door lining

Splayed rough ground

Panel

Hanging stile

Bolection moulding

PLAN OF DOOR JAMB

Moulding out of stile

Moulding out of stile

Moulding out of stile

Moulding out of stile

Fielded panel

(cupboard door with no moulding at back)

Moulding out of stile

Planted moulding

Beaded panel

Bolection moulding

Moulding out of stile

Bolection moulding

Planted moulding

DOOR MOULDINGS

Top rail

Frieze rail

Muntin

Closing stile

Locking rail

Hanging stile

Panel

Bottom rail

INTERNAL DOOR

FIRE PRECAUTIONS

When dwellings are upgraded, or converted into hotels or offices, the Building Control Division of the Department of Property Services and/or the fire authority may demand that fire doors be installed at certain openings. Requirements to comply with the Building Standards (Scotland) Regulations 1990 are set out in Section E2 of the supporting *Technical Standards*. When the upgrading of fire resistance is necessary in historic buildings, this should be achieved if possible by fire-proofing the existing door assemblies, rather than by replacing them with proprietary fire doors and frames. It must always be remembered that the new use may only be temporary and therefore even fire safety work should be reversible.

The most common requirement is for a half-hour fire resistant door (the FD 30 door) which may also be required to control smoke (FD 30S door). All fire doors must be self-closing. The performance of leaves, along with their frames and stops, when tested must comply with BS 476. The decision on whether doors require upgrading to higher standards should always be referred to Building Control and the local fire authority.

The least damaging way of upgrading the fire resistance of an existing panelled door to the FD 30 standard is by the application of an intumescent paint system. Paint systems normally require two or three coats and are available in clear, matt and gloss finishes, in a wide range of BS colours. The existing door must be in good condition and fit snugly within its frame. Stiles must be a minimum of 35 mm thick and panels be no thinner than 7.5 mm. Door stops must not be less than 25 mm thick or, where this is not possible, an intumescent strip must be set into the edge of the frame of the door. The preparation of wood surfaces and the application of paint itself must accord strictly with the manufacturer's instructions in order to achieve the correct loadings. The manufacturer must certify that the quantity of paint supplied and its application will achieve the required fire resistance, and the painting contractor must certify that the treatment is in accordance with the manufacturer's instructions. Ideally, doors which have been treated should be identified with a small, surface-fixed plate so that it is taken into account in future redecoration work.

The technique for upgrading a door assembly is a matter for negotiation with Building Control. Where intumescent paint is not acceptable the FD 30 standard can be achieved in the following manner, although this considerably alters the appearance of the door and should be avoided wherever possible:

1   The door must be in good condition, with the stiles and rails having a minimum thickness of 45 mm. It must be true and lie straight against the stops when closed. The gap between the door edge and frame should not exceed 3 mm.
2   The door stop should not be less than 25 mm deep. An existing stop can be built up, using a hardwood strip glued and screwed in position.
3   A panel of fire resisting material should be fitted closely over the existing panel, unless the door construction consists of loose panels secured by planted mouldings, in which case the timber panels can be replaced by fire resisting board held in place by the original planted mouldings. If this is done, the original timber should be labelled and laid aside for possible reinstatement at a later date, or for use elsewhere.
4   The insert must be fixed (e.g. by gluing) to the original panel to prevent it falling away during a fire.
5   The joint where the insulation board and frame abut must be covered by a new hardwood beaded bolection moulding at least 12.7 mm × 25 mm, glued and pinned into position.
6   The hinges must be of non-combustible material.
7   A self-closing device capable of closing the door from any angle should be added, either self-closing or check action type (see DOOR FURNITURE for further information).

If stiles and rails are less than 38 mm thick the door may only be retained if the face of the panels on the room or risk side are entirely covered by fire resistant board of appropriate thickness to be relatively flush with the framing of the door, and the whole of that side of the door then covered by fire resistant board, securely screwed to the door. In such a situation it may be better to set the door aside for reuse elsewhere within the building, and to construct a new fire door to match the original pattern.

DOOR FURNITURE

INTRODUCTION

Door furniture is a distinctive feature in the New Town. Within an overall uniformity there are subtle variations in the design and quality. For instance, the finest materials were used on internal doors in the most important rooms, less high quality on external main doors and cruder cast iron on external basement and cellar doors.

Much of the original door furniture remains but a considerable amount, particularly from internal doors, has been lost over the years. When restoring a door, the original furniture should always be retained. As in all matters of restoration, when replacing missing or inappropriate door furniture, examination of similar properties which have retained original features will be invaluable, though care must be taken to avoid copying inappropriate examples. Most of the commercially available architectural door furniture described as 'Georgian' is not acceptable for use in the New Town because it is based on English models. Good reproductions of some Edinburgh door furniture are available or can be made to specification. Advice is available from the ENTCC, where samples of external door furniture are displayed. If one cannot find authentic replacements, second-hand but excellent Victorian door furniture is appropriate and infinitely superior to modern 'Georgian' reproductions. Sometimes door furniture has been replaced in the past with good quality contemporary items and, in this instance, retention of these is preferred to replacement with poor reproduction fittings.

MATERIALS AND MAINTENANCE.

New Town brass fittings have a high content of zinc. Fittings of aluminium or plastic are inappropriate and should be replaced if possible. Brasswork which has been painted or electroplated can be restored and the ENTCC can advise on specialist help for restoration.

Regular polishing will keep brass looking attractive and in the best condition, although chemicals for cleaning brass may harm adjoining paintwork and stone if used carelessly. Covering the adjoining stone or timber with a plastic or cardboard template can make cleaning easier.

Door furniture in Edinburgh is usually left in position when painting doors; the side faces of brass backplates and numerals are painted to avoid the need for these to be cleaned, thereby avoiding marking the painted surface of the door.

DESCRIPTION

Numerals

The earliest numerals are likely to have been the small brass plates on which Roman or Arabic characters were engraved and filled with black wax. For restoration purposes, filling with paint is preferable because wax crazes in the sun. Brass plates were superseded by cast brass numerals, and six complete and different sets have been identified in the New Town, two in Roman characters and four in Arabic. New brass numerals specially cast to match those most commonly found in the New Town are available from the ENTCC. Painted numerals were probably standard on basement and some common stair doors. Sometimes the numeral was painted on stonework above or adjacent to the door and formed part of a painted panel, for example circular or shield shaped with a shaded border. Numerals painted on fanlights were probably a later introduction.

Numerals, backplates and other items of door furniture should be secured to doors by a threaded rod screwed into a ferrule on the reverse side of the door and held in place by a nut and washer sunk below the inner face of the door so that the method of fixing is invisible, or secret.

Nameplates

Nameplates should be brass, with incised lettering in Roman characters. These were originally filled with black wax, but are now filled with black paint. Small plates for residents' names are screwed to the backplate of the bell pull outside the door to the common stair or the flat. Larger brass plates, used by advocates and other professionals, are usually screwed to the muntin of the door. When large new nameplates are required by companies these should be placed on the wall beside the door and should exactly cover the face of the stone to which they are fixed. Fixings should always be brass or stainless steel either secret fixed or with brass covers, and must be well clear

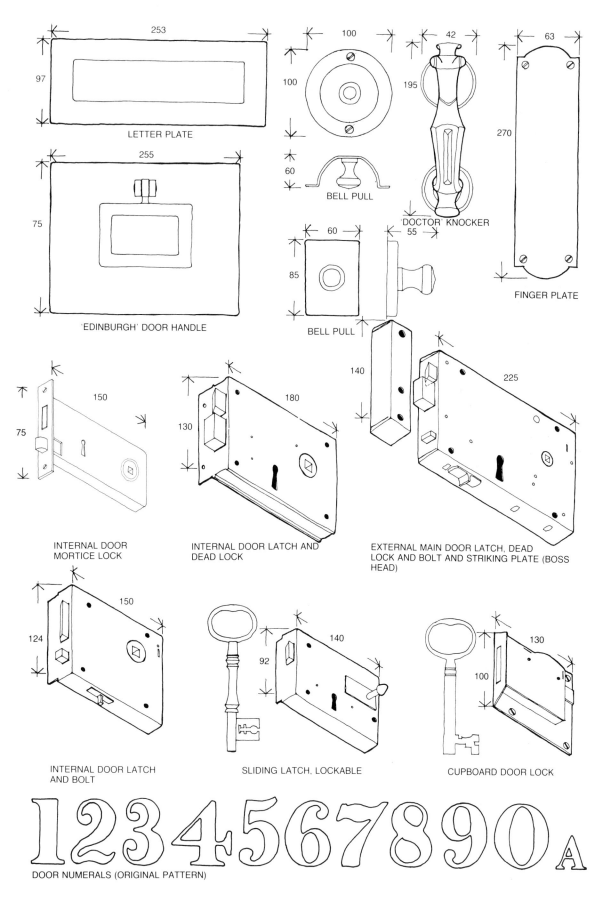

253

97

LETTER PLATE

100

100

60

BELL PULL

42

195

'DOCTOR' KNOCKER

63

270

FINGER PLATE

255

75

'EDINBURGH' DOOR HANDLE

60

85

BELL PULL

55

140

INTERNAL DOOR
MORTICE LOCK

150

75

180

130

INTERNAL DOOR LATCH AND
DEAD LOCK

225

EXTERNAL MAIN DOOR LATCH, DEAD
LOCK AND BOLT AND STRIKING PLATE (BOSS
HEAD)

150

124

INTERNAL DOOR LATCH
AND BOLT

140

92

SLIDING LATCH, LOCKABLE

130

100

CUPBOARD DOOR LOCK

1234567890A

DOOR NUMERALS (ORIGINAL PATTERN)

of the stone arrises. The plates should be of cast brass or of sheet brass covering a timber former, to resemble a cast plate. The plate may incorporate the bell pull, bell push or entryphone.

External handles and knobs

Several designs of door handle can be seen in the New Town, the most common being the Edinburgh handle, which is a nineteenth century design frequently used from the 1830s onwards. This design is found on main doors only and has a raised rectangular backplate with a rectangular drop handle. Earlier types of original main door handle include a decorative 'french' drop handle on a cast rectangular or occasionally oval backplate, and plain oval, round or octagonal brass, fixed knobs.

Second-hand cast brass handles can usually be found in Edinburgh's antique shops and new brass reproduction handles can be purchased in Edinburgh.

Bell pulls

These are usually fitted on or recessed into the stone at the side of the door, the pull knob being mounted on a flat brass backplate or recessed into a cup-shaped backplate. The bell is operated by wires running in conduits built into the internal plaster with cranked junctions. It is still possible to have the bellwire system repaired or renovated, but where an electrical system is preferred, the original manually operated pull can be retained and converted to electricity. Very often bell pull knobs have been replaced by bell pushes, retaining the backplate.

Stair door crank and lifting apparatus on common stairs

Tenements originally had a manual security system at the common stair door. Each flat had a bell, usually located in the hallway, which was activated by the bell pull located at the common entrance door. Using a system of wires and cranks, similar to that described for bell pulls, the common stair door was opened by pulling a lever on the stair landing which lifted the latch on the common stair door. Some of these original levers on stair landings still work, but where the system has broken down restoration to full working order is possible and further information on this is available from the ENTCC. Care should be taken not to damage the zinc conduit during any replastering works and it is important to avoid exposed wiring and cranks becoming paint-bound during the redecoration of common stairs.

A similar system of wires and levers was used to open gates for the access to basement areas. A metal plunger operated by foot is often still found on the third step down below pavement level.

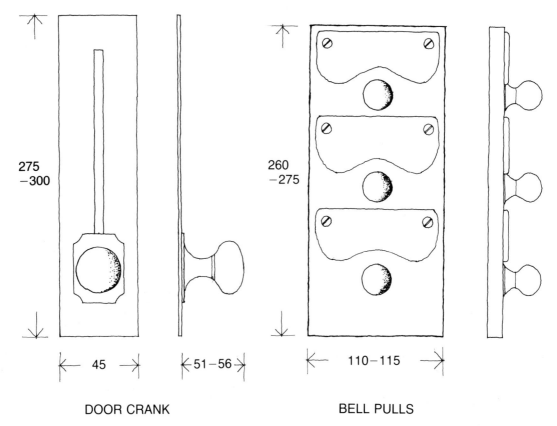

275
−300

45

51−56

260
−275

110−115

DOOR CRANK

BELL PULLS

104

Entryphones

Increasingly, electric entryphone systems are being installed at the common entrance doors of flatted tenements. The advantages of restoring the original stair crank and lifting apparatus should be considered, but where an entryphone is preferred care should be taken to choose and install a robust but unobtrusive system. If possible, it is best to fix the entryphone plate into the timber door stop or facing. Alternatively, it can be fixed into the stone jamb, care being taken to ensure that fixings and sinkings are kept well clear of the corner. The ENTCC can advise. Bell pull and entryphone systems have recently been developed by the ENTCC using flat or dished backplates of traditional pattern and a small speak/listen perforated brass plate fixed on the stone jamb, which is preferable to the combined push button plate. This permits the reuse of the original bell pulls suitably adapted to the system. Flats can be identified by name in the customary manner or by relative position in the stair.

Knockers

Originally, it is said, only doctors had door knockers; other houses and flats had bells. They were introduced later as decorative elements of brass or, on basement doors, iron. The temptation to add new knockers should be resisted, particularly on doors to flats from the common stair, where they are inappropriate and noisy for adjoining occupants.

Letter plates

Letter plates would probably not have been fitted until the introduction of the penny post in 1840. The earliest were of plain cast brass, without any decoration, often fitted vertically on the centre line of the door. There was also a narrow roll-edged type. Early examples usually have a small opening which is unsuitable for modern mail. The configuration of the door should dictate the size and location of the letter plate but the post office currently recommends that letter openings should be a minimum of 250 mm × 38 mm, positioned 1070 mm from the ground. Letter plates can be fixed horizontally on the centre line of the lock rail of a single door. Where there is a single door with central beaded muntin, or a pair of doors, the letter plate should be fixed in the centre half of the lock rail opposite to that of the door knob. A letter plate should never be placed across the central beaded muntin because this weakens the door. Replacement letter plates should be of cast brass. Aluminium, bronze, BMA (bronze metal alloy) or anodized finished plates are not appropriate.

Rim and mortice locks

Initially, most doors in the First New Town were secured by rim locks, those to principal rooms being of brass with moulded edges and sliding concealed fixings, only able to be locked by key from the outside and secured by a sliding bolt and snib from the inside. Locks on secondary rooms were of cast iron, able to be locked from both sides and with a locking snib. Occasionally timber-boxed rim locks are found on cellar doors and in some situations there is just a simple latch. Later, by the mid-1820s, mortice locks were in common use for main rooms, often incorporating a mortice bolt operated by a small knob similar to the main knob.

When a lock is not working every attempt should be made to repair rather than replace it because cast iron locks are no longer available and pressed steel rim locks are inappropriate. Most reputable lock makers do produce horizontal mortice locks but the original locks have a larger backset 152-178 mm (6-7 inches) than replacement locks, which have a 127 mm (5 inch) backset at most, and the door will need to be altered to accommodate this discrepancy.

Internal knobs

As with locks and external door handles, the hierarchy of knobs within a household depended on the relative importance of the rooms. The most common knobs were made of pressed brass and there are two predominant designs (as illustrated) in the New Town, both made in two sizes; the larger is used on main room doors, the smaller on lesser doors, such as for bedrooms, servants' quarters and cupboards.

At the top of the scale, in very wealthy households, the principal or reception apartments may have had ormolu (gilded bronze), ceramic (china or porcelain with a transparent glaze and occasionally gold lustre decoration, or red earthenware with an opaque blackish glaze), ebony, cocoawood, bone or horn knobs. The choice depended on the overall decorative scheme for the room; for example, black wooden knobs are found in dining rooms, where the marble chimneypiece is usually also black.

Roses were generally made from the same material as their knob, with the exception of wooden knobs, where brass was used. Brass roses were either pressed or cast and generally seem to have been between 45 mm and 60 mm in diameter, with a range of moulding profiles and often a milled edge. Normally, roses were surface fixed with screws although occasionally screw fixings were hidden at the junction between the rose and door knob shank.

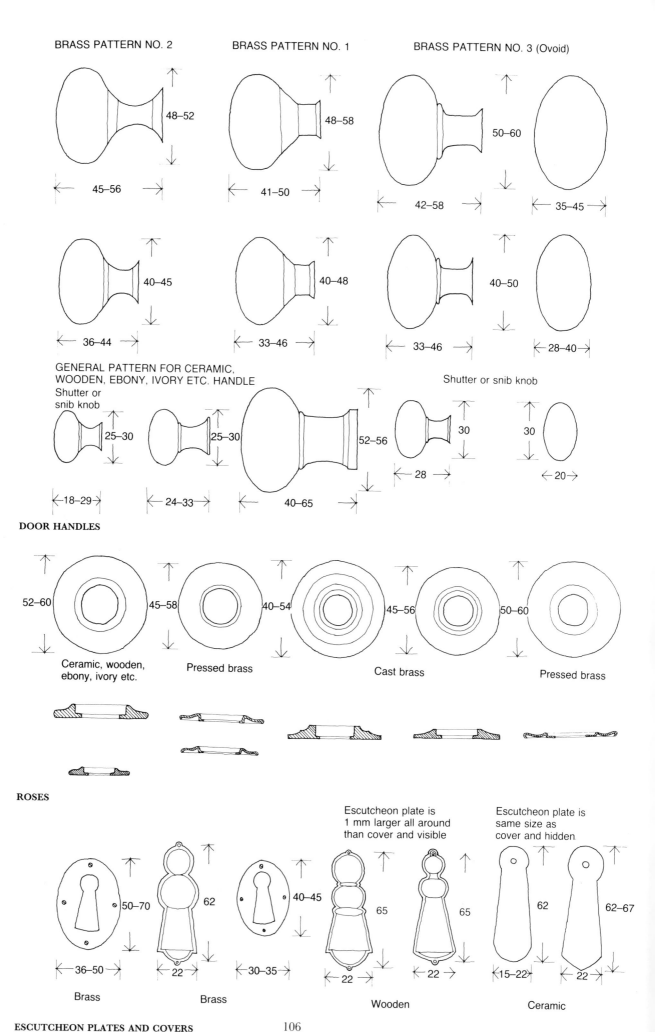

BRASS PATTERN NO. 2

BRASS PATTERN NO. 1

BRASS PATTERN NO. 3 (Ovoid)

48–52

45–56

48–58

41–50

50–60

42–58

35–45

40–45

36–44

40–48

33–46

40–50

33–46

28–40

GENERAL PATTERN FOR CERAMIC,
WOODEN, EBONY, IVORY ETC. HANDLE

Shutter or snib knob

Shutter or
snib knob

25–30

25–30

52–56

30

30

18–29

24–33

40–65

28

20

DOOR HANDLES

52–60

45–58

40–54

45–56

50–60

Ceramic, wooden,
ebony, ivory etc.

Pressed brass

Cast brass

Pressed brass

ROSES

Escutcheon plate is
1 mm larger all around
than cover and visible

Escutcheon plate is
same size as
cover and hidden

50–70

62

40–45

65

65

62

62–67

36–50

22

30–35

22

22

15–22

22

Brass

Brass

Wooden

Ceramic

ESCUTCHEON PLATES AND COVERS

106

Escutcheons

An escutcheon plate is a flat piece of metal fitted round the keyhole in a door. Escutcheons were always provided in the New Town except on very minor cupboard doors or where the keyhole was cut into the backplate of a door handle. Main doors had cast brass oval escutcheons, sometimes with a top-hinged brass cover. Internal minor room and cupboard doors generally had simple oval sheet brass escutcheons while more important doors often had ornately shaped cast brass escutcheon plates, sometimes with decorative covers in brass or a material to match the door knob. Cast iron oval escutcheons were fitted to external cellar and basement doors.

Hinges

Main doors and internal doors to rooms (about 50 mm thick) were usually hung on a pair of five-knuckle steel butt hinges. Lighter doors (about 32 mm thick) for cupboards or water closets were hung on a pair of three-knuckle steel butt hinges. Original butt hinges are distinguishable from their modern counterparts because they are generally larger with fewer knuckles per hinge. The external common stair door was usually self-closing, with a pin and offset crank hinge, while crook and band hinges made of wrought iron were common on heavy outward-opening external and cellar doors.

Fingerplates

It is thought that fingerplates were not used originally in the New Town but they have often been added at a later date and are commonly made from brass, glass or ebony. New fingerplates should be as discreet as possible.

Door closers

Door closers are required by the fire authority and the Department of Property Services on new and upgraded fire and smoke-check doors. Surface-mounted units will cause the least damage to the door but are very obtrusive and are not always aesthetically acceptable – if used, the older-style closers are best. Alternatively, concealed door closers can be used, which can be fixed either into the hinge stile or the top rail of the door (see DOORS).

ORIGINAL DOOR FURNITURE FOR EXTERNAL DOORS

Main doors to houses: bell pull; separate escutcheon and rim lock; Edinburgh door handle; letter plate and numerals (fitted later).

Common stair door: usually a round or octagonal knob on a raised backplate, pierced with a key opening in the shape of an inverted 'T', known as a French latch (key for this latch is known as a lift – see sketch); numerals. Bell pulls with a nameplate were usual.

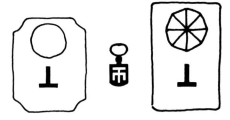

COMMON STAIR LATCH AND KEY

Basement door: A cast iron knob on a small rose or backplate; cast iron door knocker; no bell pull (sometimes a pull was fixed on the gate to the street). Letter plates and numerals were not original but were added more recently when basements became houses separate from the main door at ground floor.

Cellar door: cast iron or timber-boxed rim lock only; a pair of wrought iron strap hinges and pins (crook and band hinges), with the crooks or cranked pins built into the masonry.

Some shop doors and, occasionally, basement doors had loose iron bars secured by a padlock; one end of the bar was fixed into the door jamb, while the other end fitted over a large iron eye on the opposite jamb or on the door. Glazed shop doors, as in St Stephen's Street, had loose timber security panels bolted through the door (see SHOPS AND SHOP FRONTS).

ORIGINAL FURNITURE FOR INTERNAL DOORS

Entrance doors to flats: usually a knob on a raised backplate, though sometimes a handle of Edinburgh or French Marot pattern; bell pull; cast iron rim lock and security chain; brass escutcheon, usually without a cover. Very occasionally an iron security bar (see above).

Main room doors: door knob; an escutcheon made from the same material; often a small knob to activate a mortice bolt; a cast iron mortice or brass rim lock.

Minor rooms: brass knob; brass escutcheon; lock (as above).

Cupboard doors: commonly, a thin rim lock with a key in the keyhole to act as a pull; escutcheon; escutcheon cover in main rooms; dummy pull knobs in brass or in a material matching the decoration of the room.

FANLIGHTS AND OVERDOOR LIGHTS

INTRODUCTION

The term fanlight, derived from the semi-circular shape of a fan, tends to be applied to any glazed opening above a door, but rectangular openings are sometimes called overdoor lights. In either case, they were generally placed above solid unglazed doors to admit light into the hallway. The arcs are not always semi-circular or concentric, and the glazing bars or astragals are not always at right angles to the circumference. It is possible that they were subtly corrected to improve the appearance of the fanlight when seen from the pavement.

A wide variety of glazing patterns is found in the New Town, reflecting the taste of the first owners. It was fashionable in the second half of the nineteenth century to remove astragals and to fit a single sheet of plate glass, but this is now being reversed and the original patterns are being restored. If there is no evidence in old drawings or photographs of the original design, authentic patterns can usually be found elsewhere in the same street. The ENTCC can supply a list of firms which make or repair fanlights and overdoor lights.

CONSTRUCTION

Most fanlights and overdoor lights have a timber frame with timber astragals, but some were of cast or occasionally wrought iron, and astragals had a copper or timber web with a decorative cast lead flange. The original crown glass gives an attractive sparkle and if one of the panes is broken, salvaged or new crown glass should be used if possible. Unlike windows, fanlights and overdoor lights have mouldings on the outside and are glazed from the inside, except for dummy lights, which can be reglazed only from the outside. Sometimes single panes were hinged to give ventilation, and a few lights have a recess for an oil lamp. If the first half landing in a tenement stair was built just above the front door, the fanlight was often a single panel of Hailes stone on which astragals were painted, or a dummy glazed fanlight painted black on the inner face of the glass was constructed to conceal the stairs behind.

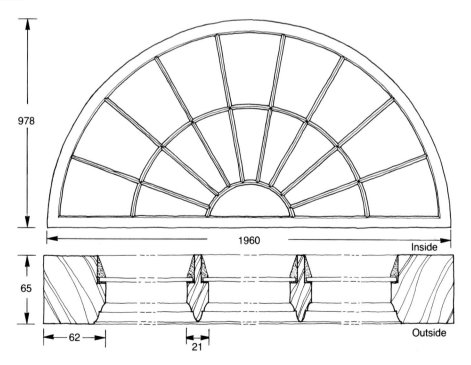

DIMENSIONS AND DETAILS OF A TYPICAL TIMBER NEW TOWN FANLIGHT

109

48 QUEEN STREET

Photo: Desmond Hodges

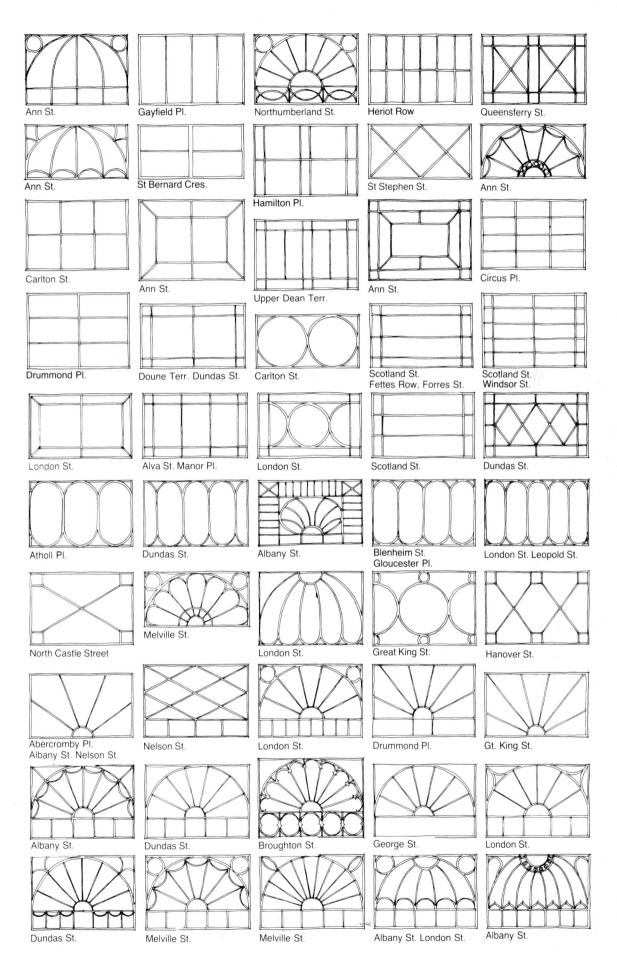

Ann St.

Gayfield Pl.

Northumberland St.

Heriot Row

Queensferry St.

Ann St.

St Bernard Cres.

St Stephen St.

Ann St.

Hamilton Pl.

Carlton St.

Ann St.

Circus Pl.

Upper Dean Terr.

Drummond Pl.

Doune Terr. Dundas St.

Carlton St.

Scotland St.
Fettes Row, Forres St.

Scotland St.
Windsor St.

London St.

Alva St. Manor Pl.

London St.

Scotland St.

Dundas St.

Atholl Pl.

Dundas St.

Albany St.

Blenheim St.
Gloucester Pl.

London St. Leopold St.

North Castle Street

Melville St.

London St.

Great King St.

Hanover St.

Abercromby Pl.
Albany St. Nelson St.

Nelson St.

London St.

Drummond Pl.

Gt. King St.

Albany St.

Dundas St.

Broughton St.

George St.

London St.

Dundas St.

Melville St.

Melville St.

Albany St. London St.

Albany St.

111

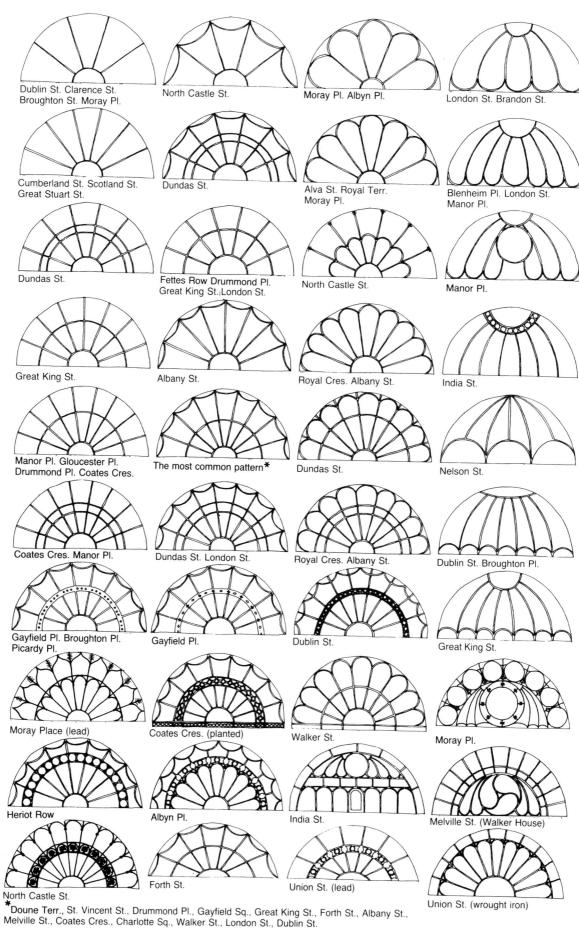

Dublin St. Clarence St.
Broughton St. Moray Pl.

North Castle St.

Moray Pl. Albyn Pl.

London St. Brandon St.

Cumberland St. Scotland St.
Great Stuart St.

Dundas St.

Alva St. Royal Terr.
Moray Pl.

Blenheim Pl. London St.
Manor Pl.

Dundas St.

Fettes Row Drummond Pl.
Great King St. London St.

North Castle St.

Manor Pl.

Great King St.

Albany St.

Royal Cres. Albany St.

India St.

Manor Pl. Gloucester Pl.
Drummond Pl. Coates Cres.

The most common pattern*

Dundas St.

Nelson St.

Coates Cres. Manor Pl.

Dundas St. London St.

Royal Cres. Albany St.

Dublin St. Broughton Pl.

Gayfield Pl. Broughton Pl.
Picardy Pl.

Gayfield Pl.

Dublin St.

Great King St.

Moray Place (lead)

Coates Cres. (planted)

Walker St.

Moray Pl.

Heriot Row

Albyn Pl.

India St.

Melville St. (Walker House)

North Castle St.

Forth St.

Union St. (lead)

Union St. (wrought iron)

*Doune Terr., St. Vincent St., Drummond Pl., Gayfield Sq., Great King St., Forth St., Albany St., Melville St., Coates Cres., Charlotte Sq., Walker St., London St., Dublin St.

MAINTENANCE

Fanlights are more susceptible to water penetration than normal windows because the moulding is on the outside of the light. It is important that the glass is fully bedded in putty and the joint between the glass and the moulding kept sealed to prevent water penetration and decay. Paintwork should be inspected annually and repainted when necessary to protect the timber from decay, or iron from rust.

Reglazing must be done from the inside of the light and can often be a specialist operation because templates may be needed for the shaped panes of glass.

When repainting astragals on the stonework of dummy lights, care should be taken to copy the original design which can be traced beneath later layers of paint. The dimensions of real astragals and their three-dimensional effect should be taken into consideration. Dummy glazed fanlights should be maintained as described in DUMMY WINDOWS.

Similar specifications to all other external paintwork should be followed (see EXTERNAL PAINTWORK). The finishing coat of paint for real and dummy astragals should be to BS 4800, Off-white 10 B 15 possibly mixed with a little White 00 E 55. The finishing coat for glass in dummy lights should be Black 00 E 53 .

FIRE PRECAUTIONS

When a door has to be upgraded for fire performance, glass fanlights and overdoor lights must also comply with the fire regulations (see FIRE SAFETY and DOORS). This usually applies to flat entrance doors off common landings; street doors do not normally require upgrading. Clear flame and heat resistant glass should be fixed dead with a solid wood bead of not less than 13 mm in cross-section. The effect of this is not always acceptable, and the glass should therefore be fitted into a separate frame on the inside face of the fanlight allowing the original window to be left intact. A discreet frame can be achieved by using a metal angle frame and fire-rated glass set into intumescent mastic.

SASH WINDOWS

INTRODUCTION

The sash window, used widely throughout Edinburgh's New Town, consists of a pair of glazed sashes which slide vertically in a case or frame. Each sash is balanced by a pair of weights contained within the case; these are attached to the sashes by cords or chains which pass over pulleys set into the top of the case. The window cases are wedged in position behind the stonework of the opening and the junction with the external stone jambs is pointed up with a mastic mix.

In conjunction with internal timber shutters, this efficient design is well suited to combat Edinburgh's climate. The majority of windows have withstood the test of time remarkably well, and there are no inherent defects in their traditional construction which regular maintenance cannot cure.

DESIGN AND CONSTRUCTION

The sashes themselves were usually constructed of 50 mm (2 inch) or 38 mm ($1\frac{1}{2}$ inch) good quality pine, although oak was occasionally used. Mouldings varied considerably, as did the construction of the case. In the larger, more important houses, pulley boxes would have a back lining to protect the cords and weights.

In the New Town, and indeed in most of Scotland, the members of a sash window are such that the width of the meeting rails is equal to the visible width of the stiles and top rail; the bottom rail is usually larger. The size and number of panes and the arrangement of astragals vary widely depending on the date and position of the window, the relative importance of individual rooms, the improvements in glass manufacture and subsequent changes in fashion. For example, in the 1820s it became fashionable to have floor-to-ceiling windows in drawing rooms on the first floor (the 'piano nobile') and the cills were lowered accordingly, as can be seen in Northumberland Street and Heriot Row. Uniform design was important, particularly on street frontages, where windows with six panes in each sash are most common, but considerable variation does occur, especially on rear elevations. Buildings in the Victorian New Town have plate glass windows without astragals.

Most New Town windows with original astragals were glazed with either crown or cylinder glass rather than the more modern cast or sheet glass. The slight imperfections and convex planes create interesting reflections and give depth to the facade and original glass should always be retained wherever possible.

INSPECTION AND MAINTENANCE

Simple inspections can be carried out during cleaning, but windows should also be checked annually by a tradesman and quinquennially by an architect. Both inspection and cleaning are made easier if Simplex hinges are fitted so that the lower sash can be swung inwards (see WINDOW IRONMONGERY). It can be difficult and dangerous to clean the outer face of the upper sash. However, the simple provision of properly secured and regularly tested eyebolts to which a safety harness can be attached will provide complete security. Eyebolts should be located in a suitable position within the room, avoiding damage to original joinery; the floor is often a good position. Where this is not possible, fixing eyebolts into the external masonry jambs may have to be considered. This is unlikely to be allowed on important front elevations but may be acceptable on the rear of the building provided that non-ferrous fixings are carefully positioned and inserted, and very well secured.

Decay in timber windows is usually caused by moisture penetration, which can be prevented by thorough painting, regular maintenance and prompt attention to necessary repairs. Repainting should normally be carried out every five years, but cills may have to be painted more frequently, particularly on elevations exposed to the south and west. Old paint should always be removed in a well ventilated area and vapour masks should be worn when removing lead based paints. Great care must be taken if paint is burned off because there is a danger of cracking thin panes of crown glass. If a paint stripper is used, the chemical must be properly neutralized before painting.

The outer edge of the sash stiles and the portion of the pulley stile with which they are in contact when closed should never be painted because this prevents smooth running. However, the extreme base and top of the pulley stile is painted so that bare timber is not exposed when the sashes are partially opened. Brass fittings, such as sash lifts and rings, can be removed before internal painting, but the removal of sash fasteners is not recommended because it can be surprisingly difficult to refix them in their original alignment (see WINDOW IRONMON-GERY).

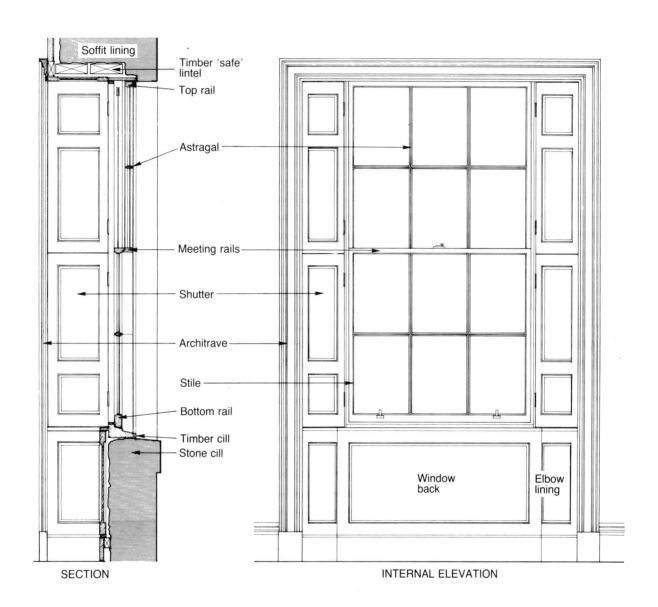

Soffit lining

Timber 'safe' lintel

Top rail

Astragal

Meeting rails

Shutter

Architrave

Stile

Bottom rail

Timber cill

Stone cill

SECTION

INTERNAL ELEVATION

Window back

Elbow lining

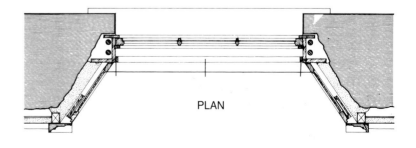

PLAN

TYPICAL NEW TOWN SASH WINDOW

A high-specification paint system should be used, the additional cost being outweighed by its longevity (see EXTERNAL PAINTWORK). The Planning Department requires that all external window joinery be painted white for visual uniformity (white paint also reduces thermal movement most effectively). Off-white (BS 4800, 10 B 15) or a mixture of this and White (BS 4800, 00 E 55) is preferable to a modern brilliant white. Originally, most windows were probably painted dark brown or bottle green, or grained; several still survive in these colours. Within the New Town, Listed Building Consent and planning permission are required for any change to the existing colour. Graining may be allowed in some exceptional circumstances; advice on this should be sought from the Planning Department.

DEFECTS AND REMEDIES

Being exposed to the elements, timber windows are particularly prone to timber decay unless they are well maintained and protected. Of the two most common types of timber decay, wet rot usually affects cills, the bottom of the cases and the bottom rails of sashes, whereas dry rot may grow unnoticed behind the window case.

Wet rot in windows is recognisable by cupping and discolouration of the paintwork, the underlying timber being soft, friable and offering no resistance to a penknife. It is caused by moisture penetration (usually the result of contact with wet stonework), defective putty or poor paintwork. Damage caused by wet rot can be repaired by cutting away the affected timber and piecing-in new sections, taking care to reproduce the original profile and mouldings accurately. At the same time, it is essential to remedy the causes of moisture penetration. Any new timber used in repair work should be pretreated with preservative by pressure impregnation; *in-situ* treatment of existing timber is no longer recommended unless essential (see TIMBER DECAY). Open joints in sashes are also vulnerable to the weather, particularly at bottom rails, where moisture accumulates. Weak joints can be strengthened by using small non-ferrous angles; mild steel angles which are more readily available will rust and should therefore be well painted and maintained.

Timber cills are particularly susceptible to decay, and if it is very rotten, the full cill may need replacing. Timber cills were originally made from good quality timber and new full cills can sensibly be made of a durable native hardwood such as oak, which should be protected by thorough priming and painting. To achieve a successful full cill repair it is usually necessary to remove the whole window, under which circumstances the opportunity should be taken to inspect and repair any damage to the window as a whole. In many cases, however, it is only the front exposed section of the cill which has decayed and a half-cill repair may be sufficient. This can be carried out *in situ*, but care must be taken to seal the joint between old and new timber properly. Unfortunately, this type of repair is rarely successful in the long term because it introduces a joint at a very weak point. The butt joint between existing and new half-cill should be covered by the bottom rail of the sash when it is shut. When repairing cills it is always worth considering whether it is possible to improve the design by inserting a damp-proof course, or introducing a drip on the underside, if the profile of the stone cill makes this possible. The mastic pointing below the timber cill should be recessed.

The width of the outer bead varies from window to window and, when it is only just checked in behind the ashlar face, particular attention needs to be paid to the pointing. Faulty joints between the stonework and the window frame should be pointed with a traditional mastic made from buff-coloured burnt sand and boiled linseed oil, which is still available commercially. Polysulphide mastics and silicone seals are not generally recommended. Voids behind the mastic pointing, between the outer lining and the masonry, need to be tamped with lime mortar, which is then keyed to receive the mastic and primed with a coat of boiled linseed oil. Traditionally, rolled-up wet newspapers were used to pack very deep voids and to provide a backing for the mastic. However, there is a danger that because newspaper contains cellulose, it may encourage rot growth if in contact with damp masonry. Consideration should therefore be given to filling voids with modern inert fillers or strips of solid insulation although care must be taken not to foul the pulley boxes. If the gap between the masonry and the frame is very large, it may be difficult to point without cracking occurring. One method used to avoid this is to wedge in a stainless steel expanded metal lath or a plaster bead to provide reinforcement for built-up pointing in lime mortar or mastic.

Draughty windows are a common complaint. It is difficult to draught-proof a sash window completely, because allowance must be made for easy movement, shrinkage and expansion. Properly functioning sash fasteners and the use of shutters and heavy curtains will eliminate most draughts. Modern weather-stripping systems are now available which can effectively cut down both draughts and sound intrusion. Draught excluders should never be fitted where they will be visible externally. However, some form of ventilation is always desirable, and essential if there is a gas fire in the room, and over-zealous draught-proofing may only increase the risk of condensation. Where condensation is a real problem, more drastic measures may be necessary to increase both heating and ventilation (see INSULATION for further guidance). Double glazing involving replacing the existing glass, or whole sashes, is not permitted by the Planning Department, although secondary glazing, which involves the provision of an independent internal window in addition to the existing sash window, may be acceptable if it

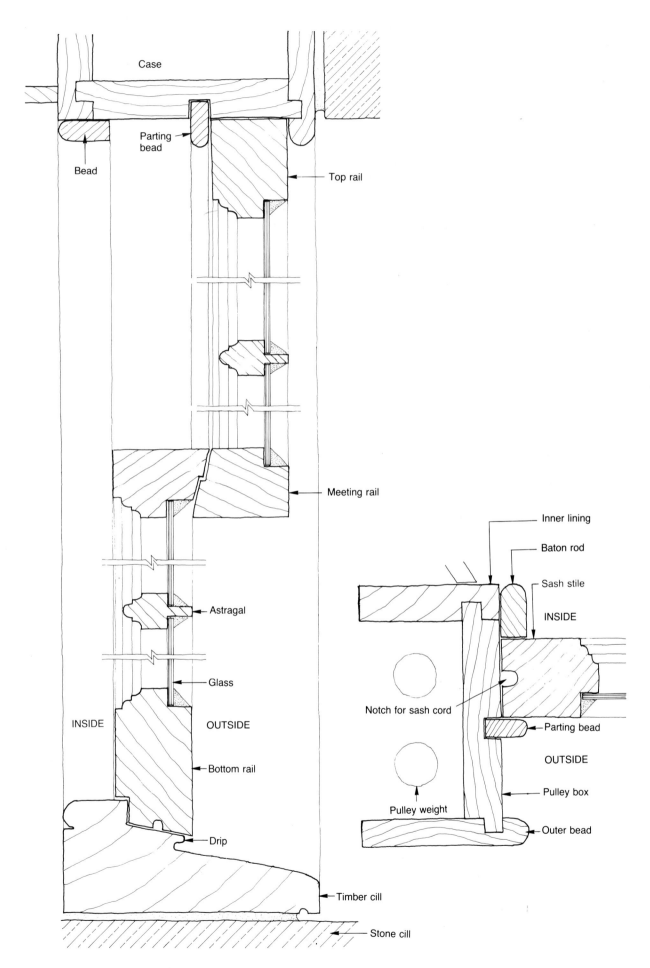

Case

Parting
bead

Bead

Top rail

Meeting rail

Inner lining

Baton rod

Sash stile

INSIDE

Astragal

Notch for sash cord

Glass

Parting bead

INSIDE

OUTSIDE

OUTSIDE

Bottom rail

Pulley weight

Pulley box

Drip

Outer bead

Timber cill

Stone cill

SASH AND CASE WINDOW DETAILS   117

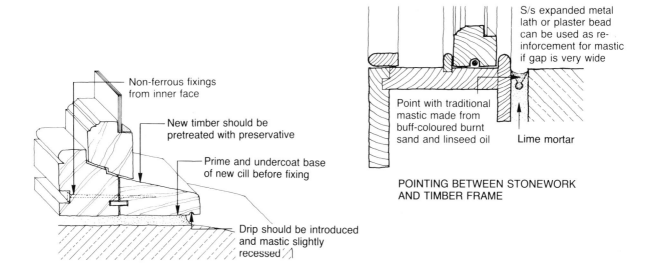

Non-ferrous fixings from inner face

New timber should be pretreated with preservative

Prime and undercoat base of new cill before fixing

Drip should be introduced and mastic slightly recessed

HALF-CILL REPAIR

S/s expanded metal lath or plaster bead can be used as re-inforcement for mastic if gap is very wide

Point with traditional mastic made from buff-coloured burnt sand and linseed oil

Lime mortar

POINTING BETWEEN STONEWORK AND TIMBER FRAME

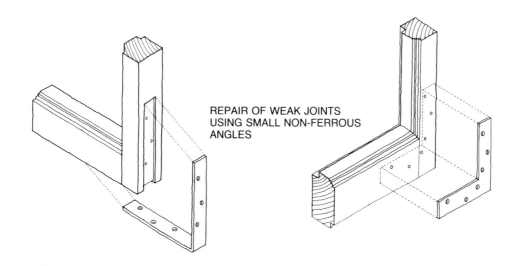

REPAIR OF WEAK JOINTS USING SMALL NON-FERROUS ANGLES

does not damage original sashes, shutters and panelling. Historic Scotland have recently produced a Technical Advice Note on this subject.

While draught-proofing and energy saving are both important issues, the maintenance of adequate ventilation is vital in old buildings. The Building Standards (Scotland) Regulations 1990 require the provision of permanent trickle ventilation in some cases when converting an existing building. This requirement will only apply very rarely to existing or replacement windows in the New Town. Where necessary, this standard can be achieved by carefully forming a vertical slot between the meeting rails, though this can weaken the sash. Simpler methods may include the use of a smaller pane of glass in an appropriate position creating a small air gap, or the fixing of permanent blocks to the pulley stiles so that the top sash is kept marginally open to meet the necessary ventilation requirements. Proprietary trickle ventilators are unlikely to be visually acceptable, and in any case cannot usually be accommodated in an existing sash. It is also possible to devise discreet forms of ventilation through the pulley stiles. Additional permanent ventilation may be required after fitting a gas-fired appliance; this may have to be provided through a wall, ceiling or solum if a large vent is necessary. Extractor fans fixed within the panes of glass are strongly discouraged for aesthetic reasons (see Historic Scotland's Technical Advice Note 6).

Crown and cylinder glass is difficult to replace and should always be protected from damage while work is in progress, particularly when scaffolding is being erected or dismantled. Probably the most common damage to windows is broken glazing. Original glass from derelict sashes is ideal for reglazing but it is very brittle and so great care must be taken when using it in repair work. The expense involved in salvaging old glass can usually be justified when compared to the high cost of new crown glass. It is recommended that if glass is removed in the course of repair work, it should always be reset where possible. When reglazing with crown glass, the convex side should face outwards. Small panes of true crown glass are now being manufactured and cylinder blown glass is available from France and Germany (the ENTCC has information about suppliers). Horticultural glass, which

118

INTERIOR, 11 DUNDONALD STREET   119

used to have imperfections, has now improved in quality and is therefore no longer a reasonable substitute for crown glass. Poor quality sheet glass from Germany and Eastern Europe, which has many distortions, may be a suitable alternative because its two surfaces are never perfectly flat and parallel and there is always some distortion in the reflections, as with crown glass. Samples should always be seen and approved before specification or acceptance on site. If a whole elevation has to be reglazed, there are various grades of proprietary reproduction glass which are more suitable. (A list of suppliers is available from the ENTCC.)

Glazing should be fixed using traditional linseed oil putty. Glazing beads and modern glazing compounds are unsuitable. Reglazing or reputtying will be necessary if regular painting has been neglected and the putty is brittle, cracked or missing. This should be carried out with extreme care to avoid damaging the original glass, which should be retained *in situ* wherever possible. Old putty should be carefully removed by inserting a glazier's hacking knife between the putty and the timber and tapping down with a hammer. If the putty is old and hard it may be necessary to soften it either with a paint stripper or by dipping the sash in a bath of non-caustic solvent before scraping it off. Dipping can take the life out of timber, but if the sashes are beyond repair it is the simplest way of salvaging the crown glass. A putty lamp is now available from Sweden which uses infrared heat to soften the old putty, making this task considerably easier. (The ENTCC can provide details of suppliers.)

New putty needs to be protected by a full paint system, which must be applied within twenty-eight days of glazing. Failure to provide this protection may lead to early deterioration of the putty, loss of adhesion and cracking. Glazing bars and sash members should be primed before being glazed so that the oil in the putty will not soak into the timber, causing the putty to dry out and perish.

REPLACEMENT WINDOWS

Windows are an important feature in the elevation of any building and their replacement by different designs can seriously affect the character of the building. A repair can often achieve a much-extended life without the need to replace the complete window. In the New Town, traditional timber sash and case windows should be used in all elevations and any change requires Listed Building Consent from the Planning Department.

If a window is in such a bad state of repair that it needs to be replaced, its modern substitute should exactly match the original in terms of materials, size, pattern and construction, with particular attention paid to the mouldings. Reference to WINDOW MOULDINGS AND ASTRAGALS shows that there is a wide variation in mouldings and that standard modern sections are not acceptable. Sashes with horns should never be used in the New Town except on late Victorian buildings with strengthened sashes designed to take the heavier weight of cast glass.

Timber for new windows is usually Baltic redwood which should be pretreated with preservative by pressure impregnation. New windows should also receive one full coat of primer before being delivered to site (see EXTERNAL PAINTWORK).

SASH WINDOW
RE-CORDING

The vertical sliding double-hung sash window as found throughout the New Town has an upper and a lower sash, both counterbalanced on each side by weights concealed within the pulley boxes of the window case and hung on cords or chains which run over the pulleys. A properly balanced and corded window will open and close smoothly, with the minimum of wear and strain on the timber sections. The dangers implicit in the untimely snapping of sash cords are only too obvious!

The size, shape and material of the sash weights are determined by the weight of the sashes themselves and the available space within the pulley boxes. Most weights were and still are made of cast iron, but lead can be used for exceptionally heavy sashes or where space within the pulley box is limited. Access to the weights is gained by removing the timber pocket pieces at the base of the pulley stiles. Some windows in the New Town were constructed with a stop pocket piece which prevents the upper sash from being lowered to the cill. Pocket pieces fit tightly into the bottom of the pulley stile and need to be prised out and replaced with great care. Whenever this is done, the opportunity should be taken to remove any loose mortar or debris which has accumulated behind the window case.

The weights are generally strung on sash cords and should be hung at least 75 to 100 mm clear of the bottom of the pulley box when each sash is fully raised. To prevent the weights bottoming or fouling the pulley wheels, the top sash cords should be kept 75 mm shorter than the measured length and the bottom sash cords 75 mm longer. Cotton cord was used traditionally and is still recommended for repairs or replacement. Rot-proof nylon is also available but its appearance is not considered acceptable. Sash chains can be found in some Victorian New Town buildings. Cords should not be painted because this stiffens them and impedes their smooth passage over the pulley wheels. Cords should be checked annually for signs of wear and tear; if one cord has broken, it is advisable to renew the other three if they are of the same age.

Cotton cord is manufactured in various gauges for use with different weights of sashes. Gauge nos 6, 8 and 10 are the most commonly used and the table below gives details of their relative sizes and strengths.

Sash cord	Diameter (mm)	Breaking strain (kg)
cotton no. 6	5	105
cotton no. 8	6	155
cotton no. 10	8	225

Any alteration to the sashes, such as a change in the type of glass or the restoration of astragals, will require changes in the size of the weights and in the strengths of the cords. Each sash should be separately weighed by spring balance to calculate the weight required. The combined weights should be heavier than the top sash in order to hold it up securely, but lighter than the bottom sash in order to keep it down. For example:

Top sash weighs 14 kg – weights for top sash should weigh 14 kg + 1 kg = 15 kg, i.e. each weight = 7.5 kg.
Bottom sash weighs 14 kg – weights for bottom sash should weigh 14 kg − 1 kg = 13 kg, i.e. each weight = 6.5 kg.

The procedure for re-cording a sash window is described below. Although this appears to be fairly simple, the weight of large sashes can be considerable and the re-cording, particularly of windows in upper storeys, should be carried out only by a competent tradesman.

PROCEDURE FOR RE-CORDING A SASH WINDOW

1   Raise bottom sash and engage Simplex hinges (see WINDOW IRONMONGERY).
2   Release removable section of baton rod.
3   Pull down sash cord on this side, securing the weighted cord in the clutch.
4   Swing bottom sash on hinge and remove knot-holder (or cord grip) from sash stile; swing sash fully open. This procedure is also used for window-cleaning purposes. Windows not fitted with Simplex hinges require to have the baton rod on one side unscrewed.
5   Remove pocket piece adjacent to broken cord and withdraw the dropped weight.

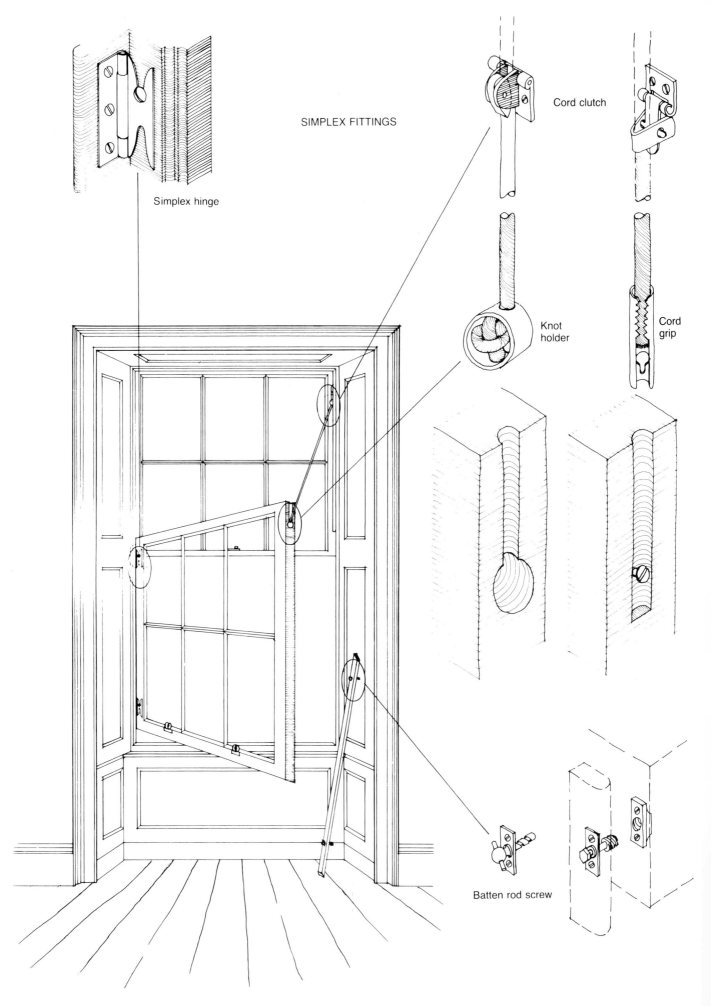

Simplex hinge

SIMPLEX FITTINGS

Cord clutch

Knot holder

Cord grip

Batten rod screw

122

6   End of broken cord on bottom sash can now be removed. If the cord for the top sash has broken, the parting bead on the same side as the removed baton will have to be carefully prised out to allow the top sash to tilt inwards and the end of the broken cord to be removed.

7   Use tack lifter to remove nails securing cord set in groove in sliding face of sash stile.

8   Use both sections of broken cord to mark off length of replacement cord but do not cut length yet. The cord may stretch through time when weighted and should be cut to leave 75-100 mm clear at the bottom of the box. Dust and lime mortar debris which may have accumulated in the pulley box should be removed.

9   Tie light string to a bent nail or similar weight and feed this over pulley and down to bottom of box behind pulley stile.

10 Tie other end of string to new sash cord and pull through over pulley to pocket piece where the weight is now attached, knotted as previously.

11 Nail the pre-marked length of new cord to the stile (or fit to knot-holder if appropriate) and then cut off excess cord.

12 Manoeuvre sash back into position and check that it operates in balance before refitting parting bead and baton rod.

ASTRAGALS AND WINDOW MOULDINGS

GENERAL

Crown glass and cylinder glass were only made in small sheets and the size of the panes was therefore limited, hence the need for rebated glazing bars, or astragals, in large windows. Strictly speaking, the term astragal refers to a particular moulding which was often used for the simplest glazing bars, the profile being similar to that below the capitals on classical columns. In Scotland, this word is used to describe a glazing bar, probably because the 'astragal and hollow' glazing bar which succeeded the 'fillet and ovolo' in the 1760s became the dominant type in Scotland and was almost universal in the New Town until the 1830s. The profile of the astragal moulding is carried onto the stiles and rails of the window sashes.

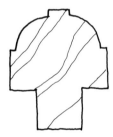

FILLET AND OVOLO

ASTRAGAL

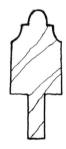

ASTRAGAL AND HOLLOW

The earliest astragals were heavy in section, but from the mid-eighteenth to the early nineteenth centuries, when the New Town was designed and built, they gradually became finer. Astragals may occasionally be of different timber from that used for the other window sections – generally, the finer the astragal, the harder the timber. Some of the finest astragals were reinforced with brass strips, as for example in Manor Place.

After 1845, when the weight tax on glass was abolished, larger sheets of heavier drawn glass came into use and astragals once again became thicker to support the extra weight. Soon afterwards even larger sizes of panes became available and astragals were no longer required because a complete sash could be glazed without the need for sub-division. Many of the later New Town houses had plate glass on the front elevation but retained the cheaper crown or cylinder glass with astragals at the rear. In most streets, however, the visual rhythm of the astragals forms an essential element in the character of Georgian Edinburgh and astragals should therefore always be retained together with the original glass. During the late nineteenth and twentieth centuries there was a trend to remove astragals and reglaze windows with larger sheets.

Within the New Town, the ENTCC offers grants towards the cost of replacing missing astragals. When astragals are being restored to an elevation, the fanlight should always be restored at the same time.

RESTORATION

Sound old sashes should always be retained in preference to new replicas often of inferior timber. It is possible and preferable to insert astragals into an existing sash but, because they need to be morticed into the frame, the old sash will have to be knocked apart and reassembled and this should be undertaken only by a skilled professional. If it is too frail to survive this operation, or if the joints are rotten, a new sash may have to be substituted. The existing heavy glass may be cut into smaller panes and reused to avoid changing the sash weights.

Astragal mouldings can vary from house to house and even from front to back within the same house. Patterns for the correct mouldings in each case can usually be found by studying and measuring neighbouring windows, or even from the mouldings of the stiles and rails of existing sashes. It should be noted that the existing sash may not be original, and existing mouldings may be unreliable evidence, unless signs of mortices suggest the earlier

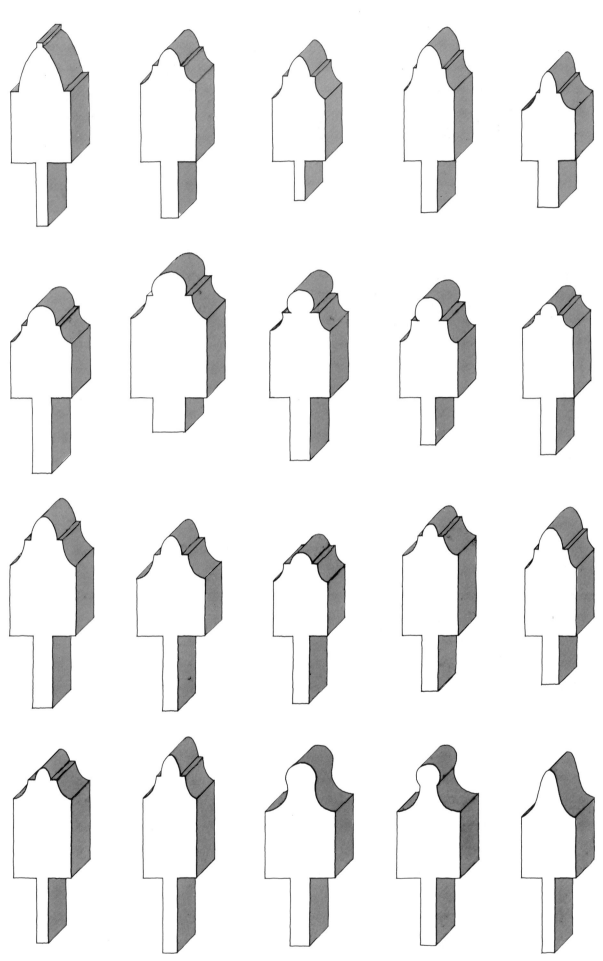

VARIOUS ASTRAGAL MOULDINGS IN THE NEW TOWN

presence of astragals. On no account should astragals be inserted where there is no visual, comparative or documentary evidence of their original existence or subsequent removal.

Skill is needed to measure an astragal moulding, because layers of paint can obscure the detail and distort the dimensions. If an old sash is to be discarded it is easy to cut through the astragal to get a clean cross-section which can then be given to the joiner to use as a pattern; otherwise, an accurate, full-size drawing must be produced. Even so, it is always wise to ask for a sample of the new astragal moulding for approval before the whole window is made up. Where window mouldings are concerned every millimetre matters!

A number of joinery manufacturers will supply ex-stock sash windows with modern standard astragals, mouldings and lugs, but these are not appropriate for use in the New Town.

The practice of gluing dummy astragals onto the glass is not recommended because it is usually short lived, visually unacceptable and is not approved by the Planning Department.

WINDOW IRONMONGERY

Traditional fittings for sash and case windows, such as sash fasteners and sash lifts, are generally simple and effective. These were originally made of cast brass and later of bronze. Pulleys and weights were usually made of cast iron.

When repairing or upgrading windows care should be taken not to discard old ironmongery; a lot of sound items are lost through carelessness and most can be cleaned and reused. There is an established market in second-hand ironmongery and a number of antique shops in Edinburgh stock good quality brass fittings, particularly brass sash lifts, for reuse on new sashes. The ENTCC can supply a list of stockists.

SASH FASTENERS

The *quadrant arm fastener*, made of brass, brings together the meeting rails of the top and bottom sashes to secure the window and to reduce draughts. It is only suitable for close-fitting and properly aligned windows. Variations of this type of fastener are available, but the type illustrated is preferable.

The *fitch fastener*, also made of brass, although not original, is better for windows whose rails do not meet properly. A disadvantage of this type of fastener is that the catch does not spring back. If it becomes loose after the top sash is lowered it can damage the meeting rail when the sash is raised again.

The spring-loaded *Brighton sash fastener* is often preferred because it provides better security. However, this fastener may be difficult for the elderly to unscrew and is not really appropriate for the New Town. *Sash screws* which link the meeting rails together, or *Week's Patent sash stops* which are fixed to the sash stiles and only allow the window to open partially, are common security measures.

Fasteners should not be painted, nor should they be removed during painting, since it is difficult to refix them in their true alignment. However, removal may be necessary if the timber rots due to condensation and requires repair, or if accumulated dirt and dust make painting difficult with the fastener *in situ*. Steel screws can corrode and should therefore be replaced with brass or non-ferrous screws.

SASH LIFTS

Hook and *ring lifts* are available in cast brass, and are screwed in pairs to the inside of the bottom rail of the bottom sash. Modern pressed or extruded sash lifts can be sharp-edged and awkward to use and it is always preferable to retain the existing cast fittings if possible. Small *sash knobs* in brass or timber are known to have been used in some of the early Georgian houses. *Bar lifts* in brass, which were introduced after 1850 to cope with the heavier weight of plate glass, are also suitable.

The *sash eye* is a single brass ring fitted flush to the top rail of the top sash so that the sash can be opened and shut with the aid of a window pole. These, together with corded *top sash mountings*, came into use in the mid-nineteenth century to help cope with heavier glass and larger sashes.

PULLEYS AND WEIGHTS

Each sash is balanced by a pair of weights concealed within the window case. *Sash cords*, which pass over axle pulleys on which the weights are hung, should be cotton or jute rather than nylon. *Chains* were used on heavier Victorian sashes instead of cords.

Axle pulleys were originally made of cast iron. The pulley wheel must be of sufficient size to allow the weight to hang clear in the pulley box, and the cord to hang clear of the pulley stile. The cover plate of the pulley is usually painted. Modern nylon pulley wheels in cast aluminium cases are not suitable for use in the New Town, nor are spring-loaded balances. New cast iron pulleys are still manufactured, although brass pulleys, which are also acceptable, are more readily available.

Sash weights were made of lead or cast iron; the former should be used for sashes above 7.26–8.17 kg (26–28 lb) in weight. The weights are concealed behind removable pocket covers set into the pulley stile of the window case. The method for sizing new weights to achieve the proper balance is described in SASH WINDOW RE-CORDING.

Early 'country' or less important windows such as dormers had sashes without cords. They were kept open by a small pivoted stay which engaged into a notch cut in the baton rod.

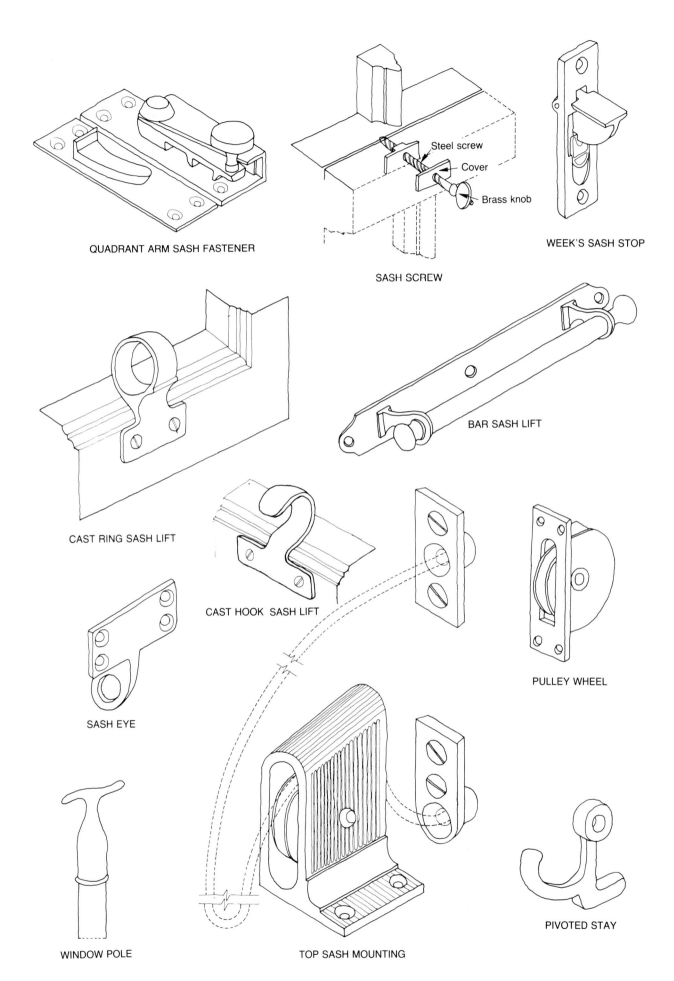

QUADRANT ARM SASH FASTENER

Steel screw

Cover

Brass knob

SASH SCREW

WEEK'S SASH STOP

CAST RING SASH LIFT

BAR SASH LIFT

CAST HOOK SASH LIFT

SASH EYE

PULLEY WHEEL

WINDOW POLE

TOP SASH MOUNTING

PIVOTED STAY

SIMPLEX FITTINGS

Simplex fittings enable the lower sash to be swung inwards, which greatly facilitates cleaning, external painting, inspection and repair of upper windows. Although they were not fitted originally, their use today is required in order to meet statutory requirements for the safe cleaning of windows more than 9 m above the adjacent ground level (see WINDOW IRONMONGERY). The use of Simplex fittings seems to be peculiar to Edinburgh. They consist of a number of separate elements:

- A pair of *slotted butt hinges*, which are usually fixed to the lower half of the baton rod to support the weight of the sash when the cords are disconnected. Mild steel hinges tend to rust even if painted. However, the use of a lubricating oil on unpainted hinges will inhibit rust formation. Brass hinges are preferable and are now available;
- A *baton rod screw*, which enables one of the baton rods to be removed so that the bottom sash can swing inwards on the hinges mounted on the opposite baton rod. *Baton rod hinges* are also available but are slightly less convenient to use;
- A *cord clutch*, fixed to the pulley stile below the pulley prevents the disconnected cord from being drawn into the pulley box. However, this cannot happen if a cylindrical *knot-holder*, morticed into the side of the sash, is used, as it enables the cord to remain in a set position during disconnection from the bottom sash (see SASH WINDOW RE-CORDING). These items are now available in cast iron.

SHUTTER IRONMONGERY

Shutters have their own ironmongery which usually consists of:

- A pair of small *shutter knobs* with roses, usually of brass, but occasionally of timber or ceramic to match door furniture in the same room;
- *Butt hinges* on the shutters and surface-mounted *backflap hinges* on the shutter backflaps which fold in; originally iron but later of steel;
- A shutter fastener, which may be a small brass *shutter hook* or, where security is more important, an iron *shutter locking bar*.

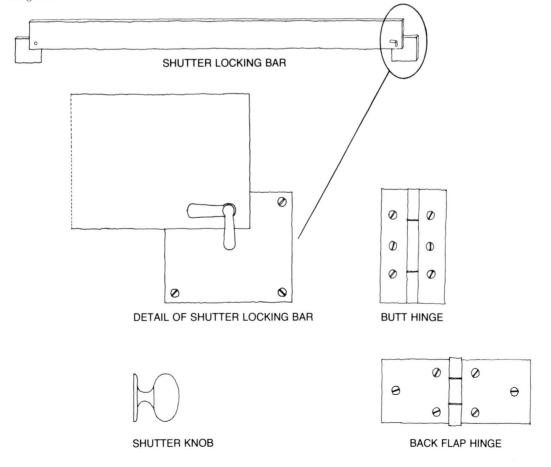

SHUTTER LOCKING BAR

DETAIL OF SHUTTER LOCKING BAR

BUTT HINGE

SHUTTER KNOB

BACK FLAP HINGE

DUMMY WINDOWS

INTRODUCTION

The aesthetic purpose of a dummy window in the New Town is to maintain the rhythm of openings in a facade in positions where a real window could not be built, such as in front of a flue or partition wall. Sometimes there is a cupboard or a fireplace situated behind the dummy window, effectively reducing the thickness of the wall. It is a myth that dummy windows in the New Town were inserted to avoid paying window tax.

Inaccessible dummy windows in a tenement are often neglected, the proprietors regarding them as part of the whole facade and feeling no personal responsibility for them. Under the law of the tenement, each real or dummy window in the wall of a flat should be maintained by the proprietor of the flat on that level.

Various methods of construction were used, and each has its own requirements for maintenance and repair. A rare type of dummy window containing a dummy Venetian blind made from a solid sheet of wood can be seen at 87 Great King Street.

DEFECTS AND REMEDIES

The most common type of dummy window consists of four ashlar spandrel panels 50-75 mm thick, recessed about 150 mm from the face of the wall. The lower panels are set back slightly to give a shadow line to simulate the overhang of a top sash. It is not uncommon for the stone panels to be face-bedded and if they have delaminated they may need to be replaced. Where a flue is situated directly behind the dummy window the inside face of the stone may have been eroded by flue gases. Repairs to dummy windows should not alter their original appearance and replacements should be constructed in the same manner as the original. Renewal in stone is the ideal solution even though this may be awkward where the panels have been checked in behind the ashlar of the wall. A less satisfactory repair is to replace the stonework with a render on a stainless steel lath. The render should be left plain and not lined to represent the coursing of ashlar. Some dummy panels were originally painted black with meeting rails and astragals picked out in white.

Another type of dummy window had fixed, glazed timber sashes built into the opening like true windows, but without pulley boxes or mouldings on the inside. Maintenance should be as for normal windows with repainting every five years; on upper floors this will require access from a scaffold tower or hoist. The Planning Department recommends that the woodwork and putty of fixed sashes be painted white to match the rest of the street.

The inner face of the glass was often painted black or brown to hide the brick or rubble stonework behind, but oil paint has a tendency to craze in sunlight. The inner face is inaccessible and cannot be repainted *in situ*. Furthermore, the unventilated cavity behind the dummy window can become so hot that the glass may crack. Replacement with black glass and the provision of trickle ventilation is a good solution. Plastic laminates, which will break down after long exposure to ultraviolet light, are not suitable.

The opening up of dummy windows is to not to be encouraged; it may upset the interior arrangement of the room inside.

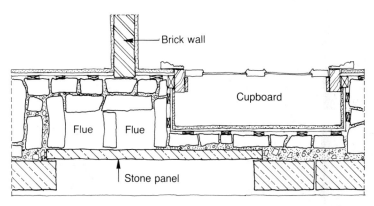

PLAN OF DUMMY WINDOW

Stone Dummy window at 2 Dundonald Street

Painted stone dummy window at 86 Great King Street

Dummy glazed sash at 87 Great King Street

Dummy glazed sash with dummy venetian blind at 87 Great King street

131

DUMMY WINDOWS

Photos: The Royal Commission on the Ancient and
Historical Monuments of Scotland

EXTERNAL PAINTWORK

GENERAL

In the New Town, only two materials must be painted – wood and ironwork. Apart from the aesthetic aspect, wood is painted to prevent decay and to reduce movement due to moisture. All ironwork, cast or wrought, should be painted to protect it against corrosion.

The Planning Department of the City of Edinburgh District Council has recommended colours for external paintwork on doors in the New Town; otherwise all windows should be painted white, and railings and balconies painted black (see SASH WINDOWS). When a gold finish is needed, the surface should be gilded rather than painted. For any proposed change of colour, the Planning Department should be consulted because Listed Building Consent and planning permission will be required (see LISTED BUILDINGS AND CONSERVATION AREAS).

WOODWORK

Inspection and maintenance

The paintwork should be regularly examined for blistering, flaking or crazing. Joinery will normally need to be repainted every five years and in some exposed situations, more frequently. Neglecting paintwork, especially after the breakdown point, causes the timber to deteriorate, particularly at the joints, and will mean that all the old paint will have to be removed before repainting; for further guidance see SASH WINDOWS.

Preparation and painting

Painting of surfaces which have not been properly prepared is a sure recipe for early failure. Existing paintwork which is still intact should be washed with non-alkaline soap or detergent and water, and rinsed thoroughly with clean water. Sugar-soap is good for removing grime. The surface should then be rubbed down with waterproof abrasive paper, rinsed clean and allowed to dry completely before applying undercoats and gloss coats. Any defect in the existing paint should be spot-primed and brought forward with undercoat after washing is complete.

Paint which is seriously defective or is so thick that it obscures the detail of the timber mouldings should be burned off, taking care to keep the flame clear of glass. Areas near glass should be stripped with a solvent-type paint remover, care being taken to wash off all residue. The surface should then be prepared and painted as for a new surface (see below).

New joinery will require the following complete paint system:

- *Knotting:* It may be necessary to seal knots in the timber to prevent resinous material from damaging the paint system. This is best done before pretreatment and priming, or after paint removal.
- *Pretreatment:* It is common to use Baltic redwood for external joinery, and this will normally be pretreated with an organic solvent preservative which can be painted over. However, if the timber has not dried out sufficiently, the paint may be affected by stains, wrinkling or peeling.
- *Priming:* Two coats of primer are best, especially on wood surfaces nearest to masonry. It is important that all end grains, edges and faces should be properly coated, whether visible or not. Primers should be in accordance with BS 5082: *Water-thinned Priming Paints for Wood*, or BS 5358: *Specification for Low-lead Solvent-thinned Priming Paints for Woodwork.*
- *Stopping and filling:* After priming, defects such as cracks, joints and nail holes should be made good with best-quality, exterior-grade filler (not the water-mixed kind). Loose or defective putty in the glazing rebate should be hacked out, and the rebate primed and filled with a traditional linseed oil putty before repainting.
- *Undercoats:* Oil-based undercoats are applied over the prepared surface to increase adhesion, build up a paint film, strengthen the colour and improve the appearance of the finishing coat. On all new exterior work two undercoats are required.
- *Finishing coat:* Oil-based finishing paints are available in varying degrees of gloss, eggshell and matt. Full gloss is the only type to be used for exterior finishing. It may be advantageous to give horizontal surfaces such as window cills an extra coat.

Lead paints may still be used in Grade A listed buildings, but only after approval from Historic Scotland. Acrylic paints should not generally be used out of doors but, when properly applied, they do give outstanding durability to external woodwork which can be very useful in more inaccessible areas.

Microporous paints have been developed for external use and the manufacturers claim that they combine the advantages of alkyd-based paints, which penetrate the surface giving a good key for subsequent coats, with those of acrylic paints, which have a very durable finish. Microporous paints are not suitable for horizontal surfaces however, and are therefore not recommended for painting woodwork in the New Town.

After applying each coat to sash windows, the sashes should be moved up and down to prevent sticking. The top surface of the top sash and the bottom surface of the lower sash should not be painted. The width of paint overlapping onto the glazing should not exceed 1.5 mm; otherwise the overlapping paint may curl away from the glass in the heat. Paint should not be removed from the glass by scraping because there is a risk of scratching the glass.

Fire doors can be painted with an intumescent coating to help achieve required standards of fire safety. Further information can be found in DOORS.

IRONWORK

Inspection and maintenance

Decorative cast iron should be inspected annually by the owner or tradesman for signs of corrosion and deterioration of the paintwork system. Particular attention should be paid to joints and welds – the points where corrosion usually begins (further information is given in RAILINGS). The inspection of rhones, downpipes and soil and waste pipes should form part of the annual inspection by a roofing contractor. All ironwork should be included in an architect's quinquennial inspection.

Preparation and painting

Repainting should be carried out every five years or at the first signs of rust, whichever is the sooner. No painting should be done in damp or humid conditions or extremes of temperatures, as this prevents proper adhesion and can delay drying.

The inside surfaces of rhones should be painted with two coats of bituminous or tar-based paint. It is important that adequate access for repainting be provided behind rhones and rhonepipes where corrosion can so easily remain undetected until more serious damage from dampness or leakage has occurred. Where access to the base of gutters cannot be provided, e.g. behind box-section conductors and ogee gutters, it is essential to prime thoroughly (at least two coats) or paint with bituminous or tar-based paint before erection.

Where proper maintenance has been neglected for a number of years, and if the metal has corroded, the architect will need to decide whether it is preferable to replace the relevant sections with new castings, or to strip them down to the bare metal and apply a new paintwork system. However, stripping and repriming can be an expensive procedure.

Manual or power-driven wire brushing, blast cleaning, hand scraping or mechanical grinding in conjunction with propane gas burning are all used for the removal of old paintwork from cast iron and for preparation before priming. This work is laborious and the use of blast or flame cleaning is often considered too expensive on small items. The only alternative is a chemical stripper, preferably of the spirit solvent type, the residue of which must be removed by white spirit or plenty of water, according to the manufacturer's instructions.

Red lead, the traditional primer, is still probably the best; quick-drying versions of this paint are now available and will dry within hours rather than days. Zinc phosphate is a suitable alternative, but it is advisable to use a high build type, preferably reinforced with micaceous iron oxide, as this will provide a highly impermeable coating. The importance of applying two coats of primer is often forgotten. With one coat of primer it is almost impossible to produce a continuous film of even thickness and free from pinholes. A second coat of primer, rather than an extra top coat, can extend the life of the whole paint system. The primer should then be protected with two layers of undercoat and a finishing coat of exterior gloss paint. No undercoat should be allowed to weather for more than a few days on site and, as with woodwork, acrylic paints should not be used out of doors.

The cleaning, priming and painting of ungalvanized mild steel should be even more thorough.

ROOF STRUCTURE

GENERAL

Most roofs in Edinburgh's first New Town are steeply pitched, with a high central ridge containing one, or occasionally two, attic floors within the roof space, as in Dublin Street, York Place, Charlotte Square and Castle Street. Roofs in the second and later developments of the New Town, built after 1800, characteristically have two parallel ridges making a double-pitched 'M' roof with a centre gutter. The rear wall may be a storey higher than the front, so that the attic floor is lit by dormer windows or skylights at the front and by full-height windows at the rear, with the ridge correspondingly offset. A few roofs have a mansard behind a balustrade on the front elevation as seen at Henderson Row and Royal Terrace, although mansards and flat roofs such as in Douglas Crescent were built mostly towards the end of the nineteenth century. Flat roofs are usually the result of later alterations and extensions, but small flat areas covered by lead were required in earlier buildings for centre gutters and cupola platforms.

The ownership and maintenance of roof and roof space is usually shared by all the proprietors but if the title deeds are silent on this issue the responsibilities may rest with the top-floor proprietor(s). The subject is explained in more detail in the OWNERS' GUIDE.

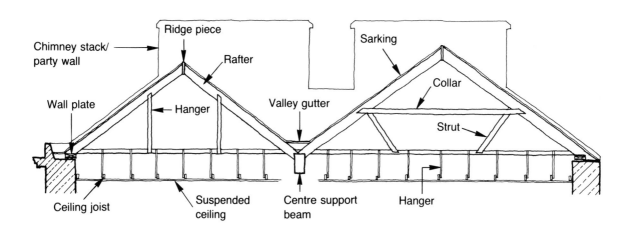

CONSTRUCTION

Pitched roof structure

Pitched roofs are triangular in section with ceiling joists forming the base of the triangle and small rafters the other two sides. This basic configuration is stiffened by collars which tie across between the mid-points of the rafters and hangers or struts which prevent deflection in the ceiling joists; trusses or purlins were seldom used originally.

The feet of the rafters are notched or birdsmouthed onto timber wallplates carried on the inner edges of the front and back walls, and into timber beams which support the central gutter. Lateral support is given by timber sarking boards and ridges. The rafters vary in size but are often about 250 mm × 63 mm (10 inch × $2\frac{1}{2}$ inch) in cross-section and are usually set at 400 mm (16 inch) centres, meeting at a ridge member.

Flat roof structure

Flat roofs are generally formed of simple joisted construction, with firring pieces and sarking to form a deck supporting the lead sheet above and sometimes with timber hangers supporting the ceiling below.

DEFECTS AND REPAIRS

Structural distortion, sometimes leading to failure, is most commonly caused by timber decay at points where rainwater collects, affecting wallplates, gutter soffits and beam ends but occasionally it is caused by ill-considered alterations such as the insertion of dormers and rooflights (see TIMBER DECAY and GUTTERS AND DOWN-PIPES).

Wallplates which are usually bedded straight onto the stonework of the wallhead and rafter feet are particularly susceptible to rot. Rafter feet can be repaired by splicing on new ends or, if only isolated joists are affected, by bridling across to sound rafters on either side. A structural engineer should be consulted for all but the most minor of repairs, because the types of joints, numbers of bolts, and so on will vary with the particular loading. Similar techniques are suitable for the joists of flat roofs.

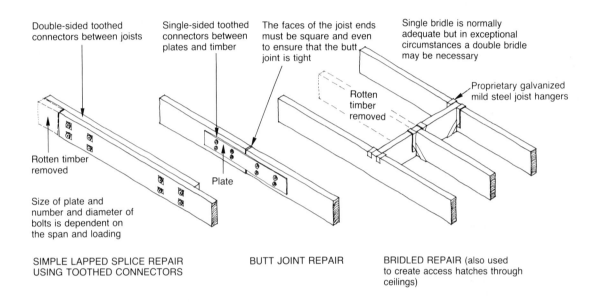

Double-sided toothed connectors between joists

Single-sided toothed connectors between plates and timber

The faces of the joist ends must be square and even to ensure that the butt joint is tight

Single bridle is normally adequate but in exceptional circumstances a double bridle may be necessary

Rotten timber removed

Proprietary galvanized mild steel joist hangers

Rotten timber removed

Plate

Size of plate and number and diameter of bolts is dependent on the span and loading

SIMPLE LAPPED SPLICE REPAIR USING TOOTHED CONNECTORS

BUTT JOINT REPAIR

BRIDLED REPAIR (also used to create access hatches through ceilings)

Fixings for splices should ideally be of stainless steel, tightened with a torque wrench but galvanized mild steel is acceptable because there should be little danger of corrosion in the roof space. Bolts should be of high-tensile steel, at centres and of sizes recommended by the structural consultant. All new timber should be pretreated or treated with preservative during repair, and new sections of wallplate should be laid on a bituminous damp-proof course to isolate them from the masonry.

The beams under centre valley gutters are particularly vulnerable to rot, especially at abutments with party walls.

Under certain conditions, condensation within a pitched roof can cause damage, especially where roofs are covered with underslating felt. Methods of introducing additional ventilation to help combat this are discussed in INSULATION. Condensation within a flat roof construction is more difficult to remedy. Warm, moisture-laden air passing through the ceiling is unable to escape from the small voids within the roof structure and moisture can condense on the soffit of the boarding above. Even lead, the most durable of roof coverings, will corrode under attack from condensation. There is limited opportunity for the introduction of additional ventilation into the void beneath a flat roof, but lead ventilators and suitable details are given in ROOF COVERINGS.

Deflection in the timbers may cause ponding on a flat roof. This should be remedied by strengthening the structure, for example the introduction of deeper or additional joists. It may also be necessary to increase the amount of fall, either by raising all the joists at one end or by fitting firring pieces on top of each joist.

Where a dormer window has been inserted into a pitched roof, distortion may occur if the rafters on each side have not been doubled up to take the extra weight from the trimmer and in this situation an engineer should be consulted.

Dirt and debris accumulates in roof spaces, particularly around hatches. All refuse should be cleared for easy inspection and to reduce the risk of fire and rot. Industrial vacuum cleaners are particularly useful for removing dust and dirt.

135

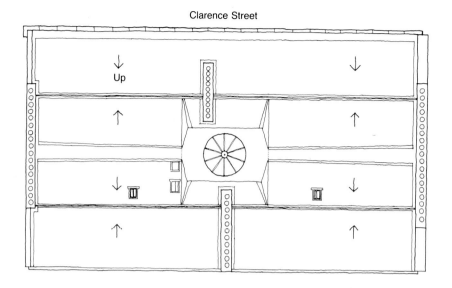

Clarence Street

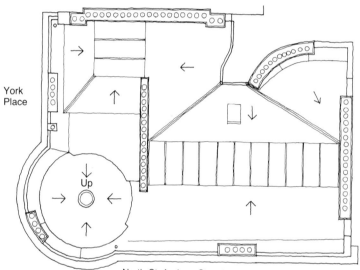

York Place

North St Andrew Street

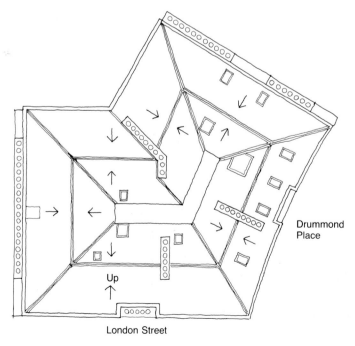

London Street

Drummond Place

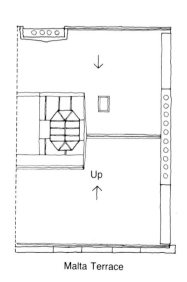

Malta Terrace

ROOF PLANS

136

ROOF COVERINGS

INTRODUCTION

Most of the pitched roofs in the New Town are covered with slates fixed to timber sarking, with lead gutters and flashings, forming an efficient and durable roofing system. Seen from the high viewpoints in the city, these slated roofs are an essential part of the visual character of the New Town. The original Georgian houses possessed only small areas of flat roof around the cupola which were covered with lead. Some Victorian additions or infill buildings have flat or mansard roofs, which were usually zinc clad, but many have since been re-covered with asphalt or bituminous roofing felt.

MATERIALS

Slates

Coarse-textured blue-black West Highland slates were used throughout the New Town, but unfortunately the last quarry closed in 1955. Demolition sites now provide the best available source and some builders' merchants and roofing contractors hold stocks of second-hand slates, although these are becoming increasingly rare and of ever poorer quality and size. When a roof is carefully stripped, 60 per cent or more of the slates should be reusable. Cracked or damaged slates can usually be redressed and shortened for use higher up the pitch or on small pitches. Ballachulish slate is preferred for its dark blue-grey colour, rough texture and durability. Easdale slate is also suitable but contains more iron pyrites, is rougher than Ballachulish, and is more susceptible to damage by foot traffic. Of the non-Scotch slates, blue-grey Cumbrian (Burlington) are probably the nearest match to Ballachulish, although they are smoother, thinner and lighter in colour. This slate can be seen on new housing in Jamaica Street (1981). Although Cumbrian slate is not a perfect substitute for Scotch slates on outward-facing slopes, it is acceptable behind parapet walls or in a central valley. Welsh and Spanish slates are not a good match for Scotch slates and are not recommended.

Slates are laid in diminishing courses, with the largest at the bottom and the smallest at the top, giving a pleasing perspective to buildings; the use of slates of varying sizes is a traditional device for minimizing the amount of wastage at the quarry. The length of the slate is measured from the centre of the nail hole to the base. In Edinburgh an 11 inch slate is called a No. 11; a $10\frac{1}{2}$ inch slate a No. 10; a 10 inch slate a No. 9, and so on down to a 6 inch slate which is a No. 1. Similarly, a $11\frac{1}{2}$ inch slate is a No. 12, a 12 inch slate a No. 13 and so on. The vertical lap, or cover, is usually 3 inches from the nail hole, and the distance between the twills (the horizontal lines of the courses) equals the length of the slate less the cover, divided in two. It therefore follows that because slates are sized in half-inches, the twills on a well-slated roof will diminish by a quarter of an inch per course from eaves to ridge. In practice, the cover is slightly less at the ridge to allow the smaller slates to lie properly. Side laps are determined by the width of the slates but, generally, should not fall below 2 inches.

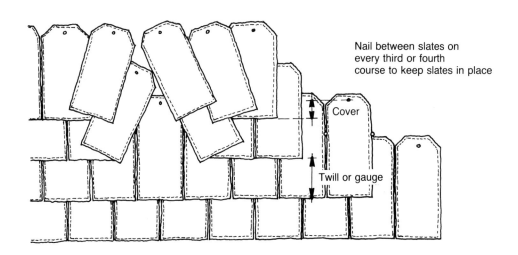

Nail between slates on every third or fourth course to keep slates in place

Cover

Twill or gauge

ROOFSCAPE

Photo: Graham Metcalfe

Nails

Slates in the New Town were fixed with iron nails, which are strong until they rust; galvanized mild steel nails have commonly been used by slaters during the last fifty years for reslating roofs, but because they also rust eventually, they are not ideal. Large-headed copper clout nails with jagged shanks, at least 50 mm long, complying with BS 1202: Part 2 *Copper nails*, are recommended by the ENTCC. Stainless steel nails are also suitable where cost is not a major consideration.

There are two ways of fixing slates to a roof, by either single or double nailing through the head of each slate. Single nailing, which is common practice in Scotland, allows the slate to pivot easily about its fixing, thereby making repair work easy. Double nailing secures the slate more firmly in position and is often used at ridges, verges, dormers and other positions where the slates are exposed to wind suction.

Sarking

In Scotland, slates are traditionally nailed directly to sarking, which is rough-sawn, butt-jointed softwood boarding, usually about 200 mm × 25 mm, fixed on top of the rafters. All new sarking should be impregnated with preservative under pressure to BS 5268: Part 5 *Preservative Treatment of Constructional Timbers*, and should follow the original construction. Replacement of sarking with plywood or tongued and grooved boarding should be avoided because they both inhibit ventilation of the roof space.

Sarking felt

Although when the New Town was built it was not the custom to cover the sarking with felt, it is now common practice when roofs are reslated. The felt has certain advantages: it acts as a second layer of defence against water penetration, it prevents dust and dirt blowing into the roof space and it acts as a temporary roof covering while slating repairs are in progress. However, underslating felt can restrict the free passage of air around the slates and in extreme cases it can cause condensation within the roof space, leading to softening and delamination of slates and decay in the roof timbers.

When underslating felt is used it should be permeable to moisture. Alternatively, building paper (breather type) to BS 4016 should be used. A number of other underslating membranes are now marketed which provide good protection, durability and ventilation.

Ridges and flashings

Ridges in the First New Town were almost certainly covered with lead, often replaced in zinc from around 1800. Abutments are traditionally weatherproofed by mortar fillets. Further information on ridges, mortar fillets and other flashings is given in FLASHINGS.

Flat roofs

Between 1860 and 1914 new infill buildings and extra storeys were often built with flat roofs, covered in zinc. Zinc only lasts for about forty years in polluted city atmospheres and much of it has been replaced by bituminous felt which is even less durable. Lead is still the most aesthetically pleasing and, in the long term, economically viable covering for flat roofs. Its malleability and ease of working means that it can be readily adapted to most roof structures.

INSPECTION AND MAINTENANCE

Roofs should be inspected for damage and deterioration at least once a year, preferably in late autumn after the leaves have fallen and before the severe winter weather. The inspection should be both internal and external. Even small leaks can cause major defects like dry rot and it is better to avoid future expense and disruption by correcting defects as soon as they are discovered. Both inspection and maintenance will be greatly facilitated if adequate and safe means of access can be provided to all parts of the roof.

Roofing contractors are normally willing to inspect a roof on an annual basis and to make running repairs as part of a routine maintenance programme. It will reduce costs if these inspections are carried out on a shared basis, either with adjoining owners or through a street association.

Pitched roofs

The easiest way to check for slipped slates is to look along the twills, rather than from the eaves or the ridge, although on external pitches this should only be done with the help of appropriate safety equipment. In a single nailed roof, the slates can easily be pulled aside to check the fixings. All rainwater drainage should be examined for blockages. Internally, the roof space should be examined for mould growth, daylight, decay in the timbers and so on. Any visible deflection of roof members, dampness on the ceiling below or damage to recent decoration should be investigated.

Flat roofs

The inspection of flat and shallow pitched roofs of built-up timber construction should follow the same basic principles as for a pitched roof. The waterproof coverings should be examined for signs of stress, cracks or blisters, paying particular attention to the condition of the roof coverings at upstands and outlets. The roof should also be checked for ponding, especially for ponds more than 25 mm deep which may signal rot in supporting timbers or structural movement. In dry weather, evidence of regular ponding can normally be seen from staining and minor plant growth.

DEFECTS AND REPAIRS

Pitched roofs

Slates are generally very resistant to decay, although they tend to delaminate in stagnant or polluted air and can be damaged by frost action if constantly wet. A sound slate will ring when struck with the knuckles or a hammer and any slates which produce a dull thud should be discarded.

Slates become loose when the fixings rust or fail under wind pressure. Water retained between the slates by capillary attraction can soften the slates or cause iron nails to rust, although the rough texture of West Highland slates is less likely to trap moisture than smooth Welsh slates. Old hairy felt underlay used under slates holds moisture, also causing nails to rust.

Single nailed slates can easily be replaced by sliding apart the overlapping slates. If the slates are double nailed at the head the damaged slate has to be removed with a ripper, and a copper or stainless steel clip or tingle is then nailed between the slates below, the new slate inserted and the end of the clip bent up to hold it in place. This type of repair is unsightly, not very durable and has to be inspected regularly. If a large number of clips are visible in the roof, the condition of the remaining fixings must be suspect. When fixings are subject to frequent and widespread failure the roof is said to be nail sick and all the slates should be stripped off and renailed. Sarking boards should be checked over at the same time and any damaged or decayed boards renewed. It is often wise to completely renew the sarking while the slates are stripped.

Slates are very easily cracked under foot traffic, especially if they are loose. Duck boards and roof ladders are a menace when rotten and neglected, but on balance they are worth having, to protect surfaces and to give easier and safer access (see ACCESS TO ROOFS).

Moss growth between slates should be raked out, because it encourages moisture retention; it is also the source of organic acids, which will corrode valley and gutter linings. Pitches subject to these growths should be sprayed thoroughly with a suitable fungicide. In extreme cases, copper strips can be inserted into the roof pitch to combat growth (see CONTROL OF VEGETATION).

Flat roofs

Poor design and bad workmanship will shorten the life of lead or zinc. Where insufficient allowance has been made for thermal movement, the stresses in the metal may be very high and will be increased if the water is lying in ponds on the roof, exposed to the heat of the sun. Sometimes the lead may be sound but is no longer securely fixed to the timber rolls. However, lead usually fails because it has been laid in panels which are too large or too thin to accommodate normal thermal expansion; the steps and drips do not allow for movement; or the fall is insufficient. Therefore, when renewing lead it is often necessary to re-form the substrate to satisfy current practice. If it is impossible to provide the number of steps required, lead sheets can be joined using a proprietary system which consists of two metal strips connected by a neoprene expansion joint. The neoprene should be protected from ultraviolet rays by covering it with a light (code 4) lead flap fixed at one side only. It is possible to repair cracks in lead sheet by burning in patches of lead, but it is essential to remedy the cause of cracking before attempting what may otherwise be only a temporary repair. This will often involve reducing the size of sheets and introducing extra drips. Care in the burning operation is essential because, unless the correct nozzle is selected for a particular situation, undercutting may occur and this will result in a new defect within a short time. Defective lead is sometimes repaired with bituminous patches or coatings; these repairs are rarely successful and only hinder the identification of subsequent leaks. For details of laying new lead, refer to GUTTERS AND DOWNPIPES.

Flat roofs are also particularly susceptible to damage by foot traffic. Traditional hollow rolls, used to join two lead panels, are not only readily crushed under foot but inhibit the movement of lead at the point where the roll is carried over the edge of a drip joint, and should be replaced with wood-cored rolls. Lead flat roofs are also particularly vulnerable to attack by condensation because of lack of ventilation below the lead or because insulation has been incorrectly positioned. The provision of roof ventilation is discussed in INSULATION.

Zinc has a shorter life expectancy than lead and is prone to attack by the sulphur products of air pollution; small perforations in areas made thinner by oxidization indicate that it has reached the end of its useful life. If zinc is being repaired, antimony-free solder of 50/50 or 60/40 tin and lead should be used, the zinc being cleaned first with spirits of salts. However, zinc sheet is unlikely to be worth patching unless it is fairly new and thick and replacement in lead should be considered.

Bituminous felt is not a durable material, even when guaranteed, and is therefore not recommended for historic buildings. Although it is used extensively for roof repair and replacement because of its relatively low cost, it has a limited life, generally no longer than fifteen years. It will be subject to the same problems of thermal movement and the effects of ultraviolet light which cause failure in other sheet materials. It also suffers from lack of adhesion often caused by water vapour expanding within the roof void and this results in bubbles in the membrane which crack underfoot in frosty weather.

FLASHINGS

GENERAL

Flashings protect the vulnerable joints between different materials and elements of a building. They are normally associated with roofs, and usually made of metal, such as lead or zinc, but mortar fillets often protect the junctions between wall and roof. Copper is not used in the New Town. In this century zinc, being cheaper, has replaced lead on many ridges and piends (hips).

All flashings should be checked during the annual roof inspection, preferably in late autumn. Any defects must be repaired immediately because flashings are a critical part of the roof fabric and leaks may lead to serious defects in the building below. It is recommended that a professional survey be carried out every five years.

DEFECTS AND REPAIRS

Flashings generally fail because they have deteriorated through age or because of structural or thermal movement. They often cover junctions between materials which have different coefficients of expansion and contraction, and the flashing material will crack if it is unable to move independently of at least one of the materials which it protects. This often occurs if original fixing details were inappropriate or where lengths of metal flashings were too great. Geotextile material is now laid under the lead to allow it to move freely.

When renewing or extensively repairing flashings, fixings and details must follow current practice in order to overcome types of lead failure caused by the inadequacies of traditional detailing.

Mortar fillets

Mortar fillets, used to protect junctions between roof and wall, are a traditional detail but they have only a limited life. Cement-rich mortar may crack, allowing water to penetrate into the building. Patch repairs are rarely successful and fillets should be completely renewed as soon as defects occur, using 1:4 hydraulic lime/sand or 1:2:9 cement/lime/sand mortar built up in at least two layers, the lower being keyed to receive the upper layer. A strip of bituminous felt between the fillet and the slates will allow the fillet to adhere to the wall but not to the slates which can then move more freely with the timber structure beneath. Stainless steel lath, fixed to the wall, will provide a key for the mortar.

A degree of moisture movement at mortar fillets is normal, and acceptable where there is a well-ventilated roof space below, but dampness can be a problem in attic rooms, and in this case a lead watergate detail is better (see later). In cases of extreme damp penetration it may also be necessary to introduce a lead damp-proof course below the skew cope or across the exposed wallhead.

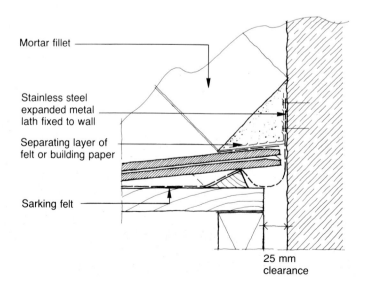

Mortar fillet

Stainless steel expanded metal lath fixed to wall

Separating layer of felt or building paper

Sarking felt

25 mm clearance

Watergates

Mortar cannot be compared in durability with lead, but before changing from a mortar fillet to a lead watergate it is important to remember that this may change the visual character of the building. A watergate is a flashing which forms a channel or gutter between a slated verge and adjoining masonry; it is constructed by dressing lead over a tilting fillet below the slates and across the sarking to form an upstand at the wall. This upstand is covered by a lead cover flashing. The edge of the lead beneath the slates is turned into a single welt forming a secondary check against wind-blown rain. Rain blown over the tilting fillet is channelled down the welt, and so it is important to ensure that water can discharge freely and safely at the bottom of the watergate.

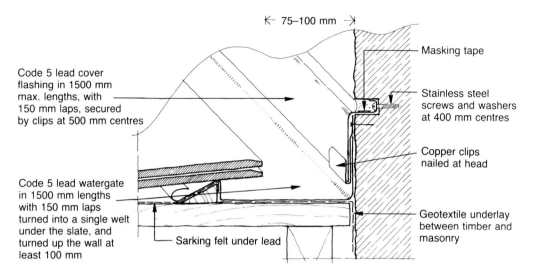

Code 5 lead cover flashing in 1500 mm max. lengths, with 150 mm laps, secured by clips at 500 mm centres

Code 5 lead watergate in 1500 mm lengths with 150 mm laps turned into a single welt under the slate, and turned up the wall at least 100 mm

Sarking felt under lead

75–100 mm

Masking tape

Stainless steel screws and washers at 400 mm centres

Copper clips nailed at head

Geotextile underlay between timber and masonry

Pitched valley gutters

These occur at the internal angle where two roof pitches meet. A lead lining, in code 5 or 6 lead sheet, should be laid over a geotextile underlay placed on top of the sarking, in maximum 1500 mm lengths. It is important that underslating felt is not used because summer heat can soften the bitumen content of the felt causing the lead to stick to the substrate, inhibiting its movement and resulting in cracking. The lead is fixed at the head of each undercloak sheet using two rows of copper or stainless steel nails and is covered by the 150 mm laps of the overcloak. The sides are dressed over a tilting fillet and turned into a single welt, as a secondary defence against wind-blown rain, as discussed in the section on watergates above.

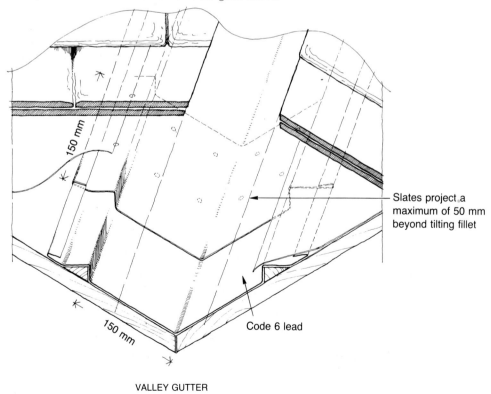

150 mm

150 mm

Slates project a maximum of 50 mm beyond tilting fillet

Code 6 lead

VALLEY GUTTER

144

Cover flashings

Cover flashings are strips of lead sheet used to cover the junction between a wall and an abutting material or element. The top edge of the lead sheet is fixed with lead wedges, or preferably stainless steel screws, into a raggle or a joint cut into the stonework and is then pointed up with a 1:2:9 cement/lime/sand mortar. A layer of masking tape should separate lead from mortar to prevent the movement of the lead causing the mortar to crack and fall out. Code 5 or 6 lead should be used, laid in maximum lengths of 1500 mm and retained along the bottom edge with lead or stainless steel clips at 500 mm centres. Laps between panels should be a minimum of 150 mm, also secured with a retaining clip. The vertical lap over the abutting material should be a minimum of 75 mm, but preferably 100 mm.

Cornice overcloak

Generally, cornices in the New Town were not covered with lead and, provided that the stone is sound and the joints are well filled, there is usually no need to alter this. However, where moisture penetration through the cornice has become a problem or where the top surface is delaminating, it is advisable to flash the upper weathered surface with lead. The fixing of a lead overcloak into new raggles on the blocking course above the cornice or the use of copper clips on the face of the cornice can be awkward and unsightly. It is better to use a lead apron with separate cover flashing, secured into a raggle or, better still, if the rear of the cornice stone has a raised joint as shown on the sketch, to insert the overcloak into a raggle at the base of the blocking couse. If the joint is not raised or if there is no fall to the front of the cornice, the raggle should be cut higher on the blocking course, if possible not more than 50 mm; cover flashing should not be necessary. The front edge of the overcloak should be turned down about 25 mm over the cornice and doubled back to be secured by continuous or spaced lead clips fixed with stainless steel or brass screws. Individual panels should be no more than 1500 mm long, with double-welted joints. If there is no overcloak, and it is necessary to protect the cornice from local penetration, i.e. beside an overflow, a wide strip of code 7 lead can be laid on top of the cornice and fixed with lead dots.

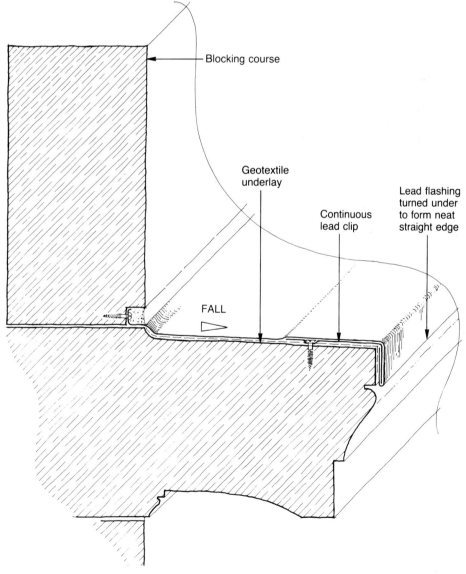

145

Ridge and hip flashings

Ridge and hip flashings were originally made in lead, but have often been replaced with zinc secured with galvanized steel clips or screws. Galvanized steel soon rusts and such fixings need to renewed regularly. Lead creates a visually softer profile than zinc and, when possible, the ridge and hip flashings should be reinstated in code 6 lead flashings, laid in maximum 1500 mm lengths over the timber roll and extending down the slope for 150 mm on either side. The lead should be nailed to the ridge roll at the underlap end only, using copper clout nails covered by 150 mm laps, and secured against wind uplift with lead clips at 300–500 mm centres.

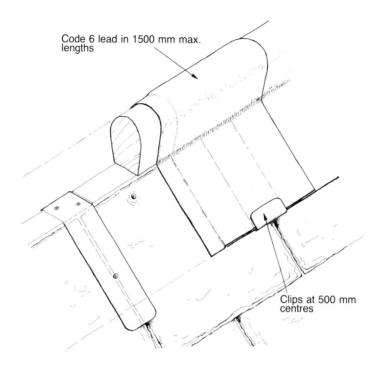

Code 6 lead in 1500 mm max. lengths

Clips at 500 mm centres

Sacrificial flashings

Leadwork is susceptible to attack from acidic run-off, for example at valley and parapet gutters where the lead is directly beneath lichen-covered slates (see GUTTERS AND DOWNPIPES and CONTROL OF VEGETATION). An extra layer of sacrificial flashing is sometimes provided by burning lead clips at 500 mm centres onto the vulnerable lead and laying an extra flashing over the top. This flashing is allowed to deteriorate – hence the term sacrificial – and protects the underlying lead. When its useful life is finished it can easily be replaced by bending back the retaining clips, removing the sacrificial lead and inserting a new length of flashing.

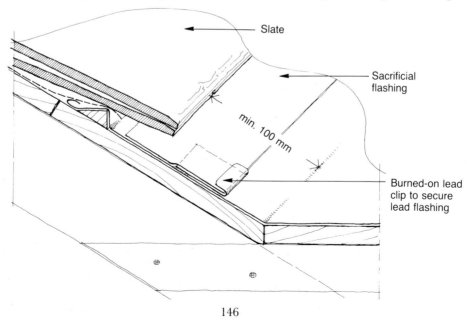

Slate

Sacrificial flashing

min. 100 mm

Burned-on lead clip to secure lead flashing

146

GUTTERS AND DOWNPIPES

GENERAL

Many roofs in the first New Town have high, single ridges on trusses shaped like the letter 'A'. Later buildings of the same depth have two lower ridges and a central valley – 'M' shaped in cross-section – so that the roofs, although still pitched, are largely hidden behind the front parapet, balustrade or blocking course.

Gutters collect the water from a roof and discharge it into rainwater pipes. These pipes, usually vertical, carry the water down to the drainage system at ground level. It is typical throughout most of the New Town to find wallhead or parapet gutters at the front of the building, a half-round, cast iron rhone at the rear and, in the majority of cases, a centre gutter between the two pitched roof sections. External rainwater pipes are often tucked into corners of the front wall or avoided altogether by conducting rainwater from the front gutter back through the roof space to a central stack in a duct on the common stair.

Parapet and wallhead gutters are concealed from street level by parapet walls and blocking courses, respectively. Occasionally the wallhead gutter is formed in the top of the blocking course or cut into the top of the stone cornice, and in Dundas Street the gutter is a channel cut into the top of the projecting band course at the head of the wall. Such gutters formed in stone were not lined with lead, depending instead for waterproofing on fine joints and good pointing. The simple and comparatively trouble-free cast iron rhone (gutter) is in keeping with the vernacular treatment of rear elevations.

INSPECTION AND MAINTENANCE

Water penetration is the most common cause of decay in buildings. Wallhead gutters and downpipes are often the culprits and they should be inspected for defects at least once a year and cleared of leaves or debris of any sort at regular intervals, using a wooden shovel. Roof spaces should also be inspected, particularly where centre gutter pipes or open channels actually pass through the roof space.

The terraced form of the New Town makes it possible to inspect several roofs without entering the buildings, and adjoining owners could share the cost of annual inspections, thus avoiding expensive repairs.

Any defects should be repaired immediately by a qualified professional. Do-it-yourself work on roofs and gutters is not recommended.

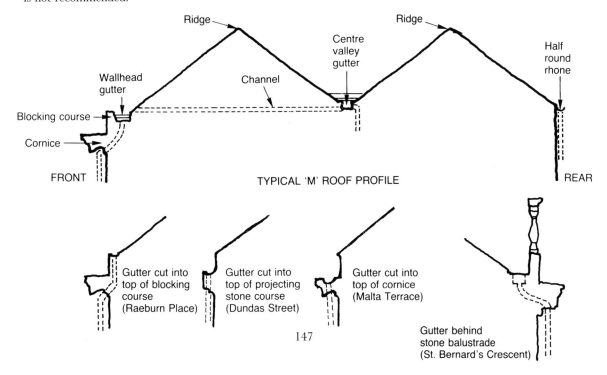

147

CENTRE LEAD VALLEY GUTTER
HENDERSON ROW/CLARENCE STREET

148

Photo: Hunter Walker

DEFECTS

Gutters and downpipes may fail due to faulty design or installation, physical or chemical damage, lack of maintenance, or merely because the materials involved have reached the end of their useful life.

1 *Gutter boarding may be deficient because:*
   - Settlement of the masonry has allowed the timber to twist, move or break;
   - The timber has been affected by wet or dry rot;
   - The boarding has been laid with insufficient fall;
   - The boards are inadequately supported.
2 *The gutter may fail because:*
   - The size of the metal sheets is too large, resulting in lines of fatigue and cracks;
   - Badly detailed joints allow water to penetrate;
   - The material has reached the end of its useful life (three-layer bituminous felt may have a life of only about fifteen years).
3 *The gutter may suffer from physical damage because:*
   - A timber duck board has not been provided for foot traffic;
   - The gutter boarding is not strong enough to carry foot traffic;
   - Dents in the lead prevent it from sliding easily over the boarding, resulting in cracks;
   - Loose slates can slip and cut the lead;
   - Sharp blows can crack an asphalt membrane in frosty weather;
   - Felt gutters often form large air bubbles and these may be cracked by foot traffic in frosty weather.
4 *Cast iron rhones and pipes can be damaged or dislocated by:*
   - Movement in the stonework;
   - Ladders leant against them;
   - Corrosion;
   - Fractures caused when trapped water freezes and expands.
5 Choked outlets cause rainwater to back up in the gutter and to flood into the building. Overflows, which were seldom fitted originally, should be provided to discharge conspicuously clear of the building. Outlets should be protected with copper or stainless steel mesh balloons, to prevent blockage from leaves and debris. All pipes must have at least one accessible handhold or inspection plate, through which a rod can be inserted to clear blockages.

REPAIRS TO PARAPET AND CENTRE GUTTERS

Lead sheet, laid correctly, should be used for wallhead and centre gutter construction. Where existing lead is in comparatively good condition and of adequate thickness, minor defects may be repaired by burning-in patches of new lead, after first checking that the cause of failure was not an oversized sheet. The lead should be welded using lead filler strips heated by gas jet. Solder, which has a different coefficient of expansion, should not be used for lead but is suitable for repairing zinc. All surfaces should be clean and the meeting edges and faces of the lead should be shaved. The thickness of the seam should be between one-third and a half thicker than the lead sheet.

Patching and burning-in will be economical only for a small number of defects and, if the gutter is in poor condition, total replacement is necessary.

After stripping the covering, the boards should be checked, replaced and, if necessary, treated with preservative. Boarding should always be laid in the direction of the fall, and tongued and grooved boards should be laid with the heart side up, cambering upwards. This is to avoid sharp edges at the joints on which the lead might wear. All nails in the boards should be punched below the surface. In work of the best quality, hardboard sheet is laid over the boards.

A suitable underlay is required to isolate the leadwork from the timber below, allowing it to move freely. Non-woven needle-punched polyester geotextile underlay is recommended by the Lead Sheet Association (LSA), particularly when there is a risk of condensation or moisture underneath the lead. This material will not rot, and it dries out very quickly. Traditional felts plus one or two layers of Class A building paper, to BS 1521, can be used but many contain too much resin or bitumen for use as an underlay. A geotextile underlay should also be used to line stone gutters cut in cornices or blocking courses where there is no timber. Uneven stone surfaces can be levelled off with a 1:1:6 screed.

Sand-cast lead was originally used in the New Town and is very durable. Sand-cast lead sheet is still made by specialists, but there is no British Standard to control the quality and the thickness which may vary across a sheet. Milled lead sheet, complying with BS 1178, is now commonly used. This is available in various thicknesses and must be laid to BS 6915 following the excellent guidance given in the LSA's manual, volumes 1, 2 and 3. Code 5

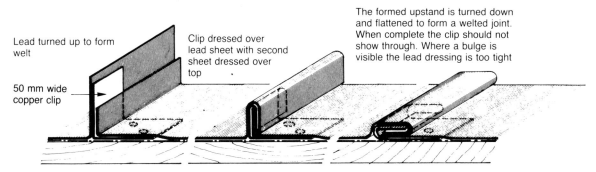

Lead turned up to form welt

50 mm wide copper clip

Clip dressed over lead sheet with second sheet dressed over top

The formed upstand is turned down and flattened to form a welted joint. When complete the clip should not show through. Where a bulge is visible the lead dressing is too tight

WELTED JOINT

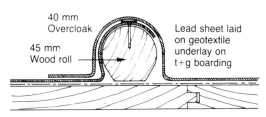

40 mm Overcloak

45 mm Wood roll

Lead sheet laid on geotextile underlay on t+g boarding

TIMBER ROLL JOINT

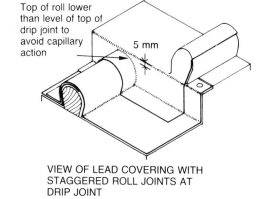

Top of roll lower than level of top of drip joint to avoid capillary action

5 mm

VIEW OF LEAD COVERING WITH STAGGERED ROLL JOINTS AT DRIP JOINT

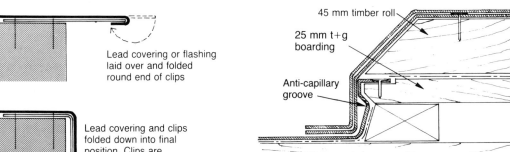

50 mm wide copper clips fixed with counter-sunk screws at 600 mm c/s

Lead covering or flashing laid over and folded round end of clips

Lead covering and clips folded down into final position. Clips are concealed and very strong

CONCEALED CLIP FIXINGS FOR LEAD COVERINGS

45 mm timber roll

25 mm t+g boarding

Anti-capillary groove

SECTION THROUGH DRIP JOINT AND END OF TIMBER ROLL JOINT

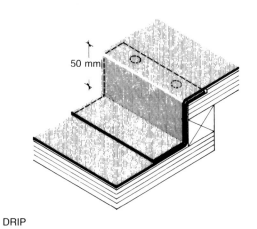

50 mm

DRIP

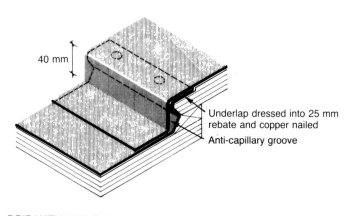

40 mm

Underlap dressed into 25 mm rebate and copper nailed

Anti-capillary groove

DRIP WITH ANTI-CAPILLARY GROOVE

150

may be used for cover flashings and code 7 for gutters, or code 8 in longer bays. Although code 8 lead sheet is expensive, it may prove cheaper than thinner sheet because there will be fewer drips, less cutting and so on.

For lead to last as long as possible, the maximum spacing of joints recommended by the LSA for flat and low-pitched surfaces are:

	Joints parallel with the fall (mm)	*Joints across the fall* (mm)
Code 5	600	2000
Code 6	675	2250
Code 7	675	2500
Code 8	750	3000
Code 9	790	3600
Code 10	790	4000

When deciding the weight of lead sheet to be used, the aspect of the gutter must be considered. Lead sheet which is seldom or never in the sun can be laid to sizes which exceed the recommendations, whereas a gutter facing south must comply with them, or be protected by duck boards. Where it is impossible to keep to the recommended lengths, the weight should be increased because this will increase the strength of the sheet. Movement in lead sheet is taken up in the joints; generally, the greater the number of joints, the less stress will be developed in each sheet.

The sole of a wallhead gutter is usually tapered and should be a minimum width of 150 mm at the lower end. If possible, the gutter should have a minimum depth of 75 mm at the top end.

The overall dimensions of lead sheet for use in centre gutters are given in the table above. If, at the highest point, the gutter is too wide to use one sheet of lead, a longitudinal joint must be introduced. This should be a timber roll, because it is less likely to suffer damage from foot traffic than a hollow roll. It is very important that the height of the roll joint is at least 5 mm lower than the height of the drip joints, to avoid capillary action from the top of the roll to the boards on the step, and that the end of the gutter terminates at a wood stop and is not rigidly fixed to the masonry.

If traditional steps of 50-60 mm and drips cannot be formed because there is not enough height, it may be possible to raise the front edge of the gutter by fixing a stainless steel angle to the blocking course. Alternatively, the sheets can be joined using a proprietary system consisting of two turnerized stainless steel strips connected by a neoprene expansion joint. The neoprene is bonded to the metal strips and these are welded to the lead at each side of the joint. The neoprene should be protected against ultraviolet light by a code 4 lead flap welded to the upper strip of stainless steel only.

On both wallhead and centre gutters a sump should be formed and the outlet from the sump dressed into the top of the downpipe. It should be a minimum of 150 mm deep and not longer than the width of the gutter; the length should be increased if the gutter is less than 150 mm wide. An overflow from the wallhead gutter should be formed through the blocking course. Code 6 lead pipe, 50 mm in diameter and welded to the lead gutter lining should be used, with the bottom of the overflow hole being 50 mm above the sole of the sump. If the overflow discharges above a cornice it should be protected by a lead flashing dressed up above the pipe. Alternatively, a gargoyle projecting clear of the cornice can be formed in code 8 lead. In centre gutters, the incorporation of a standing waste pipe with a similar margin of safety is a worthwhile precaution.

Where the lead is laid under the slates, it should be dressed over the sarking and the continuous tilting fillet, to lap the slates by at least 150 mm. Joints in lead sheet above gutter level should be fully welted. Slates should stop well above (approx. 150 mm) the highest step formed in the gutter. Cover flashings to gutters should have a minimum vertical lap of 100 mm.

It is recommended that centre gutters be protected with duck boards of treated softwood nailed with copper or stainless steel nails. Oak should not come into contact with lead, because of the action of tannic acid.

To protect lead against possible acidic run-off from lichen-covered roofs, sacrificial lead flashings may be installed. Alternatively, copper strips can be fitted at approximately every tenth slate course, the run-off from which kills the lichens. Free lime or cement from new mortar can initiate a slow corrosive attack on lead and it is advisable to protect new lead with a coat of bituminous paint until the mortar cures.

REPAIR TO RHONES

Rhones and downpipes are traditionally cast iron, and when they are well maintained they have a very long life. If they need replacement, it is preferable to use beaded cast iron rhones. PVC rainwater goods in black are available, but they are not recommended for buildings in the New Town because they have a limited life and replacement usually involves expensive scaffolding. They can easily be damaged by the weight of ladders leant against them and fixings designed for timber fascias are not appropriate for the New Town.

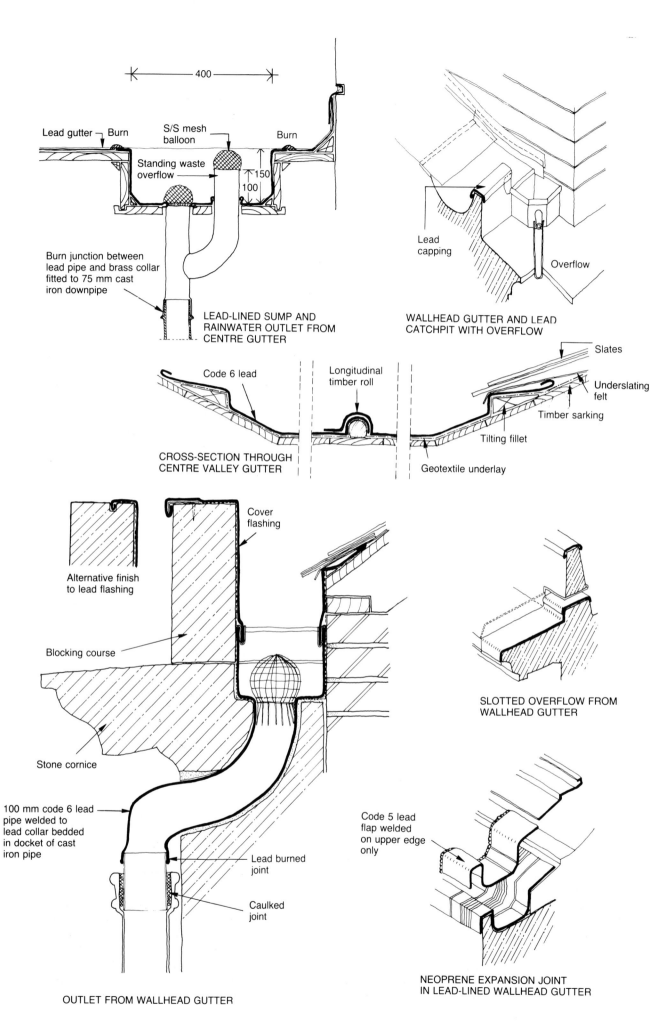

400

Lead gutter — Burn
S/S mesh balloon
Burn

Standing waste overflow

150
100

Burn junction between lead pipe and brass collar fitted to 75 mm cast iron downpipe

LEAD-LINED SUMP AND RAINWATER OUTLET FROM CENTRE GUTTER

Lead capping
Overflow

WALLHEAD GUTTER AND LEAD CATCHPIT WITH OVERFLOW

Code 6 lead
Longitudinal timber roll
Slates
Underslating felt
Timber sarking
Tilting fillet
Geotextile underlay

CROSS-SECTION THROUGH CENTRE VALLEY GUTTER

Cover flashing

Alternative finish to lead flashing

Blocking course

Stone cornice

100 mm code 6 lead pipe welded to lead collar bedded in docket of cast iron pipe

Lead burned joint

Caulked joint

OUTLET FROM WALLHEAD GUTTER

SLOTTED OVERFLOW FROM WALLHEAD GUTTER

Code 5 lead flap welded on upper edge only

NEOPRENE EXPANSION JOINT IN LEAD-LINED WALLHEAD GUTTER

DORMER WINDOWS

INTRODUCTION

Dormers, with their slated roofs and haffits (or cheeks), are distinctive features of Scottish architecture. In the New Town they were seldom part of the original design and indeed were often expressly forbidden in the original titles. However, they enable otherwise empty roof spaces to be used and many were added soon after the buildings were completed.

Dormers in the New Town usually project above the roof, although inset dormers are occasionally found. New Town dormers are mainly rectangular, bowed or polygonal in plan, and were designed to take sash and case windows of varying proportions. They normally have slated roofs and haffits, although occasionally lead was used – for example, in Manor Place. Zinc haffits are mostly Victorian.

Listed Building Consent will not generally be granted for adding new dormers to Georgian buildings in the New Town except in exceptional circumstances for instance where their addition will balance existing asymmetrically positioned dormers. The uniformity of a roofscape can also be restored by the removal of an inappropriate dormer and the insertion of a suitable rooflight (for further information see CUPOLAS AND ROOFLIGHTS). If permitted, new dormers should follow traditional designs, avoiding flat roofs and paying attention to the pitch of the dormers in relation to the pitch of the main roof and the height of the ridge. An engineer should be consulted when a new dormer is to be built. Generally, the existing rafters around the opening should be doubled and a new trimmer beam fixed between them to take the shortened ends of the intermediate rafters. Without this the common rafters may sag, making roof and dormer unsafe and unsightly. In rare cases roofs have purlins and principal rafters which must not be cut.

INSPECTION AND MAINTENANCE

Inaccessible dormers often suffer from lack of external maintenance, leading to serious leaks and major defects throughout the roof structure. All lead flashings between the dormer and the main roof, and at the gutters and aprons, should be checked during the annual roof inspection.

The Planning Department recommends that timber facings, gutter facias and other such exposed timber should be painted BS 4800, 00 A 13 (Slate Grey), while the windows should be 00 E 55 (White), although for dormers behind balusters it is preferable that all timber is painted slate grey. For further information see EXTERNAL PAINTWORK.

DEFECTS AND REPAIRS

Slated roofs and haffits can be difficult to maintain and renew. Double nailing is recommended even though subsequent repairs to the slating will be more difficult (see ROOF COVERINGS). When reslating dormers the ridges should be renewed in lead rather than zinc. A small lead flashing should be introduced at the head of the haffit to cover nail fixings on the top course of slates. To repair side watergates and flashings it is normally necessary to strip and reslate the haffit, unless the new lead can be carefully pushed up underneath the bottom course of slate, relying on a tight fit to keep the lead wedged in position.

The front lead apron flashing usually forms a tray beneath the timber window cill and it is therefore impossible to replace it without removing the window (see illustration). An apron flashing usually fails by cracking because its fixings have restricted the normal thermal movement of the lead. Once this has occurred further deterioration is unlikely and, to avoid removing the whole window, it is often more practical to insert a secondary apron flashing underneath the original, lapping the crack by 150 mm either side and securing the new lead by pushing it in as far as possible under the cill, using lead wedges if necessary, and protecting the joint with a mastic bead. Burning on a lead patch, which may result in the lead cracking elsewhere, is not recommended.

TYPES OF DORMERS IN THE NEW TOWN

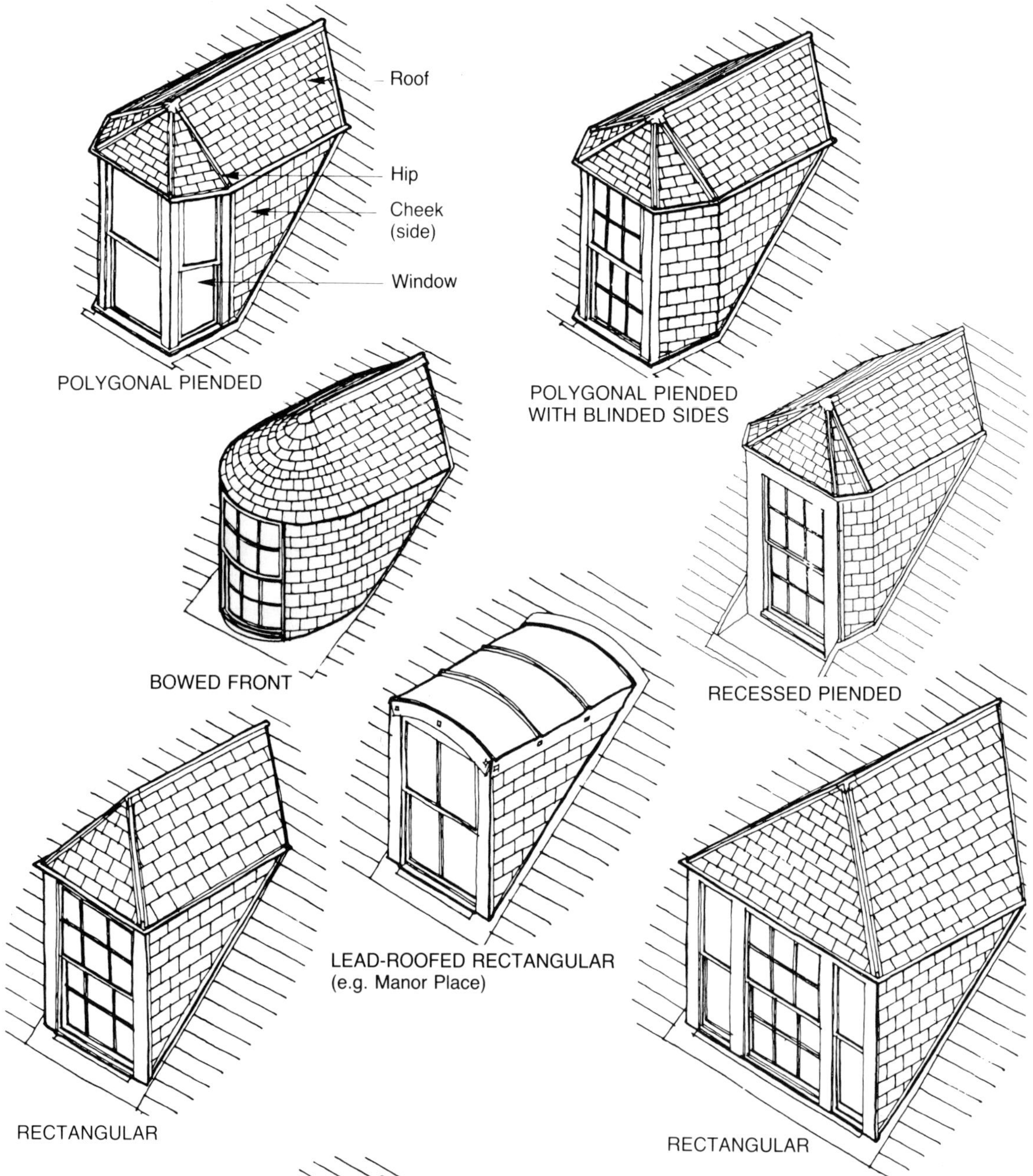

Roof

Hip

Cheek (side)

Window

POLYGONAL PIENDED

POLYGONAL PIENDED WITH BLINDED SIDES

BOWED FRONT

RECESSED PIENDED

LEAD-ROOFED RECTANGULAR
(e.g. Manor Place)

RECTANGULAR

RECTANGULAR

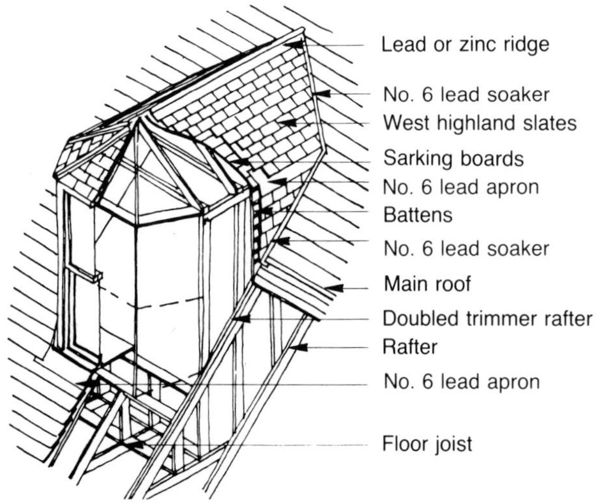

Lead or zinc ridge

No. 6 lead soaker
West highland slates

Sarking boards
No. 6 lead apron
Battens

No. 6 lead soaker

Main roof

Doubled trimmer rafter
Rafter

No. 6 lead apron

Floor joist

CONSTRUCTION OF A DORMER WINDOW

154

Timber fascias are prone to rot, particularly near the base. Depending on the extent of rot, these will either need to be completely replaced or new sections spliced on at the base. Where possible, existing timber should be undercut to protect the top edge of the new spliced repair. The bottom edge should be left clear of the apron to avoid drawing up moisture. If the fascia has to be completely replaced it is worth providing a vertical lead damp-proof course behind it, to help protect the window case and to improve the vulnerable junction between the fascia and the edge of the slated haffit (see illustration).

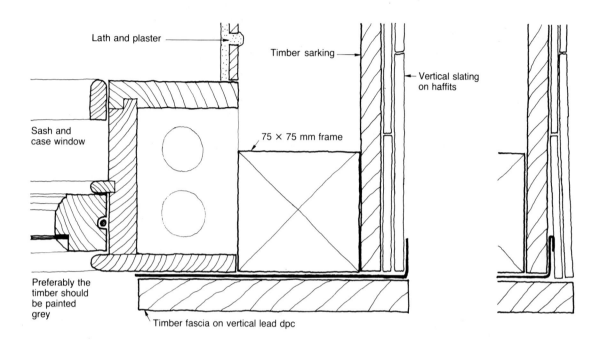

ALTERNATIVE DETAILS FOR VERTICAL LEAD DPC BEHIND REPLACED TIMBER FASCIA

CUPOLAS AND ROOFLIGHTS

INTRODUCTION

Elsewhere, the word 'cupola' describes a small dome covered with lead or copper, but in Georgian Edinburgh it applies to the large timber-framed rooflight, usually circular or elliptical on plan, above the stairwell in the centre of a building. In Victorian or later tenements, lanterns, rectangular or square on plan, are usually found above the common stair. The original type of cupola or lantern should not be replaced by proprietary preformed plastic alternatives.

Rooflights, or skylights, are windows set into roof pitches to give light to stairwells, attic rooms or roof spaces, and normally have either cast iron or timber frames, although unframed glass is also found.

Controls

In the New Town, generally both planning permission and Listed Building Consent will be required where a rooflight is being replaced or a new one introduced. In addition, a building warrant will be necessary before making a roof space into a room. If a building is converted to a new use, it may be necessary to replace existing glass with flame and heat resistant glass to satisfy fire regulations. It is essential that the Building Control Division of the Department of Property Services and the Planning Department are consulted before any conversion work is put in hand.

DEFECTS, REPAIRS AND REPLACEMENT

Cupolas

Because of their exposed position, cupolas require regular maintenance. They rarely receive all the attention they need because they are relatively inaccessible and decay is often only visible from the roof. In the case of the common stair, they are a shared responsibility which is a further disincentive to proper maintenance.

The timber kerb supporting the cupola should be high enough to accommodate the steps or drips recommended by the Lead Sheet Association for the adjoining lead platform roof and if the platform is being renewed it may be necessary to increase the height of the kerb to achieve this. If any of the timber is affected by rot, the cupola should be stripped of its glazing and all decayed timber cut out and renewed. It will seldom be necessary to renew an entire cupola; individual sections can often be repaired using pressure-impregnated redwood or a good quality hardwood such as oak. Glazing bars are vulnerable to rot at their lowest ends where the timber end grain is exposed. If not protected or regularly painted, glazing putty will crack and fall out allowing water into the joint between the glass and timber.

Cracked glass should not be taped or patched except for temporary repair, but, instead, carefully removed and replaced by a whole new pane, selected to match the remaining glass if possible. Where a cupola has to be replaced or completely reglazed it is recommended that a safety glass should be used, the type chosen being dependent on the cupola location and its susceptibility to accidental damage. The use of georgian wired cast glass (GWC) is unacceptable in an elliptical cupola because the square grid wire pattern is incompatible with the wedge-shaped panes. The installation of double glazing may also be considered for heated spaces within private houses because it is possible to increase the depth of the glazing bar section to take sealed double glazed units, but the very high cost should first be weighed against the long term benefits. Glass should be held in position by a non-ferrous metal clip (copper, zinc or stainless steel tingles) nailed to the cill and bent over the foot of the pane. These clips should be sufficiently strong to hold the glass in the event of putty failure, to prevent loose panes from falling through the cupola and causing serious damage below. A space of approximately 5 mm, generally formed by the thickness of the clips, between the glass and the timber cill should be left clear to provide ventilation in order to combat condensation which may form on the inside of the glass.

Consideration must also be given to permanent ventilation at the highest point of the cupola. This is particularly important in common stairs, many of which still have original water closet vents from flats on each floor. Cupolas usually have a lead cap with an open-ended U-shaped tube of 35-50 mm diameter, while lantern lights have louvred sections or open ridge caps, and it is important that these ventilators are kept clear.

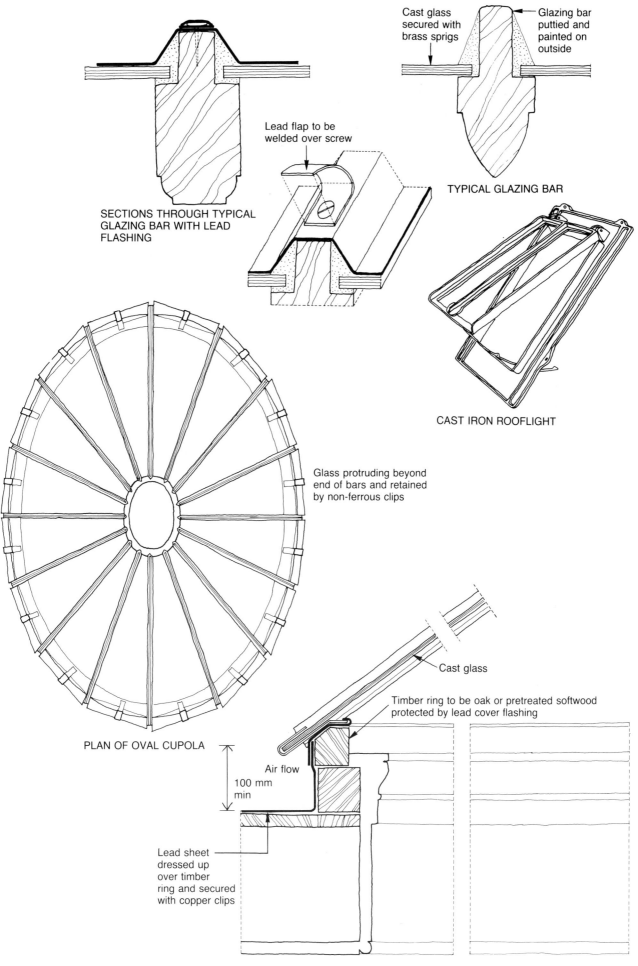

SECTIONS THROUGH TYPICAL
GLAZING BAR WITH LEAD
FLASHING

Cast glass
secured with
brass sprigs

Glazing bar
puttied and
painted on
outside

TYPICAL GLAZING BAR

Lead flap to be
welded over screw

CAST IRON ROOFLIGHT

Glass protruding beyond
end of bars and retained
by non-ferrous clips

Cast glass

Timber ring to be oak or pretreated softwood
protected by lead cover flashing

Air flow

100 mm
min

Lead sheet
dressed up
over timber
ring and secured
with copper clips

PLAN OF OVAL CUPOLA

SECTION THROUGH KERB OF CUPOLA

157

Before inserting the glass into the glazing bars a coat of primer should be applied to the rebates so that oil from the putty does not soak into the timber. Another coat should be applied immediately after glazing and, once these initial coats of paint have dried, a putty fillet and a full paint system can be applied. Ideally, glazing bars should be repainted every five years. The original detail can be improved using code 3 or 4 lead flashings carefully moulded round the glazing bar and fixed by non-ferrous screws capped with a lead dot. This is recommended where cupolas are inaccessible or where such protection will extend the life of the cupola which would otherwise require repair.

Rooflights

Rooflights, together with their associated flashings, should be included in the annual inspection of the roof. Cast iron rooflights seldom need more than thorough wire brushing and a coat of paint, but timber-framed lights may need to be repaired. Careful maintenance will ensure long life and a weathertight condition.

When replacing rooflights on the outer slopes of pitched roofs, traditional cast iron glazed rooflights should be used. Other rooflights should if possible be set inconspicuously on the inner slopes of M-shaped roofs. There are sympathetically designed replacement timber and metal proprietary rooflights available which are suitable for use in historic buildings combining double glazing, central pivoting, excellent sealing qualities and a high standard of finish. Special flashing kits are now available which allow rooflights to be set closer to the finished surface of the roof. Original timber rooflights should be repaired on site by a joiner and protected by lead flashings. Where external timber is to be painted, colour is important and in most cases it will be appropriate to use BS 00 A 13 (Slate Grey).

ACCESS TO ROOFS

Easy and safe access to roofs and roof spaces encourages efficient and regular maintenance; inconvenient or hazardous access discourages frequent inspection. Access to the traditional New Town 'M' roof is normally gained through the roof space via a ceiling hatch, either at the top of the stair or within the top flat, with a roof hatch opening out into the centre valley gutter. Some roofs and roof spaces have no access at all, in which case the formation of new hatches in the ceiling and roof are strongly recommended. A hatch off the common stair will often give access to the loft above all the top floor flats and should be securely fastened against intruders.

Ceiling hatches, when fitted between two ceiling joists, may be very narrow and difficult to negotiate. It is worthwhile and easy to enlarge them by cutting an adjacent joist and inserting a bridle. A fixed, retractable aluminium ladder is ideal for gaining access, but some modification of the hatch opening will usually be required.

Inspection of the roof space should be made as easy as possible. Dirt and debris should be cleared away and crawl boards fitted where appropriate, particularly around hatches (a foot through the ceiling below can be rather expensive!). Redundant services should be stripped out and all live cables, pipes and conduit neatly and securely fixed. Permanent inspection lights are recommended for houses and tenements if agreement can be reached over the sharing of costs.

Roof hatches are usually constructed in timber framing and boarding covered with zinc or lead sheet. Because lead is heavy to lift, zinc is strongly recommended, but after several years it tends to crack, especially at the corners. Lead-covered hatches can be fitted with gas-sprung hinges or a counterweight. Hatches were traditionally provided with iron strap hinges; if the screws are rusty, they should be renewed in stainless steel and lengthened as necessary to provide adequate restraint. It is important that hatches are fitted with a secure chain or prop to prevent them falling back onto the roof. On the whole, a prop is safer because this avoids the cover being blown shut and injuring someone using the hatch. Side-hinged hatches are easier to open and less likely to damage the slates or cause injury. Hatches should be locked internally for security.

Duck boards should be laid in the centre gutter to protect the leadwork against damage by foot traffic and to provide easy, safe access for inspection and repair. They also help to protect the gutter coverings from extremes of temperature and therefore prolong their life. Debris below duck boarding can cause water to back up and should be regularly cleared away, so duck boarding must be made in easy to lift pieces to allow for regular cleaning and inspection. Broken slats and loose nails can cut and damage the lead and regular maintenance is essential to prevent the boards being more of a hindrance than a help. Duck boards should be made of treated softwood (oak is not acceptable because the tannic acid which leaches out will corrode the lead) and slats, fixed with brass or stainless steel screws, should be no more than 13 mm apart to prevent snow from collecting beneath.

Unless there is a parapet on the front elevation, safe access down to the gutter and rhones at wallhead level (where most maintenance is required) is difficult. Some slaters will go down on a rope, but increasingly stringent health and safety requirements are restricting such casual access. This method may still be employed for short-duration work where a safety harness and rope can be secured to a substantial anchorage point, for example inside a roof hatch or at a chimney head. If front gutters are not accessible from the roof they should be inspected from a mobile hoist or scaffolding towers. The provision of permanent access ladders is recommended for internal slopes and chimneys but it is essential that these are well maintained. Roof ladders should be made of impregnated softwood and all fixings should be of brass or stainless steel. Mild steel cat ladders are sometimes fixed at party walls, parapets or chimneys where there is a change in roof level. These need to be painted regularly to be kept in good repair and to avoid rust stains on the stonework. During a repair contract, the opportunity should be taken to insert stainless steel hooks or eyes into sound masonry to provide permanent anchorages for ladders or harnesses, in order to facilitate future inspections.

ROOF HATCH, MALTA TERRACE
Photo: Ian Begg

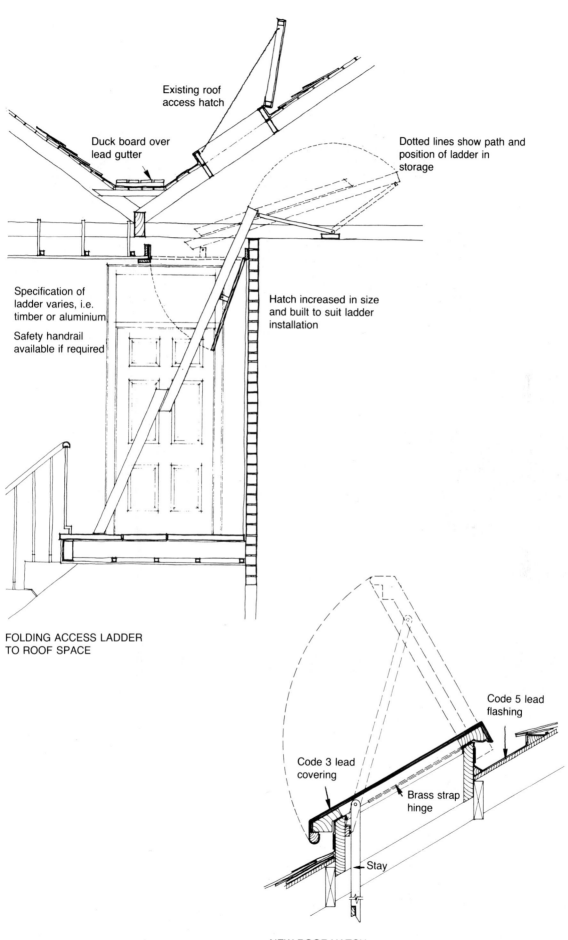

Existing roof
access hatch

Duck board over
lead gutter

Dotted lines show path and
position of ladder in
storage

Specification of
ladder varies, i.e.
timber or aluminium

Safety handrail
available if required

Hatch increased in size
and built to suit ladder
installation

**FOLDING ACCESS LADDER
TO ROOF SPACE**

Code 5 lead
flashing

Code 3 lead
covering

Brass strap
hinge

Stay

NEW ROOF HATCH

CHIMNEY AND FLUES

INTRODUCTION

Chimneys and their cans or pots are a distinctive element of the New Town skyline. Until this century, they served innumerable open fires, which were the only form of heating in Auld Reekie. There are few open fires today, but the old flues still serve central heating boilers and gas fires. They also have the useful function of ventilating the rooms and thereby help to keep the building fabric in a sound, dry and healthy condition. Under the Clean Air Acts 1956 and 1968, and the East Edinburgh Smoke Control Order 1991 (which includes the New Town) only heating appliances which use smokeless fuel are allowed. In some situations listed buildings are exempt from this order and more information is available from the Department of Environmental Health. If for no other reason, chimneys should be retained and maintained because they may well be needed for open fires in the future when the supply of other fuels has run out.

GENERAL

It is the policy of the Department of Planning that chimneys should be retained because they contribute to the architectural character of listed buildings, even if the flues are not in use, and that inferior materials, such as brick or concrete, should not be substituted for stone if the chimneys have to be rebuilt. Grants are offered by the ENTCC towards the cost of rebuilding chimneys in stone with copings and chimney cans to match the original design.

Prominent chimneys were built of ashlar while chimneys on rear roof slopes and on less important buildings were built of rubble with ashlar quoins. Divisions between adjoining flues were of thin stone, seldom more than 100 mm thick, built on cant and therefore very liable to delaminate. These divisions, known as bridges (brigs) or mid-feathers, were usually butted to the facing stones of ashlar chimneys, but were bonded through rubble facings. Sometimes the lines of flues can be traced on exposed rubble gables where the ends of these thin stones show on alternate courses.

INSPECTION AND DEFECTS

Chimneys are the most exposed part of any building and therefore vulnerable to damage by wind, rain and frost externally, and to decay by flue gases from within. The masonry, pointing, copes, cans and fixtures, such as aerials and telephone cables must be inspected every year.

Structural movement, or weakness in the original construction, such as fallen bridges may cause failure. A flue can be blocked by a bird's nest or by stones from a broken bridge between flues, and professional advice will be required. It may be possible to locate the obstruction with a boroscope, or a small closed-circuit TV camera with a lamp, lowered down inside the chimney. Once located, the flue should be opened up for cleaning and further inspection.

Defects are often due to exposure. Loose, uncramped or cracked copes with washed-out joints, loose and cracked cans and disintegrated haunching are common. The junctions between chimneys and roofs are particularly vulnerable to penetration by moisture which will soak the roof timbers and run into other parts of the building (see FLASHINGS).

Masonry and mortar are affected by the products of combustion. The first signs are dark stains on the face of the chimney or along the line of the flue. Most of the sulphur emitted by combustion is in the form of sulphur dioxide, which passes harmlessly to the outside air, but about 2-3 per cent is in the form of sulphur trioxide. This reacts with water vapour (produced when the temperature of the flue or lining falls below the dew point) to produce sulphuric acid which destroys mortar joints and pargetting. This acid also attacks sandstone and accelerates the delamination of the bridges. The top of the flue, being most exposed to the weather, will have the lowest temperature and will therefore suffer most from condensation and chemical attack. A further hazard is that of accumulated soot catching fire, a hazard made worse if coal and wood are burned together, leaving an inflammable, tarry deposit in the flue. The use of modern, high output fuels can also cause fire risk. Chimney fires can destroy pargetting, crack masonry and ignite built-in timbers. These risks can be reduced if the chimney is effectively lined.

Outward bulges on the sides of the chimneys may be caused by a number of defects. When feathers are broken, their tying action is lost and the loads imposed on the slender stone facing can cause bulging. This is compounded

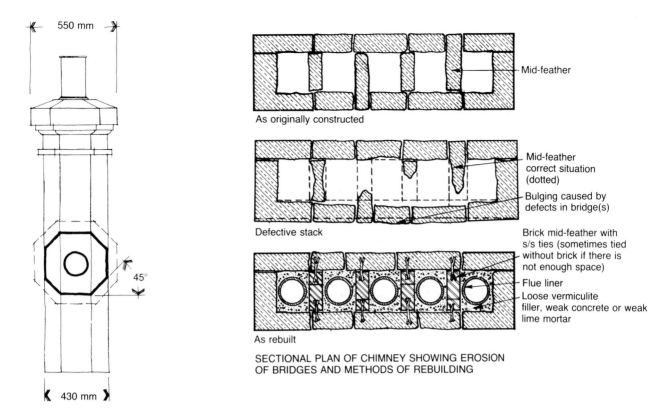

550 mm

45°

430 mm

FREESTANDING OCTAGONAL FLUE
BY WILLIAM PLAYFAIR, 1821

Mid-feather

As originally constructed

Mid-feather correct situation (dotted)

Bulging caused by defects in bridge(s)

Defective stack

Brick mid-feather with s/s ties (sometimes tied without brick if there is not enough space)

Flue liner

Loose vermiculite filler, weak concrete or weak lime mortar

As rebuilt

SECTIONAL PLAN OF CHIMNEY SHOWING EROSION
OF BRIDGES AND METHODS OF REBUILDING

SERIES OF CHIMNEY CANS COMMON TO THE NEW TOWN

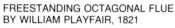

approx 600 mm

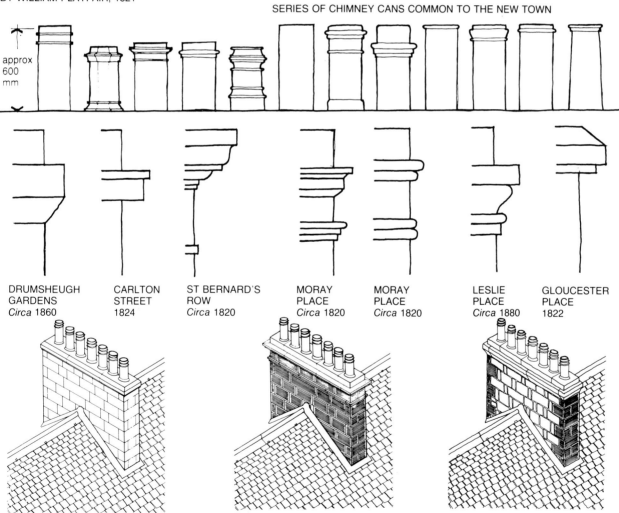

| DRUMSHEUGH GARDENS *Circa* 1860 | CARLTON STREET 1824 | ST BERNARD'S ROW *Circa* 1820 | MORAY PLACE *Circa* 1820 | MORAY PLACE *Circa* 1820 | LESLIE PLACE *Circa* 1880 | GLOUCESTER PLACE 1822 |

Polished ashlar

Broached ashlar

Broached quoins with rubble centre

TYPES OF CHIMNEY STONE FINISHES IN THE NEW TOWN

163

by erosion of the stone inside the chimney by flue gases. Increased weathering and erosion of mortar on the side facing into the prevailing rain-laden wind can cause joints to expand on that side of the chimney which will then lean away from the prevailing wind.

REPAIRS AND MAINTENANCE

Before commencing repairs to chimneys, the flues should be cleaned and temporarily sealed (for instance with sacks of wood shavings) at the fireplaces and at the top of the sound flue, to prevent debris falling into rooms.

Chimney stacks suffering from structural movement or masonry decay may have to be taken down to a sound base (usually roof level) and rebuilt. The chimney should be photographed before demolition and if the original design has been altered it will be necessary to study and measure other chimneys nearby to obtain the correct profiles, tooling patterns and texture. Once the stack has been dismantled to roof level, consideration should be given to the introduction of a lead damp-proof course tray and associated flashings to prevent moisture penetrating into the building fabric. Generally, this is not a problem with chimneys built across ridges where there is an adequately ventilated roof space allowing moisture to evaporate, but where the base of a wallhead chimney is close to the room below or where attics have been made habitable, it may be desirable to prevent the moisture movement by laying a damp-proof course.

Stone used for indenting or rebuilding should match the colour, texture and tooling of the original masonry. All eroded, laminated or soot-stained stone from the existing chimneys should be discarded. Rebuilt bridges should ideally be of stone but are more commonly built in brick with a low sulphate content such as engineering brick, which should be tied with stainless steel to the external stone. When rebuilding rubble chimneys, every second or third bridge stone should be bonded into the external face, or where reinstated bridges are of brick, the character of the bridges on the external face should be reinstated using stone. New stone for rubble chimneys should be specified to be supplied from the quarry as *cut rubble* without having been previously sawn; samples should be obtained before the stonework is ordered. The height and length of new stones, including the upper and lower limits of dimensions, should be specified. The cut rubble may require to be dressed on site to provide a size, shape and finish to match the existing stonework. To minimize wastage resulting from the site dressing, the bed size should be greater than the minimum required for the build. Where broached ashlar quoins are to be replaced, stone should be specified to be sawn, allowing an additional 10 mm on the external face of the stone for site dressing. The external face of the stone should be hewn to remove the sawn finish prior to broaching and the formation of a chisel-drafted margin.

Copings were originally of stone and should be replaced with stones dressed to the same profile, cramped together with stainless steel dog cramps and, if necessary (for example to arrest damp ingress into attic rooms below), laid on a bituminous painted lead damp-proof course. If being renewed or disturbed, adjacent skew stones on party walls should be cramped together with non-ferrous cramps and should be checked into rather than butted against the stack.

When rebuilding a chimney the old pots should be carefully taken down and salvaged for reuse if possible. A rap with the knuckles will produce a ring in good cans. Even cans damaged at their base can be cut down and reused. In most circumstances cans should be 600 mm high. Generally new cans should be double-beaded and of buff colour to harmonize with local sandstones; replacements for distinctive original pots are available to special order. It is not necessary for all old and new cans to be of the same height and in a rebuilt chimney the danger of everything being too uniform can be relieved by using cans of varying height and design. Cans are normally set down into the cope by approximately half the depth of the stone, supported on ledges cut into the cope specially for this purpose, and haunched with lime mortar, trowelled to fall to the edge of the cope. Where possible, disused flues should be kept open for ventilation. The can should be fitted with a fireclay ventilator terminal to provide an air space and to prevent rain entering the flue. Special Gas Board approved fireclay terminals should be used for gas installations. Zinc 'coolie hats' are not recommended because of their short life.

Mortar for bedding and pointing masonry should be 1:3 lime putty/sand mix, or a 1:2:9 cement/lime/sand mix and for fillets and haunching, it should be a 1:2:9 mix. In chimneys, a sulphate-resistant cement should be used. For further information on pointing refer to MORTAR JOINTS.

When in constant use, chimneys should be swept at least once a year at the end of the heating season by one of two accepted methods. The best and time-honoured method of rotating ball and brush lowered from the chimney top allows cleaning around awkward horizontal sections in rebuilt flues. The alternative technique uses a brush and rods pushed up the chimney from the fireplace. The force required to push the brushes up can damage the flue, and this method may prove impossible where flues twist or are very long.

Whenever a flue is opened up for repair or a chimney is rebuilt, the opportunity should be taken to insert a flue liner. Care should be taken to minimize the inevitable reduction of the cross-sectional area of the flue because this bears a direct relationship to the size of the fireplace opening and can affect the efficiency of the fire (see FIRE-PLACES). The space between the liner and the original flue should be filled with weak lime mortar or weak vermiculite concrete, which will stiffen the construction and insulate the liner. Generally, impervious clay liners are

GABLE CHIMNEY REBUILT AT 1-3 ALBANY STREET   165   Photo: Ian Riddell

recommended. It is essential to ensure that liners are fitted with the socket of the liner uppermost to avoid condensed gases seeping through the joints. Other types of liner include refractory concrete pipes, which are less resistant to acid attack, and flexible metal linings, either of stainless steel, which are expensive, or aluminium, which should be avoided in corrosive situations. Stainless steel is now obligatory where gas boilers are installed. When repairing a flue a proprietary liner system is available which involves pumping refractory mortar down the flue around an inflatable rubber tube former, which is subsequently deflated and removed.

ROYAL TERRACE

Photo: John Johnson

PARAPETS AND BALUSTRADES

Parapet walls, which increase the height of the front elevation without the addition of an extra storey, are rare in the New Town, although Carlton Street is a good example of the fine proportions attained in this way. Balustrades are another attractive form of architectural decoration found on wallheads and, in the later New Town, on canted bays.

INSPECTION AND MAINTENANCE

Parapet walls and balustrades are exposed to weather on both sides and so should be regularly inspected for signs of deterioration of mortar pointing or stone decay. Regular maintenance (e.g. pointing) of parapets and balustrades will ensure that joints and structure remain watertight, and prevent fixings becoming loose. Fixings are a matter of particular concern because narrow balusters can be rather unstable (and they can cause considerable damage if blown down, as happened in Carlton Street in 1973).

DEFECTS

Long unbraced sections of balustrades or parapets can suffer vertical and horizontal distortion; individual balusters may decay because they are of thin section and edge bedded and this can cause problems in the balustrade as a whole. Temperature changes and frost action can cause joints to open and stones to crack, giving rise to water penetration. Rusting wrought iron cramps, even when wrapped in lead, can burst the stone.

The lack of a damp-proof course beneath the parapet wall or balustrade, combined with badly maintained pointing, may allow water penetration through the joints and into the wall structure. This can lead to decay of timber wallplates, joist ends, gutter linings and supporting timbers, as well as staining and moss and algal growth on the cornice.

REPAIRS

There are occasions when parapet walls have to be completely rebuilt, because they contain decayed or badly distorted stones. The opportunity can be taken to introduce non-ferrous dowels and/or braces to skews and chimney heads, which will improve the stability. However, the result must have sufficient flexibility to allow thermal movement to take place and this can be achieved by avoiding rigid fixings and by setting cramps or dowels in mortar rather than epoxy resin. Since wallhead gutters are usually located behind the blocking course of the balustrade, it may also be possible to improve the lead detailing in this area and reduce the likelihood of damp penetrating through to vulnerable timbers. Traditionally, the lead flashing which protects the gutter lining is wedged into a raggle on the inner face or on the top of the blocking course and is not usually placed under the balusters. It is not recommended that lead be placed under balusters because of the problem of differential thermal movement which can cause breakdown of the lead. It is also a detail which is difficult to achieve while at the same time complying with the recommendations of the Lead Sheet Association (see GUTTERS AND DOWNPIPES).

When parapets or balustrades are renewed, the materials and detail should match the original sections. Replacement with precast concrete is not recommended and would in any case be unlikely to receive Listed Building Consent.

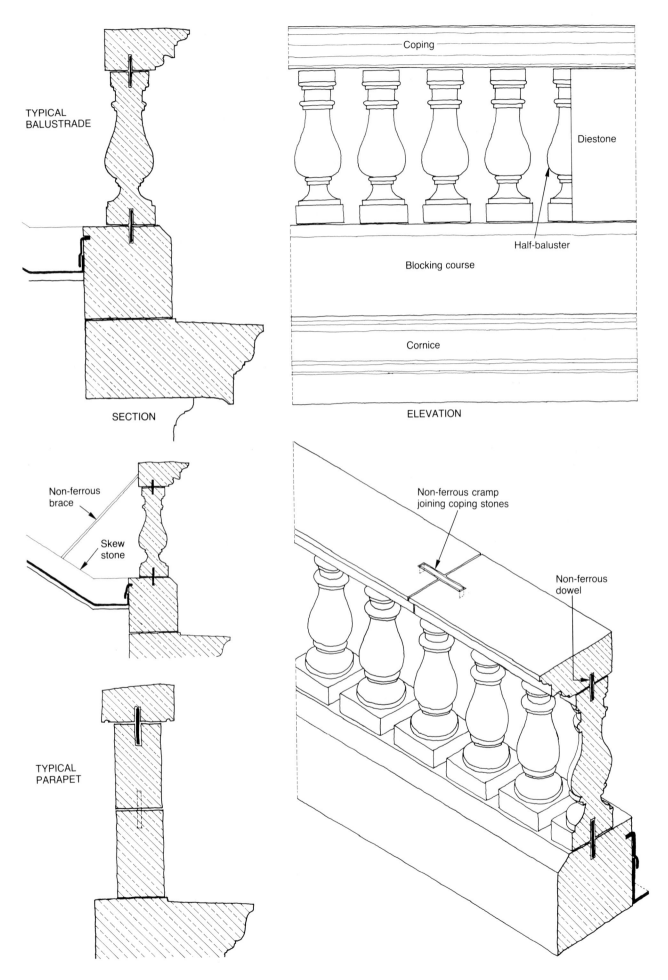

TYPICAL
BALUSTRADE

SECTION

Coping

Diestone

Half-baluster

Blocking course

Cornice

ELEVATION

Non-ferrous
brace

Skew
stone

TYPICAL
PARAPET

Non-ferrous cramp
joining coping stones

Non-ferrous
dowel

INSULATION

INTRODUCTION

This chapter examines methods of reducing heat loss and transmission of sound, and the protection of furnishings from the harmful effect of natural light. It is essential these days to conserve resources and to reduce expenditure on fuel. All new buildings must have a minimum standard of insulation but the insulation of existing buildings can cause problems, particularly if they are listed and it becomes necessary to disturb the finishes.

THERMAL INSULATION

A building loses heat through conduction, convection and radiation, the last being insignificant. Heat is lost from the elements of a typical New Town tenement in the following proportions:

Heat loss through conduction	%
Roof	15
Walls	35
Windows	15
Ground floor	10
Total	75
Heat loss through convection	25
Total heat loss	100

In older buildings draughts and natural ventilation can account for as much as 50 per cent of the total heat loss. It is therefore important that doors and windows fit well and that flues can be closed when not in use. However, it is equally important to ensure that there are some air inlets to provide adequate air movement through the building and avoid the risk of condensation, and also to provide sufficient oxygen for combustion for gas fires and boilers. When fitting new gas appliances, the current Building Standards (Scotland) Regulations require ventilation openings to comply with Part F 3.2 of the supporting Technical Standards, and Building Control officers are unlikely to accept that general building 'leakage' will provide the air required. This may seem excessively cautious, especially in historic buildings with listed facades where the existing windows are extremely leaky. The rules are intended to protect occupants against fumes and a relaxation is unlikely, but with the support of the planning authority it may be possible to obtain a waiver if the appearance of the listed building is likely to be compromised. In any case, with careful design, ventilation openings can be made inconspicuous (see SASH WINDOWS and SERVICES).

Adding extra insulation without providing adequate ventilation can result in condensation due to moisture-laden air condensing on cool surfaces. This is particularly likely in roof spaces which need good natural ventilation if the ceiling is insulated. For general advice on upgrading of thermal insulation and its associated condensation risks see the recent reports by the British Research Establishment, BRE 143: *Thermal Insulation: avoiding risks* and BRE 174: *Tackling Condensation*. A number of recent BRE Digests and Information Papers refer specifically to existing buildings.

The decision to insulate an existing building depends on its size, construction, heating system and, most importantly, the lifestyles of the occupants.

Walls

A typical external wall in the New Town about 700 mm thick, built of sandstone and lined internally with lath and plaster, has a *U*-value of about 1.45 which is good by traditional standards although the current Building

Standards (Scotland) Regulations require that walls of new houses should have a U-value of 0.45. The U-value is the rate at which heat is conducted through a material, and is measured in watts per square metre per degree Celsius (W/m^2/$^\circ$C).

It is difficult, and usually uneconomical, to improve the insulation of external walls in the New Town. The walls are finished with lath and plaster on strapping except at basement level where the plaster is applied directly to the stonework, i.e. plastered on the hard (see PLASTERWORK). It is not usually possible to add loose insulation to the external walls because the narrow cavity between timber straps varies in width and is connected to the void between floor joists. The only other practical option is to fix rigid insulation boards between the straps, but this involves the removal and replacement of all the plaster, which is undesirable and prohibitively expensive.

External walls which are plastered on the hard have poorer insulating qualities than strapped walls and, being heavyweight, they are slow to respond to heating. In rooms with little or no decorative features, dry lining can be added to these walls, which will improve the insulation and allow the rooms to heat up more quickly. This treatment is not recommended for the majority of rooms in the New Town, where the dry lining would interfere with timber finishes such as architraves around doors and windows, skirtings, shutters and soffit linings which would have to be removed, adjusted and refixed. Plaster cornices could also be obscured.

Floors

Heat loss through solid basement floors laid directly on the ground can be significant. Such floors can be uncomfortable due to cold radiation, condensation and dampness. In this situation, insulation must be combined with continuous damp-proofing or tanking. If it is possible to slightly raise the general floor level, and if great care is taken in positioning the damp-proof membrane, then insulation can be added above the solid floor without risk of condensation.

Guidelines are given in the Floors section of BRE 143: *Thermal Insulation: avoiding risks* and particular attention is drawn to the situation where the existing floor is found to be damp and to the need to avoid cold bridges at junctions. In addition, BRE Digest 145: *Heat Loss through Ground Floors* provides guidance on the calculation of heat losses from insulated and uninsulated floors, while BRE Digest 380: *Damp-proof Courses*, and BRE Good Building Guide (GBG) 3: *Damp-proofing Existing Basements* both give advice on how to prevent water penetration.

Traditional deafening consisting of ashes or sand laid between joists to reduce sound transmission between flats also provides good thermal insulation. It is not worth trying to improve the insulation between floor levels, and there is always the risk that the floor surface will be damaged if it is lifted.

Roofs

It is generally agreed that one of the most cost-effective methods of saving energy is to insulate the roof or attic space. Roof U-values can be reduced to 0.2 by using 150 mm thick insulating material. The resulting saving of energy should pay for the installation within two to three years. In houses in the New Town this can be done either within the roof space above the ceiling of the top floor or at roof level.

All insulation should be removed from the roof space before spraying with timber preservative: even so-called non-flammable chemicals become dangerous once their water content evaporates, producing a risk of spontaneous combustion.

1 *Insulating at ceiling level:* The simplest method is to lay insulating material, such as fibreglass quilt, loose fill or rigid insulation board, within the roof space loose on top of the ceiling plaster. A quilt is the easiest to handle and to lift later for repairs to services. Water pipes should either run beneath the insulation or be individually wrapped. This method of insulation results in a cold roof space, so the cold water tank must be wrapped in insulating material together with its supply pipe and valve to prevent freezing in winter. On no account should insulation be laid beneath the tank.

The roof space needs to be well ventilated to avoid condensation on the soffit of the rafters or sarking boards, or on the underside of the sarking felt. There will usually be a reasonable degree of ventilation in a roof space which is slated directly onto butt-jointed sarking boards, but when there is sarking felt, the flow of air may be insufficient to prevent condensation; therefore insulation should not be packed tightly into the eaves.

Additional ventilation can be introduced by some or all of the following methods:

● Trim back the lowest sarking board to provide an air gap (although with traditional close-slated eaves this may be difficult to achieve);
● If there is sarking felt, provide larger gaps between sarking boards;
● Insert proprietary ventilators into the inner slated pitches where they will be inconspicuous;
● Insert permanent ventilators into external gable walls in the roof space. Ventilators in party walls may allow fire to spread and should not be used.

The use of a quilt thicker than 100 mm produces a much colder roof space and with very thick insulation, condensation can even occur within the thickness of the quilt itself. The use of foil-backed quilt will help to

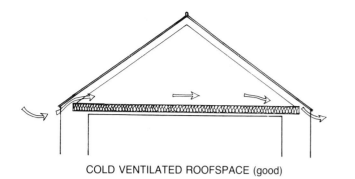

COLD VENTILATED ROOFSPACE (good)

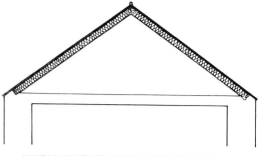

WARM UNVENTILATED ROOFSPACE (bad)

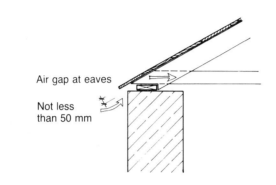

Air gap at eaves

Not less than 50 mm

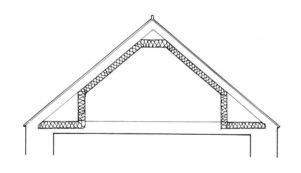

INSULATED CONVERTED ROOFSPACE

MANSARD ROOF – INSULATION BETWEEN RAFTERS

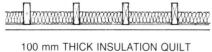

100 mm THICK INSULATION QUILT

QUILT AND POLYTHENE LAID OVER JOISTS (bad)

FOIL-BACKED QUILT (good)

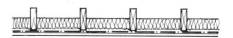

RENEWED CEILING – VAPOUR BARRIER BELOW JOISTS (good)

172

reduce the amount of moisture vapour passing through it to the roof space, although it will not provide a continuous vapour barrier. Polythene, laid tight over and around the joists, is not recommended because it will only exacerbate condensation around the timbers, which, as they cannot dry out, will decay. If a plaster ceiling is to be taken down and renewed for any reason, the opportunity should be taken to insert a more adequate vapour barrier of lapped and taped polythene sheets.

2 *Insulating at roof level:* Insulation at roof level requires limited and controlled ventilation, and is used where attics have been converted to habitable accommodation or below flat roofs, however the need for a vapour barrier on the warm side of the insulation in this case is even more critical.

Some proprietary types of rigid polystyrene insulation form vapour barriers in themselves, but it may be difficult to insert the boards into restricted spaces without opening up the roof structure. It is also difficult to provide adequate ventilation through the roof structure of an existing building. Flat roofs are particularly fraught with problems and even new leadwork can fail remarkably quickly if its soffit is subject to chronic condensation.

Windows

Ill-fitting windows are often responsible for excessive heat loss and draughts. However, the cost of suitable replacement double glazing, compatible with the building type and with neighbouring buildings, is rarely justified by the amount of energy or fuel saved. In any case, double glazing is not acceptable for the sash windows of buildings in the New Town. Most windows in the New Town are fitted with shutters, and these, together with heavy curtains, will reduce heat losses at night.

Windows account for about 15 per cent of the heat loss from typical buildings in the New Town. Despite this, Building Control are required to demand that provision is made for additional permanent fixed ventilation where new gas-fired equipment or open fires are to be installed. BRE Digest 306: *Domestic Draught Proofing: ventilation considerations*, Table 1, provides useful advice on how this might be achieved in buildings with facades which must not be altered visually. Draught sealing and regular maintenance will reduce heat loss through sash windows (see also BRE Digest 319: *Domestic Draught Proofing: materials costs and benefit*). For further advice, see SASH WINDOWS.

ACOUSTIC INSULATION

Walls

It is not normally necessary to increase the acoustic insulation of walls. Even internal timber stud partitions usually have sufficiently thick lath and plaster on both sides to be effective sound barriers.

Floors

The traditional ash deafening, up to 100 mm thick, gives good acoustic insulation, sometimes supplemented by the separation of floor structure from the ceiling suspended below. Additional insulation is seldom required, but good thick carpets help to deaden the sound. Excessive vibration can be due to structural problems and the floor may need to be strengthened or repaired. When floors have to take exceptional live loads, as might happen in studios for dancing or judo, a floating floor may have to be added.

Windows

Additional glazing is sometimes recommended to reduce the noise of traffic on busy main streets, but neither double glazing nor secondary glazing (double windows) are really compatible with the New Town. Shutters and curtains provide thermal and acoustic insulation; traditional shutters can give a sound reduction of 35 decibels which is equal to that obtained with most double window systems.

If additional glazing is unavoidable, secondary glazing should be considered rather than double glazing. Secondary glazing consists of adding a completely new window, located on the window cill, with a space of about 100 mm between new inner and old outer windows which reduces sound transmission and, to some extent, heat loss. Vertical sliding sashes are available with sufficiently slender profiles to be fairly unobtrusive, although they have the disadvantage of reflecting the front astragals and this is noticeable from the outside on sunny days. Secondary glazing renders the shutters useless, and access to cords, weights and pulleys of the outer window will be severely impeded. Fixing of the additional window or its subsequent removal can cause substantial damage to the adjacent shutters and panelling. Unless the inner window provides a complete seal, condensation on the inner face of the outer window may occur. If the outer face of the frame of the new inner window is painted black it will hardly be seen from the street. The Planning Department must be consulted before the installation of additional glazing because Listed Building Consent may be required.

PASSIVE SOLAR GAIN

Sunlight in a room can be a delight to the occupants but can also cause irreparable damage to furnishings and decoration. Light can weaken and fade materials such as paper, parchment, wood, leather, natural fibre textiles, glues, gums, resins and even synthetic dyes and plastics. The administrators of museums and historic mansions now take precautions against these harmful effects. Ultraviolet (UV) radiation which comes from the invisible part of the spectrum does the most damage. There are a number of special blinds on the market which are designed to filter out the most harmful UV rays without reducing the general light level within a room.

Blinds give privacy and protect furnishings from daylight and direct sunlight. New Town windows were fitted in Victorian times with internal roller blinds. The blinds were covered with Holland, a type of linen which is still available, and they were hung on 'Scotch' mounts, no longer available, but worth retaining if they are still in the building. New blinds and mounts of good quality should be chosen carefully and fitted by a professional. External roller awnings were also provided on the south-facing windows of principal rooms in Victorian times and these should be retained and restored.

Where the contents of a room are especially sensitive to UV light, it is possible get a specialist to coat the window with a UV-absorbent varnish or to cover it with UV-absorbent adhesive acetate film. A variety of qualities are available, in slightly differing colours, but neither the varnish nor the film need be obtrusive if chosen and installed with care. The deterioration of UV filters should be checked twice a year using a UV monitor. More information is available from the ENTCC.

The positive aspects of solar energy should not be ignored. Solar gains can be extremely useful in offsetting heating requirements particularly in large heavyweight buildings such as those in the New Town. The sun warms up the fabric during the day and this heat is later released during the evening, although adequate care must be taken to ensure that valuable furnishings are not affected by direct sunlight.

INTERNAL WALLS

INTRODUCTION

Internal walls may carry the superimposed loads of upper floors and walls, or they may be non-loadbearing. Structural movement, alterations and new uses may result in the application of unintended loads on internal walls. In old buildings, it should not be assumed that any partitions are non-loadbearing; floor and wall construction, loading patterns and beam sizes should be investigated by a survey.

CONSTRUCTION

Internal walls of constant thickness, other than those which contain flues or support common stairs, may also carry floor joists. Partitions which change their positions on plan at successive levels are less likely to be loadbearing, but it must never be assumed that they have no structural function. The thickness of internal walls varies from 150 mm to 300 mm; chimney breasts are often 450 to 700 mm thick; walls to common stairs, may be 500 mm thick; and party walls have been found as thick as 1500 mm, where two independent gables have been built back to back.

Stone

Party walls, chimney breasts and walls to stairs were usually built of stone over their full height. They were generally constructed of random rubble sandstone and plastered on the hard, although examples of lath and plaster have been found on internal walls. Other internal walls were sometimes built of stone up to ground or first floor level, above which they were generally built of brick. Internal walls give lateral support to external walls and vertical support to upper floors.

Brick

In the New Town, many internal walls are of brickwork, either a half brick or a whole brick in thickness. Many of these brick walls are built on top of 300 mm thick rubble walls which rise through the basement (and occasionally ground floors), but bricks were sometimes used to build these thicker and lower parts of internal walls (e.g. in Heriot Row). Timber studs were sometimes incorporated into slender brick walls as reinforcement against buckling. Although handmade bricks were produced in Scotland at William Adam's Kirkcaldy Brick and Tile Works from 1714 (if not earlier), large quantities of pressed bricks were produced at Portobello and were used for internal walls in the New Town. Handmade bricks are rectangular, relatively durable and difficult to cut; the length varies from 225 mm to 300 mm and the width from 100 mm to 140 mm. Variations in brick sizes were evened out by thick lime mortar beds, and the irregular vertical face of this brickwork was straightened by heavy coats of plaster on the hard, 35 to 40 mm on each side. Plaster exceeding 50 mm in thickness may indicate an attempt to cover up buckling during construction – a likely hazard for thin walls built with slow-setting lime mortar.

An opening in a brick partition was not always spanned by a timber lintel, the wall sometimes being carried on the door frame or on a soldier course of bricks forming a flat arch.

Timber

Stud partitions are common in the New Town, particularly on upper floors. It should never be assumed that a stud partition is non-loadbearing, and no alterations should be carried out without a full investigation. The size of the timber studs (usually 100 mm × 50 mm at around 400-450 mm centres) varies with storey height; Great King Street, for example, has 150 mm × 50 mm studs, and examples of 75 mm x 50 mm studs extending 3600 mm vertically have also been found. Stud partitions filled with brick on edge (bricknoggings) to provide stiffening are quite common. Where partitions run through two or more storeys, 200 mm × 75 mm or 200 mm × 100 mm full-height diagonal braces were occasionally used. Fully trussed timber partitions are also found, the whole wall acting as a trussed beam, for instance supporting the floor and ceiling over a drawing room.

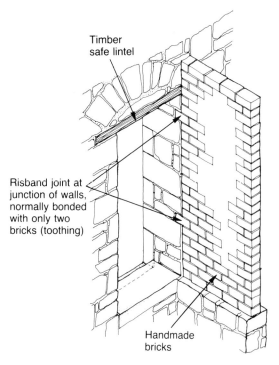

Timber
safe lintel

Risband joint at
junction of walls,
normally bonded
with only two
bricks (toothing)

Handmade
bricks

JUNCTION BETWEEN INTERNAL BRICK WALL
AND EXTERNAL STONE WALL

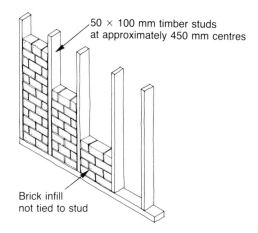

50 × 100 mm timber studs
at approximately 450 mm centres

Brick infill
not tied to stud

BRICK INFILL STUD PARTITION

COMMON DEFECTS

Internal walls will crack, bulge or bow if they are overloaded, not tied to the structure, or if there is differential settlement or other external cause of distress (see STRUCTURAL DEFECTS).

- *Loading:* In the New Town many upper floor flats had coal stores, kitchens and entrance halls paved with heavy stone flags which exert a massive load on supporting brick walls or, occasionally, stud partitions.
- *Differential settlement:* Partitions may carry loads not originally intended but now transferred by differential settlement from other parts of the building.
- *Lack of bond:* Risband (straight-butted) joints between internal and external walls are common in the New Town. The difficulty of creating an effective tie was increased by the practice of building the external walls first and the internal brick walls later. The bond provided by two or three brick tusks per storey height can be broken if the bedding mortar has weakened or if the walls have moved apart, and because of high ceilings and large rooms the internal walls are relatively slender so that the number of brick tusks may be inadequate for the strain.
- *Timber:* Rot or beetle infestation will weaken a timber member and, although a timber frame can accommodate a certain amount of movement without cracking, the connections may be inadequate for the extra strain (see TIMBER DECAY).
- *Alterations:* Instability or a change in the loading patterns on whole buildings can be caused by: the removal of the partitions which carry load or act as a structural tie; the replacement of walls with beams, because of the subsequent concentrations of load at the supports; or the removal of a structural member of a roof truss. It is essential to establish the type of wall construction before making alterations because the result of confusing a simple non-loadbearing stud wall with a loadbearing trussed partition can be disastrous. For example, cutting through a trussed partition to make a door can seriously affect the structural integrity of the partition.

  Lime plaster helps to consolidate old brick walls and its removal can occasionally destabilize an otherwise sound construction. Furthermore, the action of hacking away can cause the bond between the brickwork and its lime mortar to be loosened and this may cause eventual collapse.
- *Services:* Excavation, drilling and checks cut to accommodate pipes and conduits can weaken walls.

176

GENERAL REPAIRS

All significant repairs and alterations should be carried out under the guidance of a structural engineer and alterations will require a building warrant and perhaps Listed Building Consent.

Some cracks in internal walls will, in fact, have relieved the stresses which caused them, but their effect on the performance of the building should be determined with the assistance of a structural engineer. As a general rule, surgery should be kept to a minimum. The degree of lean or bulge, the displacement of materials, the position of the cracks and fissures, the state of the bonding, the condition and the strength of the mortar and the function of the wall must all be considered. If the movement has stopped, the wall should preferably be repaired rather than rebuilt.

It is often difficult to bond repair work to the existing building fabric, but walls can be joined and stiffened by plates, dowels and anchor ties concealed behind the plaster.

Movement resulting from failure of foundations may require extensive rebuilding of walls and underpinning (see FOUNDATIONS).

Timber partitions can sometimes be eased back to their original position using props and blocks but most structural movement is irreversible.

If a brick wall has to be rebuilt the original bricks should be reused after cleaning off all the old lime mortar. The new mortar should be as weak as possible, and certainly no stronger than the bricks. A 1:2:9 cement/lime/sand mortar will be suitable for most purposes and will allow the wall to respond to inevitable slight structural movements without cracking.

FLOORS

INTRODUCTION

There are two basic forms of flooring used in the New Town. The basement or lowest floor is usually of sandstone slabs (from Hailes or Ravelston) bedded directly onto the earth over levelling sand. No damp-proof membranes were used; regular airing and use of fires kept the rooms dry. In town houses at basement level stone slabs were used in kitchens, cellars and passageways while in rooms intended for servant accommodation there were timber floors. The joists spanned across sleeper walls covered with a wallplate. Again no damp-proof membranes or courses were used; only occasionally the space below the floor was ventilated.

Above basement level all floors are suspended using both joists and beams so that large room areas could be achieved. Usually the floors were boarded with Scots pine (redwood) but stone flags were also common in kitchens and hallways.

TIMBER FLOOR CONSTRUCTION

Most main beams run parallel to the front elevation, with the joists spanning from the front to the back walls; sometimes joists are supported by intermediate beams or internal walls, which reduce the effective span to as little as 2 m. In some cases, joists are carried on secondary beams spanning from front to back.

Boarding is usually of Scots pine, tongued and grooved and secret-nailed. Scots pine is also used for structural timbers. The slow rate of growth, high density and lower moisture content of old timber will withstand higher stresses than modern timber and sound old timber should be reused wherever possible.

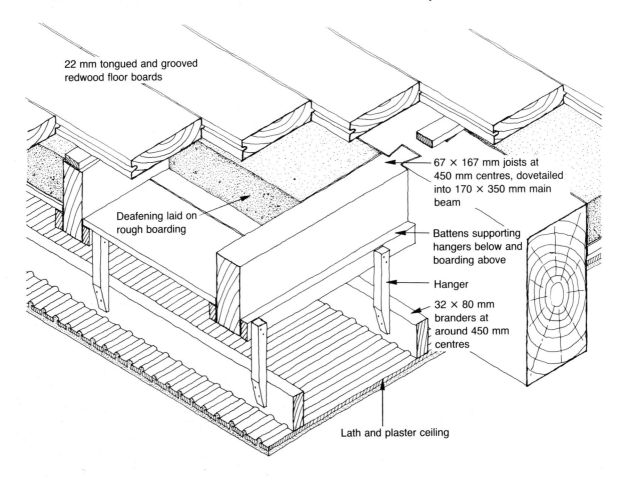

22 mm tongued and grooved redwood floor boards

Deafening laid on rough boarding

67 × 167 mm joists at 450 mm centres, dovetailed into 170 × 350 mm main beam

Battens supporting hangers below and boarding above

Hanger

32 × 80 mm branders at around 450 mm centres

Lath and plaster ceiling

TYPICAL TIMBER FLOOR CONSTRUCTION

179

Photos: Joe Rock

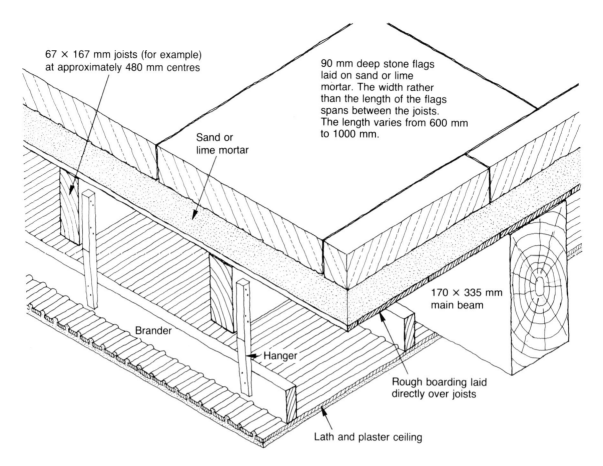

67 × 167 mm joists (for example) at approximately 480 mm centres

90 mm deep stone flags laid on sand or lime mortar. The width rather than the length of the flags spans between the joists. The length varies from 600 mm to 1000 mm.

Sand or lime mortar

170 × 335 mm main beam

Brander

Hanger

Rough boarding laid directly over joists

Lath and plaster ceiling

TYPICAL SUSPENDED STONE FLOOR CONSTRUCTION

MAINTENANCE

The only maintenance required for timber flooring is occasional washing, the water being used very sparingly and immediately wiped dry. Traditionally, floor boards were dry-scrubbed, using white sand; sometimes Fuller's earth or a mixture of herbs was preferred.

Repeated washing of stone flags with open joints can lead to the decay of the timber joists and deafening boards beneath. A small amount of clean, warm water, without soap or detergent and especially not bleach, is recommended for stone floors, although sour milk was considered to be a suitable cleaning agent.

INSPECTION

The incorporation of the deafening in suspended floors makes it difficult to inspect joists and end bearings. Usually inspection from above must suffice, taking into account the condition of the boards or flags and noting any undue deflection or vibration. During inspection, deafening should be disturbed only if the floor is suspected of being unsound. If hidden decay is suspected, it is possible to analyse the density throughout a piece of timber using a special wood decay drill, allowing the condition of joist ends to be checked without disturbing the deafening (the ENTCC can supply further information).

If it is necessary to open up the structure, this is best done at the edge of the floor to check the size and bearing of the joists. The span is not always from wall to wall and a beam will often be found at mid-span, although its exact position will only be revealed by lifting the floorboards. Secretly-nailed boards should be lifted for inspection by carefully running a feather splitter along each side of the board. When replacing the boards, there is no alternative to face-nailing and punching. It is wise to insert traps in the floor, in the form of short boards which can be unscrewed for subsequent inspection. Traps are also useful above electrical fittings and junction boxes. When inspecting or repairing a floor, deafening must be carefully lifted, put into sacks and laid aside for reuse; it can be extremely densely packed and a pick axe may be required to lift it. Due to its weight, deafening should be gently laid back onto the deafening boards and never directly on to the lath and plaster. If there is any doubt as to the safety of a floor, a structural engineer should be consulted.

DEFECTS

A floor may deflect, vibrate or collapse for the following reasons:

1 *The floor is unable to carry the load due to:*
   - An inherent defect in the supporting structure; many loadbearing partitions were built in a way which would not be considered good practice today;
   - An abnormal load; change from residential to office use can involve considerable extra weight;
   - The structure being weakened by excessive notching of joists to take pipes and conduits;
   - A reduction of the effective section of joists because of timber decay or insect attack.
2 *The bearing conditions are inadequate due to*:
   - Outward movement of walls;
   - Decay in the ends of the joists, particularly those on external walls or under defective centre valley gutters.

A wide gap between flooring and skirting is always suspect; it can indicate timber decay and the shrinkage or collapse of structural members.

It is not uncommon for a beam to run beneath the hearth, where it may be damaged by heat from the fire or from a rising flue. Such beams are particularly vulnerable if new fireplaces or boilers have been carelessly installed, or during a chimney fire. The use of modern high-rated fuels can cause excessive heat and increase the risk of flue fires.

Even if a floor is safe, the sensation of deflection can be alarming. If it is obviously at risk, it is advisable to consult an engineer, who will assess the need for remedial work.

As long as the structure is basically sound, a sloping or sagging floor is not in itself a defect and it should be left undisturbed. It is not practical to re-level a floor unless the building is being totally refurbished; of course, a new floor can be superimposed on it, but this can cause difficulties at thresholds and skirtings. Access to services must be not be forgotten.

REMEDIES

A floor may be strengthened either by inserting additional beams or joists or by adding material (commonly steel) to the existing beams and joists. Inserting new beams and strengthening existing ones are jobs more easily tackled from below. Valuable plasterwork will preclude this, and most repairs to the floors of tenements will have to be executed from above, unless the tenement is being totally refurbished.

The following examples, indicating what can be done to strengthen a floor, should not be regarded as standard methods. Any substantial strengthening will require a building warrant (see BUILDING REGULATIONS AND STANDARDS), and should be designed for individual cases under the guidance of a structural engineer.

Inserting new beams

Steel beams can be used to give extra support to joists, either at mid-span or at one-third and two-thirds of the span to avoid a flue. Before inserting a new beam, it is essential to check whether the supporting wall is capable of supporting the point load. There is usually sufficient depth to incorporate a new beam within the floor construction, but its insertion may be tricky if the ceiling beneath is to remain undisturbed. Inserting the beam in two or even three sections, spliced together *in situ*, may be easier than trying to manoeuvre a single beam into position.

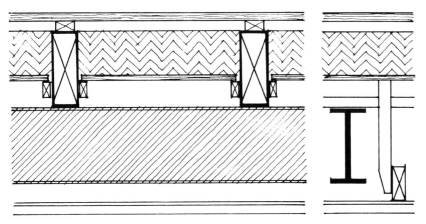

THE USE OF ADDITIONAL STEELWORK TO STRENGTHEN JOISTS

Strengthening a main beam

Where the effective strength of an existing beam has been considerably reduced by rot, notches or excessive deflection, a consultant engineer should advise on the best method of replacement or strengthening because there is no single appropriate solution to this problem.

Repair to decayed joist ends

If the decay does not extend much beyond the wall face and the work can be tackled from below, it may be possible to bolt a large mild steel angle to the wall and wedge the joists off a timber plate bolted to the table of the angle. The precise arrangement depends on the floor construction, the straightness of the rubble wall and its suitability for receiving fixing bolts. If the decay extends a short distance from the wall face, and it is preferable to cut all the joists back from the wall by 60-85 mm, bolt a longitudinal timber 50 mm or 75 mm wide to the wall and reinstate the joist bearings by means of joist hangers. This work can be carried out from above. With both these methods, it is important to maintain positive ties between the wall and the floor.

More extensive repairs may be required if the timber decay is due to extensive dry rot. If the joists need to be cut back 300 mm or more beyond the extent of the decay, new lengths can be bolted on, using toothed timber connectors (see ROOF STRUCTURE). The lap should be at least six times the depth of the joist and the bolts staggered. The decayed joist may also be reinforced with mild steel angles or channels bolted through sound timber. New timbers should be pressure-impregnated with a suitable preservative, and built into the wall with provision for circulation of air all around (for instance, by building up with stones or bricks which are in contact with the wood but keeping the mortar back 65 mm from the surface of the wood). The size and design of bolted connections should be calculated by a structural engineer.

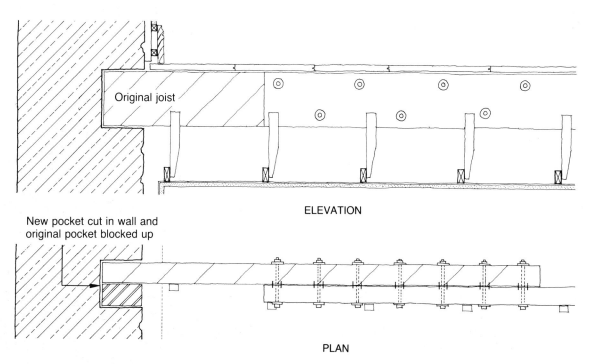

ELEVATION

New pocket cut in wall and
original pocket blocked up

PLAN

TYPICAL REPAIR TO JOISTS FOLLOWING AN ATTACK OF DRY ROT

Where a floor has been subjected to flood damage or extensive damp, it is important to lift some of the boarding to allow the deafening to dry out. If need be, the deafening can be lifted, spread out to dry on polythene sheeting and replaced. Traditional lime and ash deafening is undoubtedly still the best for sound insulation although there are various proprietary alternatives.

When repairing a timber floor, each board should be cut away and new timber, preferably of the same species, grade and size, pieced in. Even if two adjoining boards are pieced, the new lengths should still conform to the original pattern of staggered butt joints. Salvaged boarding to match the existing floor may be available from demolition or architectural salvage firms.

Private house

Common stair

Private house

Common stair

Photos: Joe Rock

STAIRS

INTRODUCTION

This chapter deals with the main stairs in New Town buildings, either common or private stairs, which are invariably made of stone. Secondary service stairs are very rare in the New Town, even in the largest houses.

Most private houses in the New Town have central stairs naturally lit from above by a cupola or lantern. Common tenement stairs are lit either from above or, if they are at the front of the building, by normal sash and case windows. In the Scottish and French tradition, tenement stairs are plain and simple in design to the point of austerity, in contrast to the elaborate decoration of the flats which they serve. This austerity is a feature of Scottish tenements which should not be modified by using modern floor finishes, some of which are extremely difficult to remove. The common stair in a tenement is considered to be a continuation of the public footpath, and so the electric stair lighting is usually maintained by the local authority. The upkeep and cleaning of the stair, however, is the responsibility of the owners and tenants of the flats.

The steps, usually of Hailes stone laid on their natural bed, are interlocking. They were built at the same time as the supporting walls, so that part of the load is carried by the side wall and part is transferred through the treads to the landing below. This is known as a pen-checked stair. Landings are often made of solid stone slabs usually supported on timber joists and trimmers but sometimes just built into the walls. The ground floor is commonly of stone flags. Stair soffits are generally left as plain stone and landing ceilings are plastered.

House stairs are usually similar to common stairs but above ground floor level have nosings to the treads, with more ornate cast iron balusters and moulded timber handrails. Tread and railing details usually become simpler on flights down to the basements and back greens or gardens. The majority of handrails in the New Town are made of a softwood core, veneered on the sides with mahogany or other hardwoods with a solid matching crossbanded top and beads, and finished with beeswax or French polish; a few are solid mahogany. They are often intricately shaped to match the bend of the stair and have decorative palm ends. Simpler pole handrails of solid pine are found in some tenements.

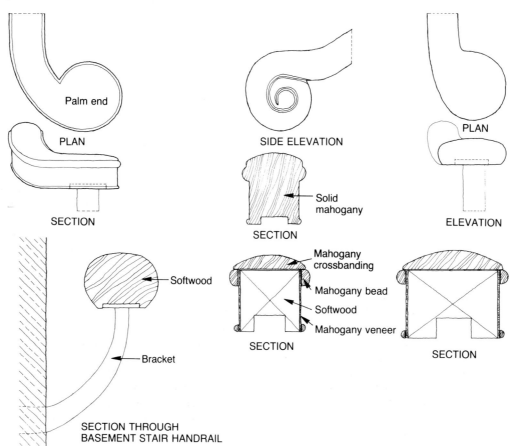

184

INSPECTION AND MAINTENANCE

The major areas for inspection are balusters, handrails, step edges, nosings and supporting walls. The balusters should be checked to see whether they are loose; the nosings should be checked for excessive wear which could be dangerous.

When washing stone, a little soft soap may be added to the water to remove grease and to disinfect but detergents and bleaches should be avoided. It is better to wipe the surface with a damp cloth or mop than to use a lot of water which may penetrate the joints; regular sweeping is best. Handrails should be polished and balusters painted periodically.

DEFECTS

The great strength of pen-check stairs is evident from the rarity of fractured treads; once built they are structurally very sound. However, stairs may be damaged by:

1 *Structural problems* (rare): the usual cause is differential movement of supporting walls. Analysis will require the services of a structural engineer who may recommend stabilization measures or the reinforcement of individual steps.
2 *Decay:* delamination is rarely found on internal stairs, although it is common on external steps. Decomposition due to the use of chemical cleaners or detergents is also rare. Continual stairwashing can cause rotting of landing support timbers.
3 *Wear:* this is the most common defect, especially on the lower and the most-used flights of common stairs, which can become a safety hazard. Handrails may also become worn or damaged.
4 *Mechanical damage:* the nosings and edges of steps are vulnerable and should be protected from accidental damage if, for example, heavy furniture is being moved. Balusters may become loose and excessive movement can cause the stone around the seating of the baluster to chip or the baluster to snap.

REPAIRS

Because of the interdependent construction of the pen-checked stair, it is almost impossible to replace an individual step, but it is possible to rebuild a whole stair; this has been done at No. 8 Queen Street and at No. 23 Great King Street. Strict geometrical principles must be adhered to, particularly on winding stairs, in order to maintain a consistent ramp and twist soffit.

If the end of the step is damaged around the baluster, it can be cut back and a new section of stone, matching the existing profile, can be indented and fixed with non-ferrous dowels and epoxy resin. Cement mortar is not suitable for repairs because it is likely to craze and spall. The indented repair involves cutting the balusters just beneath the coping, fixing the lower ends into the newly repaired stone and rewelding on completion. If the damage to the stone is restricted to a small area, a mortar repair may be more satisfactory and is relatively inexpensive. It is unwise to cut into the rear half of the tread, which is the thinnest section.

Some wear is to be expected on treads which are more than a century and a half old. Wear is part of the character of an old building and levelling should be avoided unless the stairs are hazardous. In grant-aided repair schemes the Housing Department requires that the treads should be levelled if the wear exceeds 25 mm. Treads can be levelled and nosings replaced with stone indents or epoxy screeds; both methods require that the original surface should be cut back to receive the new material and neither is entirely satisfactory. In any repair, feather edges are best avoided by cutting out a rectangular pocket to receive the screed (min 10 mm thick) or stone indent (min 40 mm thick).

Natural stone indents should be selected from a quarry to match the characteristics of the original steps, fixed with 6 mm non-ferrous dowels and bedded in lime mortar. The triangular cross-section of a pen-checked stair must not be weakened by cutting into the rear of the tread; only the worn area beside the nosing should be cut out to receive new rectangular stone indents. Stone indents should be confined to the centre of the tread, stopping at least 100 mm short of the balusters or wall. They may have to be buffed to sweeten in with the adjacent worn edges. The front arris of all treads should be slightly rounded by rubbing with another stone.

Roughening the surface to provide a key for a proprietary epoxy screed is less likely to weaken the stone. However the use of such a screed will introduce an alien material of unknown long-term compatibility. Screeds are also more liable to stain. If the adjoining original stone is very worn it may be necessary to extend the screed over the whole surface of the tread; care should be taken to reproduce the exact moulding profile of the original step.

If paving stones on landings and entrance halls are excessively worn it may be necessary to replace all of them with new sandstone or honed Caithness flags, because new indents seldom match the adjoining worn surfaces.

Balusters which are loose where they fit into the stone tread can be secured by staving the lead in the socket. This is a skilled job, particularly where the socket for the fixing is on the side of the stair, and should only be carried

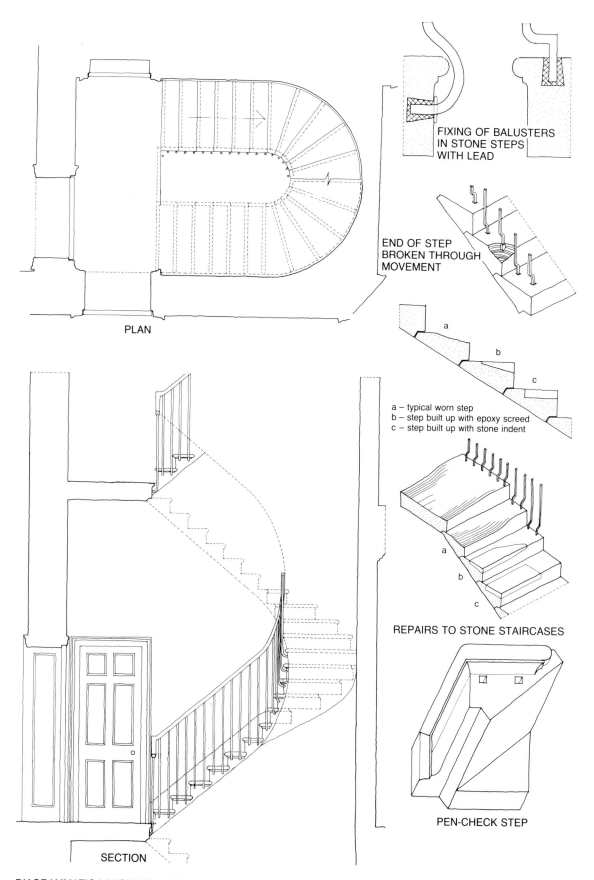

PLAN

FIXING OF BALUSTERS
IN STONE STEPS
WITH LEAD

END OF STEP
BROKEN THROUGH
MOVEMENT

a
b
c

a – typical worn step
b – step built up with epoxy screed
c – step built up with stone indent

a
b
c

REPAIRS TO STONE STAIRCASES

PEN-CHECK STEP

SECTION

DIAGRAMMATIC LAYOUT OF PEN-CHECK STAIR IN PRIVATE HOUSE

186

out by a blacksmith. When replacing a baluster it is important that it matches the existing ones and it may be necessary for special castings to be made. (For further guidance on general repairs to cast iron balusters, see RAILINGS.)

The repair and replacement of handrails is also a skilled operation, especially if the handrail ramps and twists. In such a case, much of the work may have to be carried out *in situ* by specialist joinery firms. Damaged straight sections of handrail can usually be unscrewed from the cope of the iron balustrade and repaired in individual lengths. Where the veneer or crossbanding has come off a handrail it is usually possible to repair it without having to replace the whole section.

SERVICES

Most common stairwells contain service pipes and conduits for gas, electricity and water supplies, as well as telephone and sundry cabling. These should be accessible at all times yet neat and unobtrusive. This can best be achieved by placing them in carefully designed and detailed ducts or, if exposed, by grouping them together. Ducts containing gas pipes must have at least 3000 mm^2 free area of ventilation at high level.

An interesting feature of Edinburgh tenements is the mechanical door-opening gear which allows the street door to be opened from each landing; a useful security check well worth preserving, although now being usurped by electric entryphones (see DOOR FURNITURE).

New stair lighting in common stairs must be installed, at the owners' expense, to the design and specification of the Lighting Department of the City of Edinburgh District Council, who will usually take on responsibility for the provision and maintenance thereafter. Only standard fittings are allowed, wall-mounted 2100 mm above floor level; all wiring is concealed within a protected raggle cut into the plasterwork.

INTERNAL JOINERY

INTRODUCTION

Joinery is the finishing trade which mainly involves the making and fixing of skirtings, architraves, dados, dado rails, panelling, picture rails, doors, windows, window linings and shutters, handrails and special elements such as built-in columns, bookcases and screens.

SKIRTINGS

Timber skirtings protect the base of the wall plaster and cover the junction between plaster and floor boards. As is commonly found with many joinery elements, the size and complexity of skirtings relates to the social function of the room. The simplest skirting, found in minor rooms (such as kitchens and maids' rooms), was a plain unmoulded board. Larger skirtings, found in the more important apartments, were generally made in two pieces from a base board blocked out by, and fixed to, rough grounds with an ornate planted moulding on top.

Skirtings are nailed to timber grounds fixed to bilgates (built-in battens or timber bricks) or dooks (timber plugs driven into the joints) in the wall. The top ground often also forms a stop for the plaster. These timber fixings are susceptible to rot if the wall becomes damp. (For further information see TIMBER DECAY.) Where there is dado panelling, the skirting is fixed to the panelling stiles which extend down to floor level, and sometimes there is an additional ground running between the rails at floor level. It is normal practice to scribe or fit the lower edge of the skirting to the floor. At the junction between the skirting and architrave (or architrave block) the skirting should be let into the architrave to avoid a crack appearing if shrinkage occurs.

ARCHITRAVES

Architraves are used around door and window openings, to cover the junction between plaster and door lining or window ingo panelling, shutters and their timber fixings. Normally the door and window architraves in a room follow the same pattern, but window architraves are often slightly larger.

Simple architraves in minor rooms were made from one piece of timber. Large, ornate architraves, for instance in dining rooms, were generally made in two or more pieces, either built up, one piece planted on top of another, or in two halves tongued and grooved together. The feet of architraves were either continued down to the floor or, in the case of grander rooms, finished at a plinth or block, which was slightly wider than the architrave. The architrave would be tenoned and fitted into a mortice in the block. In some drawing and dining rooms, doors were provided with an elaborate door-piece with pilasters, frieze and cornice.

DADOS AND DADO RAILS

By the end of the eighteenth century the cornices and upper parts of internal walls in important rooms were often of plaster, while a reduced form of wainscotting known as a dado covered the lower parts of the wall to a height of about a metre. In classical terms, this resembles a continuous pedestal with a skirting (base moulding), a dado rail or chair rail (cornice), with panels between. Later, the joinery was reduced to skirting and dado rail and then to a skirting alone.

The top member of the dado had a projecting moulded rail to protect the plaster from being damaged by chair backs. Sometimes, a continuous strip of timber, called a mouse moulding, was fixed to the floor at the base of the skirting to keep chair legs well away from the wall.

Full wainscotting or dado panelling is made up of a framework of stiles and rails jointed into each other and infilled with thin timber panels housed in grooves in the framing members. This construction enabled the component parts to shrink and expand independently of each other and so avoid the cracking which would occur with a rigid framework. The framework was fixed to bilgates or dooks in the wall.

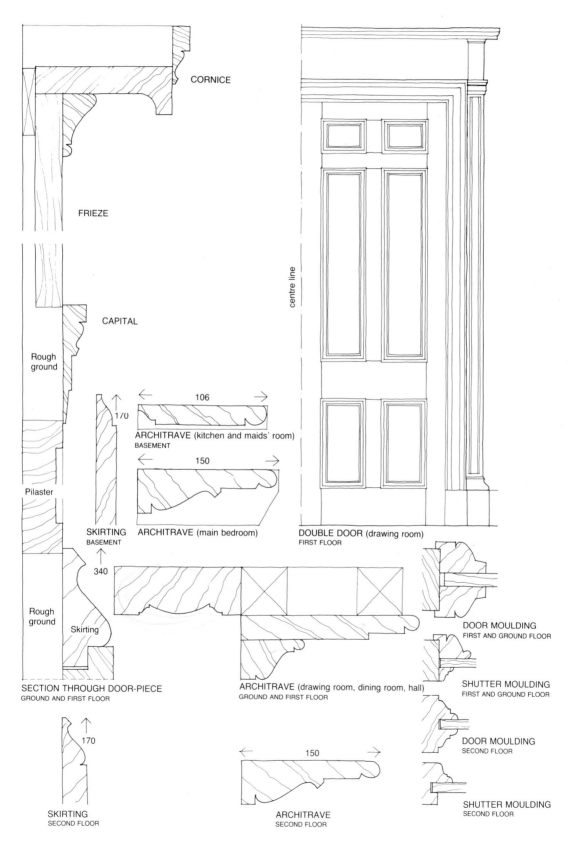

CORNICE

FRIEZE

CAPITAL

Rough ground

Pilaster

Rough ground

Skirting

SECTION THROUGH DOOR-PIECE
GROUND AND FIRST FLOOR

170

SKIRTING
BASEMENT

340

106

ARCHITRAVE (kitchen and maids' room)
BASEMENT

150

ARCHITRAVE (main bedroom)

DOUBLE DOOR (drawing room)
FIRST FLOOR

centre line

ARCHITRAVE (drawing room, dining room, hall)
GROUND AND FIRST FLOOR

DOOR MOULDING
FIRST AND GROUND FLOOR

SHUTTER MOULDING
FIRST AND GROUND FLOOR

DOOR MOULDING
SECOND FLOOR

SHUTTER MOULDING
SECOND FLOOR

170

SKIRTING
SECOND FLOOR

150

ARCHITRAVE
SECOND FLOOR

8 REGENT TERRACE

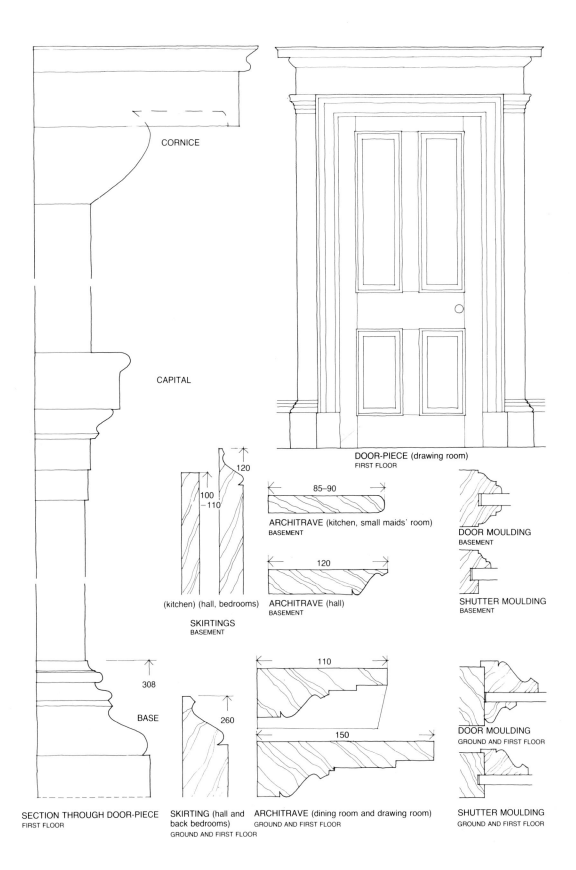

CORNICE

CAPITAL

DOOR-PIECE (drawing room)
FIRST FLOOR

120

100
– 110

(kitchen) (hall, bedrooms)

SKIRTINGS
BASEMENT

85–90

ARCHITRAVE (kitchen, small maids' room)
BASEMENT

120

ARCHITRAVE (hall)
BASEMENT

DOOR MOULDING
BASEMENT

SHUTTER MOULDING
BASEMENT

308

BASE

260

110

150

DOOR MOULDING
GROUND AND FIRST FLOOR

SHUTTER MOULDING
GROUND AND FIRST FLOOR

SECTION THROUGH DOOR-PIECE
FIRST FLOOR

SKIRTING (hall and
back bedrooms)
GROUND AND FIRST FLOOR

ARCHITRAVE (dining room and drawing room)
GROUND AND FIRST FLOOR

13 CARLTON STREET

SHUTTERS

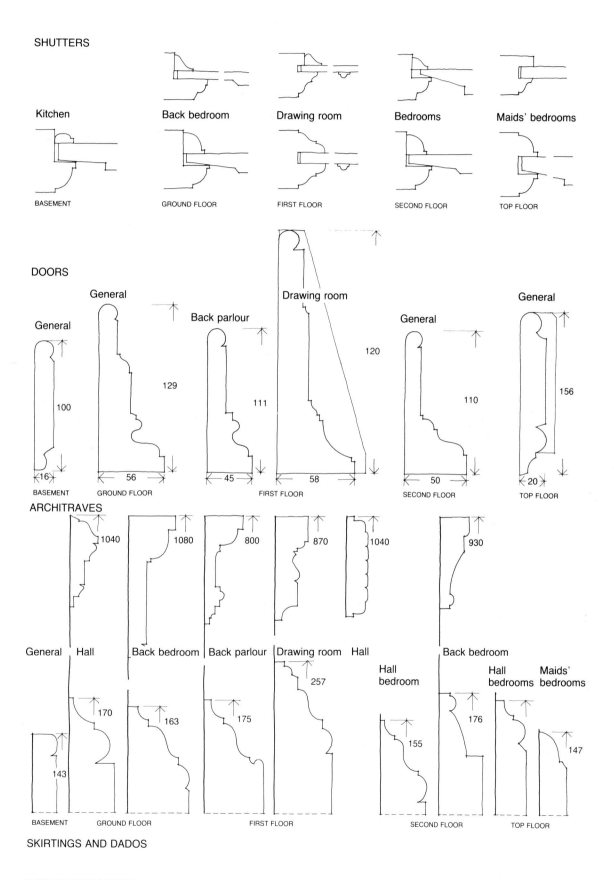

Kitchen   Back bedroom   Drawing room   Bedrooms   Maids' bedrooms

BASEMENT   GROUND FLOOR   FIRST FLOOR   SECOND FLOOR   TOP FLOOR

DOORS

General   General   Back parlour   Drawing room   General   General

100   129   111   120   110   156

16   56   45   58   50   20

BASEMENT   GROUND FLOOR   FIRST FLOOR   SECOND FLOOR   TOP FLOOR

ARCHITRAVES

1040   1080   800   870   1040   930

General | Hall   Back bedroom | Back parlour | Drawing room Hall   Back bedroom

Hall bedroom   Hall bedrooms   Maids' bedrooms

170   163   175   257   155   176   147

143

BASEMENT   GROUND FLOOR   FIRST FLOOR   SECOND FLOOR   TOP FLOOR

SKIRTINGS AND DADOS

7 CHARLOTTE SQUARE

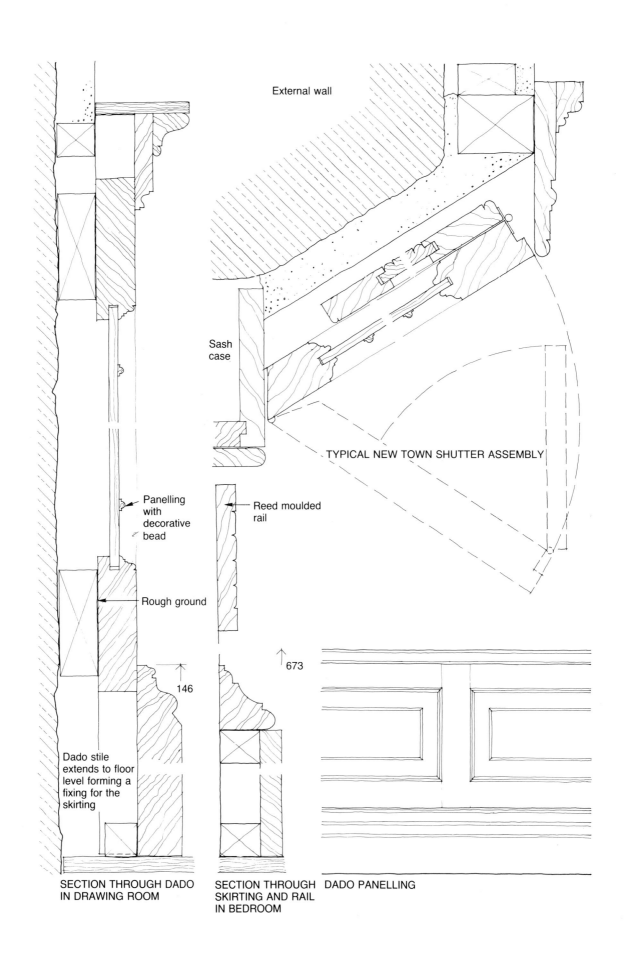

External wall

Sash case

TYPICAL NEW TOWN SHUTTER ASSEMBLY

Panelling with decorative bead

Reed moulded rail

Rough ground

146

↑ 673

Dado stile extends to floor level forming a fixing for the skirting

SECTION THROUGH DADO IN DRAWING ROOM

SECTION THROUGH SKIRTING AND RAIL IN BEDROOM

DADO PANELLING

TOP FLOOR, 72 NORTHUMBERLAND STREET

PICTURE RAILS

These were a Victorian innovation, frequently introduced into existing rooms to suit the fashion of the day. Various mouldings were used fixed to small timber plugs; metal rails with a slight inwards dip, on projecting brackets, were also used.

DOORS

The size of the door, number of panels and moulding detail depended on the importance and proportion of the room. In a room, door and shutter mouldings are generally of a matching pattern, the shutter mouldings being at a smaller scale than those of the door.

Doors were often positioned to achieve symmetry in principal rooms and frequently dummy doors were employed, to balance a door to a cupboard, or one giving access to an adjacent room. Double doors were also used for grand effect although sometimes only one leaf opened, the fixed leaf concealing structure or an element such as a staircase behind. For information on construction and repair see DOORS.

WINDOWS

The joinery of traditional sash and case windows, common throughout the New Town, is discussed in SASH WINDOWS. The internal ingoes at windows are usually splayed to admit more light and are normally lined with timber panelling and shutters. In lesser rooms, the linings may be less elaborate, such as fixed panelling rather than working shutters, or simple tongued and grooved boarding. In some cases, the ingoes were lined with plain plaster edged in a timber staff bead.

Shutters, made up from framed and panelled sections, were divided in a number of ways, both horizontally and vertically, depending on the size of the window and proportions of the room. Shutters not only add considerable elegance to a room but also cut down heat loss, prevent draughts and give extra security. They are normally secured by a flat wrought iron bar across all panels or, if security is not a problem, with a flattened hook and staple.

PLASTERWORK

PLAIN PLASTERWORK

Plain plasterwork provides a smooth and level surface to receive paper or paint. Walls in basement rooms and stairwells, structural partitions and party walls, were usually plastered on the hard, i.e. plaster was applied directly to the brick or stonework. External walls above basement level and all ceilings were strapped with timber lathing to which the plaster was applied. The lathing for ceilings was fixed to branders suspended from the joists above; sometimes the laths were fixed directly to the joists.

Lathing on structural walls was nailed to timber strapping fixed to bilgates (battens built into the wall) or dooks (timber plugs driven into mortar joints). The laths were an inch and a half (39 mm) wide by a quarter to half an inch (7–13 mm) thick and were produced from straight-grained timber, usually oak or Baltic fir, which was riven or split. These were considered less likely to warp than sawn laths. The laths were spaced about a quarter of an inch (8 mm) apart using wrought iron nails about three-quarters of an inch (18 mm) long. Often the head of the nail was turned over and left proud to improve the plaster key.

The first or scratch coat of coarse plaster was applied diagonally across the laths, penetrating between them to form wet rivets or plaster keys. After hardening, these keys supported the weight of the plaster. The first coat was scratched when soft to form a key for the second or straightening coat which was used to create a true plane. Finally, a finishing coat was applied to give a smooth, level finish.

The scratch and straightening coats would have been composed of one part mature lime putty to three parts of fairly coarse sand. Cow's hair was generally added to the mix in the proportion of half a pound to every cubic foot to reinforce the rivets and to increase the tensile strength of the plaster, making it more flexible and less likely to crack. The finishing coat usually consisted of one part sharp sand to three parts of lime putty.

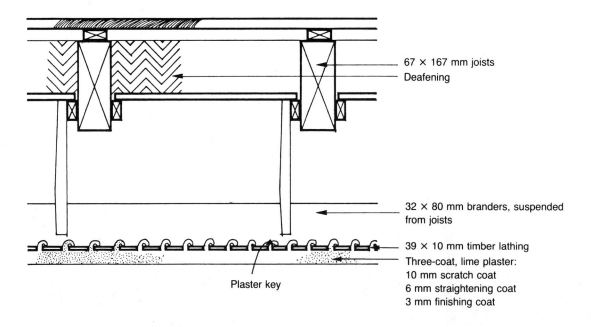

67 × 167 mm joists
Deafening

32 × 80 mm branders, suspended from joists

39 × 10 mm timber lathing
Three-coat, lime plaster:
10 mm scratch coat
6 mm straightening coat
3 mm finishing coat

Plaster key

DECORATIVE PLASTERWORK

Although the elevations of houses in the New Town conformed to an overall design, the plans and internal decoration depended on how much the first owner wished to spend.

Most decorative plasterwork in New Town houses consists of an enriched cornice, sometimes with a frieze and, in the later buildings, a ceiling rose. A few houses, mainly in the first New Town, had ceilings decorated in low relief over their whole area; notable examples still surviving are at 8 and 46 Queen Street, 1 Charlotte Square and 115 George Street.

DECORATIVE PLASTER CEILING
1 CHARLOTTE SQUARE

195

Photo: Tom Scott

The amount and detail of decoration varies from floor to floor in a typical house in the New Town, reflecting the importance of each room for which a range of motifs were available. All decorative plaster should be retained and Listed Building Consent must be obtained before any plasterwork can be removed. If missing decoration is to be restored, it may be helpful to look at plaster in other houses in the same street.

Simple cornices were built up *in situ* in coarse lime plaster on laths on timber bracketing, and the final profile was run in fine lime plaster using a metal template sliding along a level timber guiding rail temporarily fixed to the wall. The plaster would usually have been gauged with gypsum (plaster of Paris) in order to speed up the set. The template could be muffled with felt to form the base coats. Small profiles were built up directly from the straightening coat which had been scratched to form a key.

Extra enrichments such as egg and dart mouldings and acanthus leaves were cast separately using the newly developed, rapid-setting plaster of Paris which allowed lengths of elaborate ornament to be manufactured. Final modelling and adjustments to the pattern were carried out in the plasterer's workshop. The casts were then set in place on the plain lime plaster background using a thin adhesive layer of wet plaster of Paris. From the mid-nineteenth century fibrous plaster techniques were used to produce large lengths of precast mouldings.

The following materials were used for enrichments on walls, ceilings, chimneypieces and furniture:

- *Stucco:* In Scotland the terms plaster and stucco are synonymous, but generally stucco describes high relief enrichment hand-modelled by skilled craftsmen working *in situ* using pure lime putty gauged with a little gypsum. (This should not be confused with the trade name for plaster of Paris in the late eighteenth century which was Stucco);
- *Composition or Compo:* This is a mixture of resin, linseed oil, glue and whiting , which was pressed into hardwood moulds and, when set, fixed in place by glue or panel pins. It was a very flexible material and was used for delicate enrichment on timber backgrounds such as the chimneypieces in the First New Town and Gayfield Square (see FIREPLACES);
- *Gesso:* In use since the Middle Ages, this is a mixture of whiting, linseed oil and animal glue or size, built up in several layers over an armature. It is particularly suitable for rococo curls on three-dimensional wall sconces and girandoles;
- *Scagliola:* This was an effective way of imitating marble and it is seen on columns in lobbies of several houses in Charlotte Square, Heriot Row and other parts of the New Town. It was made by mixing plaster with size and pigments and could be formed *in situ* or precast into panels for invisible fixing. It was then finished by polishing to produce an effect like real marble;
- *Papier maché and carton pierre:* These are two other types of pressed ornament used for applied decoration from the 1780s and throughout the nineteenth century to represent carved wood. It was also used for ceiling decoration.

MAINTENANCE

Plain plasterwork in itself requires no maintenance as such except for periodic painting for decorative purposes. If the paint is flaking off, remove all the layers and repaint after checking for any signs of dampness.

Decorative plasterwork should be painted as seldom as possible because successive coats of paint will eventually obscure detail; if necessary, the old layers of emulsion paint should be removed. Repainting should be carried out using limewash or a soft distemper which can easily be washed off between applications. Colours should be chosen carefully to suit the original design. Further guidance can be found in INTERNAL PAINT AND VARNISH and INTERIOR DECORATION.

Traditional lime plaster depends on a high level of porosity to allow moisture caused by dampness or condensation to evaporate. Modern paint systems such as gloss or emulsion paints are impermeable and trapped moisture can cause staining and flaking. Gloss paints should never be used, and emulsion should not be applied if it has not been used before; they should never be applied to new lime plasterwork because the setting of the plaster itself will be impeded and the finish will be damaged by residual construction dampness trapped behind. Soft distemper or water-based paints are preferred because they allow the plaster to breathe and prevent moisture or salts becoming trapped behind the painted finish.

The removal of emulsion paint is an extremely difficult and laborious task and is best tackled using gentle steam and a sponge or palette knife. Methylated spirits can be applied first if the emulsion is very stable. Paint removers can damage the lime in the plaster, so a small area should always be tested first; non-caustic paint removers should never be used in conjunction with steam. For decorative work, a specialist should always be employed. Some surface decorations such as papier maché or compo will be irreversibly damaged by paint removal.

INSPECTION

The wise owner of a listed house should watch for signs of dampness, cracks, fungal attack or woodworm and take professional advice if necessary. A visual inspection can be backed up by tapping with the knuckles to detect

DRAWING ROOM CEILING
NO. 8 QUEEN STREET
ROBERT ADAM, ARCHITECT 1770-1771

197

Photos: Simpson and Brown

springiness, loose ornament or the presence of moisture. To investigate behind wall plaster, a fibre-optic boroscope or endoscope can be used. Gentle pressing by hand from below or the careful use of a broom handle wrapped in a duster can reveal sections of plasterwork which have become detached from the ceiling lath. However, the only certain way to detect loss of key is to make a visual inspection from above. This involves lifting the floorboards and clearing away the deafening, which can be inconvenient, especially in tenements. Great care should always be taken to prevent debris from falling onto the plaster, and to avoid vibration.

Opening up should allow inspection of the structural timbers and their bearings as well as the condition of rib and cornice bracketing, fixings, timber laths and the keys themselves. A litter of broken plaster rivets between the branders is sufficient indication that the ceiling is no longer keyed and may collapse; although many badly bulging and cracked ceilings remain in place for years, others collapse without warning. If a ceiling is particularly vulnerable, it should be propped before opening up takes place. If water is trapped above a ceiling, a hole should be made in the plaster to allow it to drain away and relieve the load.

DEFECTS

Defects in historic plasterwork are mainly caused by external factors, most commonly damp. They can, however, be caused by faults in the backing material or by the breakdown of the plaster itself. Inherent defects in the plaster itself are more likely to be encountered in replacement plasterwork due to poor craftsmanship or poor specification of materials. Defects may be described as:

- Physical, e.g. caused by damp;
- Chemical, e.g. caused by salt contamination;
- Mechanical, e.g. caused by accidental damage.

The following checklist of defects is based on that given in Historic Scotland's Technical Advice Note 2: *Conservation of Plasterwork* to which the reader is referred for greater detail:

1 *Cracking:* This is usually caused by structural, thermal or other movements of the building structure affecting the backing material, and cracks are usually deep and wide. Minor cracking is not usually serious and is often due to old movement or early shrinkage. In replacement work, cracking may occur if the plaster lacks tensile strength due to too little or unevenly distributed hair.

2 *Crazing:* New replacement lime plaster can craze if the sand is not sharp enough.

3 *Efflorescence:* This is caused by the crystallization of hygroscopic salts on the surface of the plaster and is commonly due to rising damp. New plaster applied to salt-contaminated masonry may show some efflorescence, creating the illusion that a rising damp problem still exists. Salt contamination can also be associated with condensation in flues or ill-advised biocidal treatments. Temporary sacrificial coats of lime plaster can sometimes be used to draw harmful salts out of a wall.

4 *Staining:* This can be caused by the rusting of the nails used to fix laths or structural timbers, by salt contamination, by the transference of dirt via water penetration, by surface mould growth due to condensation, by smoke or by the use of inappropriate paint finishes. In new work staining can be caused by impurities in the sand.

5 *Partial or total collapse:* This is usually the result of total separation of coats or separation from the backing material. The failure of the keys over an area of 1-2 m^2 can cause collapse months later due to the weight of the unsupported plaster. If there is major water penetration due to flooding, the timber lath may expand as the moisture content increases, and when it dries out the cracked keys may fail. This can happen months, even years, later. The collapse of the top coats leaving the first coat securely in place can be due to inadequate keying.

6 *Crumbliness:* This is often due to the breakdown of the plaster after prolonged water penetration and salt contamination. Another cause may be the fine roots or mycelia of vegetation or fungal growth which can destroy the integrity of the plaster. In the event of a fire, the coarser background coats may eventually disintegrate, leaving the harder top coat collapsed and fragmented. In replacement work, crumbliness is often due to insufficient lime in the mix.

7 *Separation of coats:* Excessive moisture can cause the coarse background coats to decompose behind the finishing plaster and decorative mouldings. In new work, separation of coats is usually due to the inadequate keying of earlier coats or inadequate preparation before laying down later coats.

8 *Separation from backing material:* This can be caused by the inadequate keying of a masonry backing or excessive suction due to inadequate wetting of the wall. Laths which are too closely or too widely spaced will not allow the plaster to form the ideal hook-shaped key. Other causes could be lath nails rusting away after prolonged water penetration, mechanical damage to the plaster keys, vibration or movement in the backing material, or the use of an unsuitable mix, especially lack of hair in the first coat on timber lath.

9 *Bulging:* This usually indicates an extreme case of separation from the backing material, although it can be caused by structural movement, timber decay or insect attack in timber lath. Mechanical damage caused by debris falling down behind strapping or collapsed deafening boards can cause local bulging. Bulging can indicate the separation of the top coat from the first coats due to inadequate keying of earlier coats or the use of incompatible materials.

10 *Fungal growth:* Wet or dry rot can occur after prolonged water penetration, and moulds can be caused by frequent condensation. Fungal growth may not cause any problems to the plaster itself but the underlying water penetration will almost always require more serious attention.

11 *Flaking of surface decoration:* This is caused by dampness and salt contamination. Internal finishes can be affected by changes to the exterior treatment of the building especially if this inhibits the evaporation of moisture through the wall. Flaking can be due to incompatible paint finishes or inadequate drying time.

REMEDIES

Remedial work will vary from filling small cracks to complete stripping and replastering. Very fine cracks can often be dealt with as part of the normal decorating process. The cracks should be wetted up and filler applied with a soft brush to get into all the crevices. Thin crazing may reappear later.

Larger cracks should be prepared for filling by undercutting the edges (taking care not to cut the laths) to form a dovetail key. After brushing away all loose material and dust, the surrounding area should be treated with size. The old plaster and background should be thoroughly wetted up with water and the crack filled with lime plaster, built up in coats to the depth of the surrounding plaster and ensuring that the necessary time is taken for each coat to set.

Small sections of decorative plasterwork can sometimes be reinstated by fixing them to the support timbers with long countersunk brass screws and stainless steel gauze washers. The joints can be filled as for small cracks.

If a small part of the ceiling has collapsed it may be possible to do a patch repair by replastering onto the old lathing provided that it is in good condition and all old plaster, dirt and debris are removed beforehand. Cleaning should be done with a vacuum or soft brush used only along the direction of the lath.

When there are areas of damaged or missing laths and keys it may be possible to repair from above using wire mesh and plaster of Paris. Further guidance about this technique can be found in Historic Scotland's Technical Advice Note 2, *Conservation of Plasterwork.*

It may be possible to retain and repair a sagging original ceiling, and this may be economical if it has a lot of enrichments or mouldings. A lath and plaster ceiling which has lost its key can be propped on planks supported on temporary trestles or props. Felt pads on the planks will protect the plaster from accidental damage when the ceiling is wedged back to its original position. Permanent support is then provided by metal discs countersunk into the plaster and attached by long stainless steel screws to the branders above. The countersinking is filled flush with lime plaster gauged with a little gypsum.

Where a ceiling has developed a slight bow, cradling pieces can be formed between the joists to secure the distorted profile.

After fire or flood damage, saturated plasterwork must be allowed to dry out. Holes should be drilled in the floor above and through the ceiling to allow air to circulate between the laths and to allow water to drain away. This drying-out process will take some months, and there is always the danger that the laths, expanded with water, will crack the plaster keys which will subsequently fail when drying and shrinkage has taken place. For this reason, a ceiling will often collapse unexpectedly many months after a flood has occurred. The use of dehumidifiers will reduce the drying-out time to a matter of days and at the same time reduce the risk of plaster failure.

Recently, new timber lathing has become more readily available. This should be chosen to match the original in terms of species, thickness, spacing and so on. Replacement laths should be butt-jointed at their ends and fixed by galvanized nails; the joints should be staggered to avoid long cracks. In general, lime plaster should not be used in conjunction with expanded metal lath. However, well-haired lime plaster applied very carefully to stainless steel lathing can be used where timber would be particularly susceptible to decay. Where a wall is to be completely restrapped to receive new lath and plaster it may be appropriate to remove all the original timber from within the masonry construction and replace it with stainless steel fixings. Where dooks or bilgates are to be retained or renewed in an external wall, they should first be treated with preservative. Most defects affecting plaster will also damage any decorative finish such as paint or paper.

RENEWAL OF PLASTER

Plain plasterwork

Three-coat work in traditional lime plaster is recommended in spite of the delays involved in drying out each coat. Gauging with Class A gypsum (plaster of Paris) can be used to accelerate setting, but this may cause greater

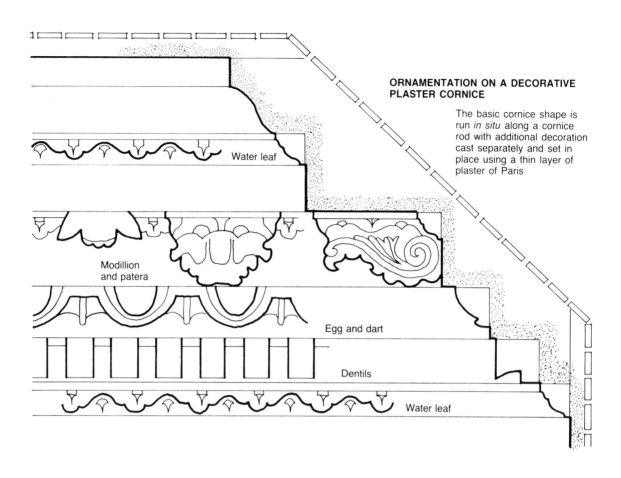

ORNAMENTATION ON A DECORATIVE PLASTER CORNICE

The basic cornice shape is run *in situ* along a cornice rod with additional decoration cast separately and set in place using a thin layer of plaster of Paris

Water leaf

Modillion and patera

Egg and dart

Dentils

Water leaf

Anthemion and palmette

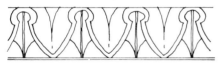

Leaf and dart

Egg and dart

Acanthus leaf

Guilloche

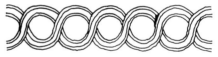

Guilloche

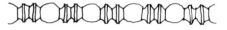

Bead and reel

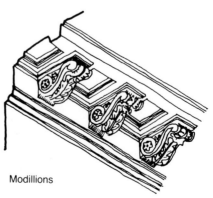

Modillions

DEVICES USED IN PLASTERWORK

shrinkage. The use of plasterboard is unacceptable because the finish is undoubtedly inferior, and the difference in thickness between the plasterboard and the traditional work may cause problems at junctions with existing cornices and so on.

After ensuring that there will be a decent key, the background should be prepared by cleaning and thoroughly wetting up. The scratch coat can then be applied with a steel trowel. When firm, but not set, the surface should be scratched to provide an undercut key for the next coat. An interval of one to three weeks (depending on the temperature and humidity) is recommended to allow the first coat to dry. This prevents shrinking, crazing and separation of coats later on.

After sweeping off the dust and further wetting, the straightening coat should be applied. When it has set, it should be wetted again and vigorously scoured or compacted with a wooden float. This will have to be done at least twice, sometimes more, to level the work and consolidate it as it shrinks on drying. This process is essential to avoid cracking and crazing in the finished work. Again, there should be an interval of one to three weeks before proceeding.

After yet further wetting, the finishing coat should be applied in two layers, and it too should be scoured (unless gypsum plaster has been used). At this stage any minor imperfections can be remedied by scouring small quantities of fairly dry finishing stuff into the surface.

When replacing plaster affected by damp it is essential to ensure that the cause of dampness has been cured, that there is no evidence of timber decay preventing a firm backing for the plaster key, and that areas which show damp stains but which are still fixed to the backing are also removed.

If there is rising damp, the insertion of a damp-proof membrane may not solve the problem immediately because the drying-out process can take several months. During this time, evaporation will lead to hygroscopic salts forming a damp patch on the plaster, which may need to be replaced to a height of 450 mm above the affected area. The new plaster should provide a dry, salt-free surface, able to resist efflorescence, residual moisture and hygroscopic salts; it should be able to breathe. A lime-based renovating plaster is recommended, applied in accordance with the manufacturer's instructions (the ENTCC can provide information on suitable products). A minimum of three days should be allowed before plastering walls which have been treated with fungicide.

Decorative plasterwork

Damaged decorative plasterwork should be repaired or replaced in the traditional manner; standard modern designs in plaster or fibreglass are unsuitable for historic buildings. If decorative plaster has been lost, reinstatement will require knowledge and skill; the names of specialists who are capable of undertaking such work are available from the ENTCC. Decorative plaster can be reproduced by taking an impression in plaster of Paris from the original shape, either *in situ* or in the workshop. Undercut enrichments, such as acanthus leaves or paterae, are usually removed for copying, but if this is impossible, an impression or a squeeze may be taken with latex, which can be peeled away from the original.

When houses were built in the eighteenth century, master casts of the ornament were taken from clay models or patterns were carved in timber. Many of these casts are kept by specialist firms, which can supply copies of accurate mouldings.

INTERIOR DECORATION

INTRODUCTION

There is no such thing as a typical New Town house. The housing stock of the New Town was created between 1770 and 1840, during a time of radical changes in taste, and their interiors reflect this variety. Many early houses were altered during the 1820s to bring them into line with the latest fashions. The greatest aesthetic change was the substitution of the Greek orders for those of ancient Rome (which had been favoured by Robert Adam). This affected the appearance of almost every decorative element, including cornices, chimneypieces, the profiles of panelling, windows and door cases and the divisions of the wall face.

During the later nineteenth century, the New Town remained a desirable residential quarter and the houses continued to be adapted to modern needs. At the turn of the century, there was a taste for re-Georgianization and many houses were renovated in a fashionable Adamitic taste which was preferred to the more austere Greek revival style. The most thorough renovation was the 4th Marquess of Bute's refitting of No. 5 Charlotte Square (now the headquarters of the National Trust for Scotland) in 1905.

During the 1950s and 60s many New Town houses suffered from ill-judged attempts to stamp a modernist aesthetic on their interiors and much damage was done through the removal of original chimneypieces and ornamental detail.

The best restoration practice will respect the various phases present in the interiors of a house and, through an understanding of the taste of the time in which these interiors were created, will aim at a sympathetic treatment preserving the character of the rooms.

Listed Building Consent is required for any internal or external alteration or extension, which could affect the architectural or historic character of a building on the descriptive lists (see LISTED BUILDINGS AND CONSERVATION AREAS). Elements which may affect the character of the interior of a building include alterations to:

- Chimneypieces; door cases, doors and their fittings; window furniture, such as shutters and pelmets; plaster, wood and paintwork; original wall coverings; reliefs and statuary; fixed tapestries; tilework and mosaics; niches and pediments; floors, plain and unspoiled or decorated in wood, marble, mosaic and so on; balusters, newel posts and tread ends; and any fixed furniture, such as bank or bar counters, or other fixtures of good quality down to about 1930;
- Any sub-division of rooms, particularly in the principal apartments and staircases, may be regarded as affecting the character.

Lack of specific references to the interior of a listed building in the descriptive lists does not necessarily mean that the interior is not important. It may just indicate that it was not possible for the inspectors to gain access to the interior when the building was inspected.

SOURCES OF INFORMATION

Archaeological evidence

Many houses preserve fragments of earlier decorative schemes which come to light during alterations or redecoration. Fragments of earlier wallpapers or decorative painting are often found behind old fuse boxes or early gas pipes and electrical conduits.

Paint sections and microscopy

Most houses retain their original decorative finishes under later layers of paint. Early wallpapers have often been papered over. These can be revealed deliberately by archaeological means. Very often decoration was not wholly or even partly removed before a new scheme was introduced, and important information can still exist below layers of paint or wallpaper. The simplest and most commonly used method of investigation is to take paint sections; literally to scrape with a knife or chemically strip a small area of decoration layer by layer. However, it can be difficult to distinguish one layer from another, and the surface colour changes of a single layer caused by the action of air and light may be misinterpreted. Microscopy is more reliable, but it can be expensive if done professionally and it is not within the expertise of most householders. It involves taking minute cores from the surface decoration and mounting them as cross-sections for microscopic examination. In this way successive layers

DRAWING ROOM   5 CHARLOTTE SQUARE AS RENOVATED IN 1905

WATERCOLOUR OF DRAWING ROOM OCCUPIED BY   203   Photos: The Royal Commission on the Ancient
COUNT ARTUR POTOCKI IN 1840,   and Historical Monuments of Scotland
POSSIBLY MORAY PLACE
(By courtesy of the Jagiellonian University, Cracow)

can be examined and the details of pigmentation revealed. Investigation by either of these methods should be made inconspicuously and with great care. It should be remembered that not all parts of a room would have been treated in the same way and that features, such as fireplaces and doors, are not always original to a room. For the best understanding of a decorative scheme, scrapes and sample cores need to be taken throughout the room, from walls (at different heights), ceilings, doors, windows, cornices, panelling and so on.

In the interests of future research it is helpful if past decoration is not removed before redecoration, unless the build-up of layers of some unevenness will seriously impair the look of a room. If removal is necessary, any research should be undertaken beforehand and a photographic record made. The ENTCC does not offer grants for internal restoration work, but information relating to the decoration of houses in the New Town and descriptions or photographs of decorative schemes as they are uncovered can be passed to the Scottish Conservation Bureau.

Documentary sources

Little research has yet been undertaken on early New Town houses, although there are a few important sources like the ledger of William Deas (*Scottish Record Office EDI/548/1*), one of the first house painters to be attracted to the New Town:

Nov 14th 1774
Mr Robertson, Architect newtown.

To 139 yds. Two coats painting in dining room outsides of doors, windows Etc at yr Lodging Newtown	£3- 9- 6d
To 21 yds Three Coats painting in Cornish finished in dead white	1- 1- 0d
To 72 yds Size yellow in walls of Dining room	18s 0d
To 33 yds whitewashing in Ceiling of Dining Rm.	2s 9d

More information is available on the later New Town, but few sets of building papers have yet been discovered. Moray Place possessed the grandest town houses in the city and appears to be the best documented. Research on individual houses can be approached through their former owners who are recorded in title deeds and the Post Office Directories (the latter are available at the Edinburgh Room of Central Library). It may then be possible to follow these individuals through family or business papers and also in their testaments and wills (the ENTCC library has an example study by Andrew Kerr, tracing the history of 13A Dundas Street). Contemporary newspapers often contain advertisements of auction sales and houses for lease which give a vivid impression of the character of later New Town houses.

The National Trust for Scotland's pioneering re-creation of 7 Charlotte Square as their Georgian House gives a good visual impression of an early New Town house. There is no equivalent for the more Grecian and opulent interiors of the later New Town. Unfortunately, there are hardly any visual records, but the few that have come to light are discussed in *The Scottish Interior* by Ian Gow. Some further information is available in the form of inventories and typescripts of lectures from ENTCC who will always be keen to learn of interesting early decorative schemes as they are discovered. The Royal Commission on the Ancient and Historical Monuments of Scotland may be able to undertake the recording of important discoveries.

NEW TOWN INTERIORS

From the outset the general arrangement of New Town houses remained standard, with the drawing room on the first floor, a dining room on the street floor and bedrooms over the drawing room. The top lighting of the staircase allowed it to be centrally located. The kitchen and other service rooms were in the basement. Early documentary information, however, suggests that there was considerable flexibility in the way a family used the other rooms in these houses.

A very high architectural standard was set both by Sir William Chambers' house for Sir Lawrence Dundas in St Andrew Square in 1770 and Robert Adam's house for Lord Chief Baron Ord at No. 8 Queen Street in 1770–1. Both houses introduced the latest metropolitan taste to Edinburgh. No. 8 provided a model for speculative builders and was the first house to have both a curved sideboard recess in the dining room and a pair of interconnecting drawing rooms.

With so many houses being built so rapidly, there was very considerable opportunity for experiment, but by the time of the Second New Town a standard plan had evolved in response to the demands of fashionable entertaining. Documentary sources show that both patrons and builders copied features that they admired in newly completed houses, with the result that a successful and almost vernacular formula was rapidly assimilated.

The entire first floor was devoted to a suite of drawing rooms connected by large folding doors to accommodate large parties and routs. A narrow lobby allowed an increased width for the dining room at the front of the house,

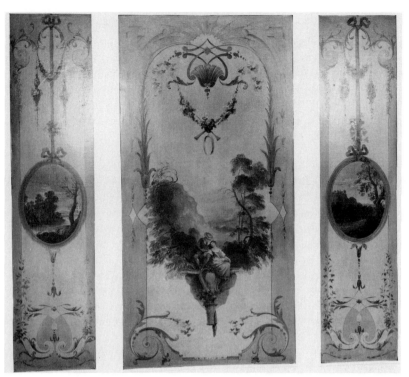

45 FREDERICK STREET

6 REGENT TERRACE

WALL DECORATION

Photos: The Royal Commission on the Ancient
and Historical Monuments of Scotland

while the back was divided into smaller apartments. With the first floor devoted to drawing rooms it became necessary for the family bedroom and dressing room to be squeezed into that part of the street floor not required for either dining room or the smaller breakfast room behind. The rest of the family would have slept on the second floor and attics (servants slept in the basements and many would have lived nearby).

The wedge-shaped plan of the Moray Place houses of the 1820s (see illustrated plans of No. 3) permitted the grandest expression of this plan type with up to three drawing rooms arranged as a circuit. Elizabeth Grant, in her *Memoirs of a Highland Lady*, recalls her mother's social calendar:

> Five or six dinners, two small evening parties, and one large evening one, a regular rout ...
> which paid my mother's debts in the visiting line each winter.

The drawing rooms were all-important, and architects treated this floor as their piano nobile. In contrast, the dining room storey was usually rusticated. By the time of Moray Place, improvements included ingenious hinges to enable the huge interconnecting doors to be folded back flat against the walls to allow for the large numbers of people invited to routs.

In general, a careful professional conformity characterized the interiors of the New Town. Many houses were built by speculative builders and let on short leases, giving tenants little incentive to upgrade their rooms. This uniformity was increased by the dominance of the city's leading cabinet-maker and upholsterer, William Trotter, who also acted as the most important house agent in the city from his premises at No. 8 Princes Street. He would also hire furniture.

From about 1810, decorative painting in the form of imitative graining and marbling began to play an increasingly important role in the appearance of New Town houses. During the 1820s and 1830s and as the city's house painters built on these newly found skills, Edinburgh emerged as a centre of innovative interior decoration. D. R. Hay, the protégé of Sir Walter Scott, founded a school of decoration based on scientific colour schemes and had many supporters among the leading intellectuals of the Modern Athens. His later experiments included ornamental painting in the Raphaelesque, Pompeiian and Watteau styles, all of which have been discovered in the New Town. Various stencilled treatments, often in gold leaf, were further specialities.

THE CHARACTER OF LATER NEW TOWN HOUSES

In comparison with the early houses, later New Town houses were characterized by the flamboyant opulence characteristic of Regency taste in Britain as a whole. The capital investment poured into the New Town encouraged the flourishing of a number of distinguished Edinburgh firms such as Richard Whytock, who invented a cheap method of printing carpets but was also capable of manufacturing furnishing textiles of the highest quality.

By pooling the documentary and archaeological evidence it is possible to give a composite impression of a typical Moray Place house during the 1820s. Its overall decorative treatment was copied rapidly throughout the New Town.

The lobby

The grained front door opened into the lobby which, in spite of its small volume, was usually treated very architecturally, often having columns and relief sculpture. The lobby was the first stage on the processional route to the drawing rooms and was usually given a hard finish. The columns were almost always marbled and it soon became fashionable to paint the entire wall-face as marble, often with fictive inlaid roundels or other ornaments in contrasting 'marbles'. The stone pavement was often treated as a geometric painted marble pavement in defiance of the real joints of the slabs. A stove, often in a niche, conveyed hot air from the kitchen range below (see FIREPLACES). The standard furniture was a set of shallow hall chairs with wooden seats and a truss-legged table, screwed into the wall, for visiting cards. In the late 1820s Pompeiian treatments were common.

The staircase

The staircase was usually separated from the lobby by glazed doors and again had a hard, exterior finish which was lined out in imitation ashlar blocks. The stone stair was usually treated as marble and the balustrade gilded, bronzed or painted in imitation of green oxidized copper. The doors were often painted to look like mahogany or even rosewood.

The drawing rooms

The drawing rooms were furnished *en suite* and accounted for over half the total expense of furnishing the house. To demonstrate their status their cornices were usually richly gilded and the chimneypieces were of pure white statuary marble. The woodwork was painted in imitation of an expensive cabinet wood like satinwood or maple. Drawing rooms were almost always papered with an expensive wallpaper, often with a gold motif and, later, gold stencilling was common. The wallpaper would be bordered by a gilt fillet. The carpet would be fitted and run through both

GROUND FLOOR
13 HERIOT ROW

Photo: Alan Forbes
by kind permission of the
National Galleries of Scotland

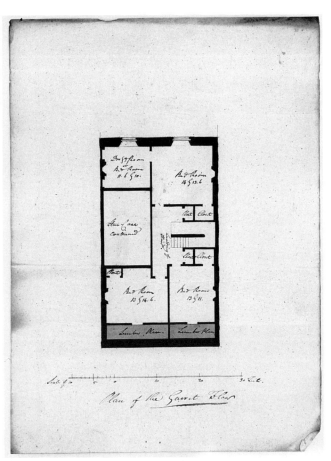

Second floor

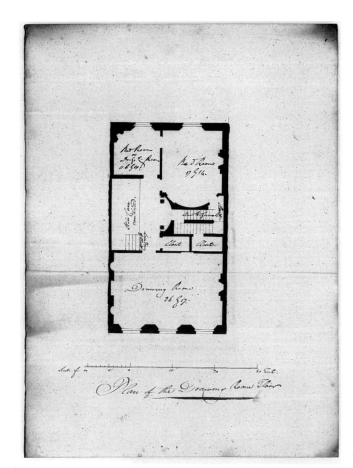

First floor

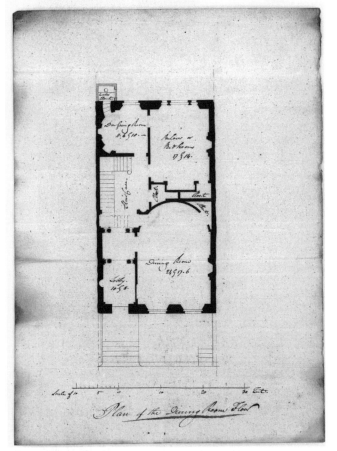

Ground floor

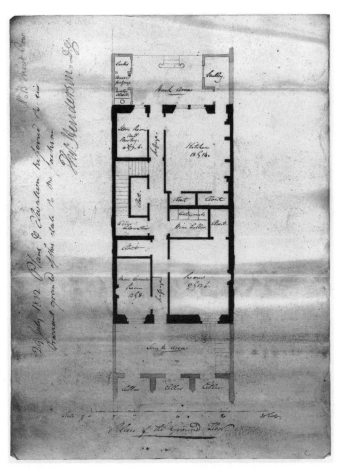

Basement

PLANS OF 13 HERIOT ROW

208

FIRST FLOOR LANDING
13 HERIOT ROW

Photo: Alan Forbes
by kind permission of the
National Galleries of Scotland

front and back drawing rooms. Curtains of silk damask at the windows were hung from a continuous pelmet with box pleats across the window wall. Furniture was often of an expensive cabinet wood like rosewood and upholstered in silk with chintz case-covers for daily use. A looking glass in a gilt and often French-style frame over the chimneypiece was usually balanced by a large gilt-framed mirror on the opposite wall above a table with truss legs.

The dining room

The dining room was treated more simply than the drawing room with plain painted walls which were frequently relieved by recessed plaster panelling. The sideboard recess was often treated architecturally with flanking pilasters. The dado would be grained in imitation oak. The chimneypiece was conventionally of black marble and this was picked up in the door and shutter handles which were often of black buffalo horn. The oak graining also frequently had fictive inlaid ebony stringing. Dining rooms always had Turkey pattern pile rugs and curtains of wool or a wool and silk mixture. The furniture was of mahogany upholstered in morocco leather. Red was the usual choice for curtains and upholstery.

The breakfast room

The breakfast room often had its walls bordered with painted lines and anthemion stencilled corner pieces. Sometimes the furniture matched that of the dining room to allow for large dinner parties.

The bedrooms

The bedrooms were frequently painted to match the ground colour of their wallpaper with a lighter tone repeated on the ceiling. The upholstery was usually printed chintz. Often each bedroom would have a set of identical chairs which could be used as a long set in drawing rooms for routs.

The standard change to the fixed arrangement of late Georgian town houses during the Victorian period was to transfer the main bedroom from the street floor to the back drawing room and it became usual to enlarge the front drawing room to create an 'L' plan. During this process the connecting double doors and the matching dummy to the stairs were often removed.

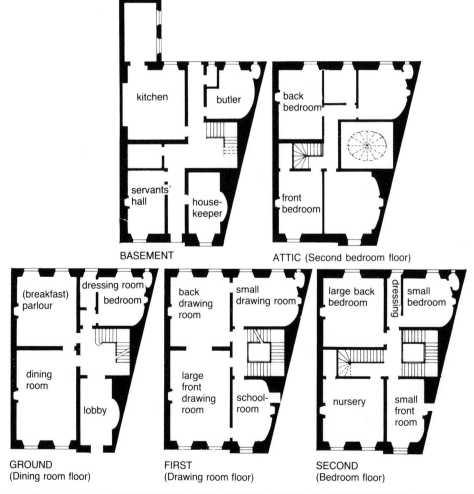

PLANS OF 3 MORAY PLACE (RECONSTRUCTED FROM DEAN OF GUILD PLANS) WITH NAMES OF ROOMS DESCRIBED ON WILLIAM TROTTER'S FURNITURE LAYOUT

WILLIAM TROTTER'S ESTIMATE FOR FURNISHING NO. 3 MORAY PLACE, EDINBURGH
FOR SIR DUNCAN CAMPBELL OF BARCALDINE, C.1825

DINING ROOM

2 Window Curtains of rich crimson 3/4 damask with plaited Valens bound with lace and finished with plummet fringe
2 large brass cornices with ornamental ends with large brass rings including brackets Pins &c
2 Roller Window blinds with patent mounting
2 Pair half sash blinds
A Brussels Carpet of best quality
A fine Wilton hearth rug
A Crumb cloth of mixture drugget
A Handsome mahogany Sideboard richly carv'd 8 feet long
Painted floor cloth for front of Sideboard
A Mahogany French fire screen on pillar and claws
A Set of very fine pillar and claw dining Tables of best Spanish wood and moulded edge, claws and legs on fine brass socket castors.
12 fine Grecian mahogany chairs with broad top rails capitals richly carved seats stuff'd in best manner and cover's w't crimson morocco
2 Arm'd chairs to suit covered & finished in same manner
A Mahogany Butlers tray
A ditto cross stand for do

PARLOUR

2 Roller blinds with patent mounting
2 Window Curtains of best crimson moreen with plated valens bound with lace and finished with fringe
2 large rich brass cornices with ornamental ends and large brass rings including brackets Pins &c
A Mahogany Pillar and claw Breakfast Table
8 Mahogany Chairs same as in dining room
A Mahogany French fire screen on pillar and claws
A Brussels Carpet of best quality
A fine Wilton hearth rug
A Crumb Cloth of mixed drugget

BEDROOM

A Roller Window blind with patent mounting
A carpet of best superfine English
A Mahogany four posted Bedstead lath bottom 6 feet wide
A Set of French Castors
A Set of Mahogany Cornices neatly finished & patent polish'd
A Set of Curtains of best watered moreen with plaited valens bound with lace and finished with plummet fringe
A ordered Straw Palliace in 2 parts in stout tick
A Bordered Hair Mattress in strip'd holland
A down bed, bolster & 2 pillows in fine linen tick
3 fine Rose Blankets
Binding do
A Marseilles Quilt
2 Bedside Tables
A Dressing Glass
A do Table
A Mahogany double wash stand table basons ewers etc
A Mahogany Linen airer
A Mahogany dwarf wardrobe with trays and shelves
4 Blackwood chairs with cane seats

DRESSING ROOM TO DITTO

A Roller window blind with patent mounting
A Carpet of best Superfine English
A Mahogany Bedet with Wedgewood pan
A Mahogany Bason Stand
A Mahogany Dressing Table
A Dressing Glass
A Chest Mahogany Drawers 3/4 Size
2 Blackwood chairs with cane seats

LOBBY

A handsome mahogany Lobby Table on truss legs 5 feet long
4 Handsome Mahogany Lobby chairs richly carv'd
2 Hemp foot matts
A Mahogany Hat Stand with brass rings & pan & brass branches
A Great Coat stand to suit wt. the above
Painted floor cloth to plan
Brass plates
Carpet for Stair cannot be calculated not knowing the number of steps
Brass Rods and butts
Drugget cover as above
A Lobby Table for Inner Lobby to fold down

LARGE FRONT DRAWING ROOM

2 Venetian blinds painted green

2 Window Curtains of Tabourett lined with durant and full plated Valens and angled dreep betwixt windows bound with silk lace & finished with silk plummet fringe

A Rich brass Cornice with ornamental ends large brass rings brass curtain bands brackets

A Brussels carpet of best quality

A fine Wilton hearth Rug

12 Very fine grained rosewood chairs partly of real rosewood highly finished and patent polished stuft loose seats in fine brown linen

2 highly finished chairs with stuft seats and backs in fine brown linen and legs on Strong brass socket Castors

A floor cover of mixture drugget

A Handsome Grecian grain'd Rosewood whole back sofa richly carv'd stuff'd in best manner with squab seat on truss feet castors

2 down pillows in white fustan

A handsome Grecian Ottoman to suit with one high end and stuft in best manner in squab seat in fine brown linen

A down pillow in white fustan

A handsome real rosewood Loo table elegantly finished with massive pillar and plinth finished with patent polish

A pair handsome pillar and claw Card Tables to suit of real rosewood finish'd with a patent polish

A handsome rosewood Sofa Table on pillar and claws to suit with the above

A pair neat Pole firescreens on elegant feet brass poles mounts finished with silk

2 handsome square real rosewood Tea Poys on thermd pillars and plinths

A Handsome Ladies Work table of fine Elm with silk bag neatly finished

A Splendid Commode of Fancy wood with truss legs highly ornamented with marble top and mirror under

A Very fine Mirror for over ditto in handsome gilded frame

A fine chimney mirror in gilded frame

SCHOOL ROOM NEXT FRONT ROOM

A roller window blind with patent mounting

A Carpet of fine Carpeting bound with lace

2 deal Tables on Square legs each with a drawer

6 rush bottomed chairs

A set hanging Bookshelves

BACK DRAWING ROOM

2 Venetian blinds painted green

2 Window Curtains same as Front room of fine Tabouret lined with durant bound with satin lace and finish'd with plated fringe

A Brass Cornice with large rings ornament &c as other room

A Brussels Carpet of best quality

A fine Wilton hearth rug

10 Grained rose-wood chairs with stuft seats in fine brown linen same as Front Room

2 French Chairs with stuft back & seats in fine brown linen same as Front Room

A floor cover of mixture drugget

A Handsome Whole back Grecian Sofa with squab seat in fine brown linen same as Front Room

2 Square down pillows in fustan

An Ottoman to suit with the above

A square down pillow in fustan

A Handsome Rosewood Loo table on therm'd pillar and plinth finished with patent polish

A Pair pillar and claw Card tables of fine rosewood on therm'd pillars and plinths

A Pair handsome Pole firescreens with silk mounts

1 Round Rosewood Tea Poys

A Handsome chimney mirror in gilt frame

SMALL ROOM NEXT BACK ROOM

2 Venetian blinds painted green

2 Window Curtains of fine tabouret lined with durant bound with satin lace and finish'd with silk plummet fringe same as other rooms

A Brass Cornice for ditto with large brass rings ornaments brackets bands &c

A Carpet of Brussels best quality

A Hearth Rug to Suit

A floor cover of mixture drugget

4 chairs same as other rooms

An Ottoman same as in other rooms

A down pillow for do

A Rosewood Card Table like those in other rooms

A Rosewood Sofa Table same as in Front Room with silk pannel and shelf

A neat grain'd rosewood French firescreen with silk pannel & Shelf

LARGE FRONT ROOM SAY NURSERY

2 Roller blinds
A Carpet of fine Carpeting
2 Eliptic roof Tent Bedsteads with lath bottoms 4 feet
6 inches wide
2 Setts printed Cotton curtains for ditto bound with
lace
2 Straw Palliaces
2 Hair Mattrases
6 Blankets
2 Binding Do
2 White Cotton counterpains
2 Bolsters & 4 Pillows
3 Crib Bedsteads with lath bottoms 2-6 x 5 feet
3 Setts printed Cotton Curtains bound with lace
3 Mattrasses in Striped holland
3 Bolsters 3 pillows
9 Blankets
3 Counterpanes
6 Cane bottom chairs
A deal Table on square legs
A deal Napery Press

SMALL FRONT ROOM

A Venetian Blind
A Window Curtain of fine Tabouret same as before
A Brass Cornice for ditto with large rings ornamental
ends bracks & bands etc.
A Brussels Carpet of best quality
A fine Wilton hearth rug
4 chairs same as in other rooms
A Rosewood Sofa Table same as in other rooms

brought forward:
Covering 32 Chairs 4 French chairs
2 Whole back Grecian sofas
3 ottomans and 7 pillows with silk cord & 3 Gimp 3
Counterpanes
Overall Slips of printed cotton bound with lace for 32
chairs, 4 French chairs, 2 sofas 3 ottomans & 7
pillows

LARGE BACK BEDROOM

A Roller blind with patent mounting
A Carpet of fine Carpeting bound with lace
3 Tent Bedsteads 3 feet 6 in wide with Eliptic roofs &
lath bottoms
3 Setts printed cotton curtains
3 Straw Palliases in stout tick
3 Hair Mattrases in striped holland
A Feather Bed bolster & pillow
2 Bolsters & 2 Pillows for the other two beds
9 Blankets
Binding Do
3 Bason stands
2 Dressing Tables
2 Dressing Glasses
2 Chests of Drawers
Blackwood chairs cane seats

ROOM NEXT BACK ROOM

A roller window blind patent mounting
A carpet of fine Carpeting
A four posted bedstead 5 feet 6 inch wide
A Set French Castors
A Set Mahogany Cornices
A Set of Moreen Curtains as before
A bordered straw palliace in two parts
A Bordered hair mattress in stripd holland
A Down bed bolster and two pillows
3 fine Rose Blankets
A Binding do
A Marseilles Quilt
2 Bedside tables
A Dressing Glass
A Dressing Table
A Mahogany double wash hand table
A Mahogany Linen airer on claw feet
A Mahogany Dwarf wardrobe same as in other room
4 Blackwood chairs with cane seats

DRESSING ROOM TO DITTO

A roller window blind
A Carpet same as Bedroom
A Mahogany shaped bedet
A Bason Stand
A Dressing table
A Chest Drawers 3/4 size
2 blackwood chairs cane seats

SECOND BEDROOM FLOOR FRONT ROOM

A roller blind
A carpet of fine Carpeting
A double Ogee roof tent bedstead 4 feet 6 in wide
A set of curtains for ditto of printed cotton
A Straw Palliace
A Hair Mattress
A Feather bed bolster & 2 pillows 3 blankets
A Bedside table
A Linen airer
A Chest of Drawers
A Dressing Table
A Dressing Glass
A Wash hand Stand
2 Blackwood chairs cane seats

BACK BEDROOM

A piece of Carpeting
2 Square Roof Tent bedsteads with lath bottoms 4 feet
    6 inch wide
2 setts check Curtains bound with lace
2 Hair Mattresses
2 Bolsters & 2 pillows
6 Blankets
2 Coverlets
A Dressing Glass
A Tripod Bason stand
2 Blackwood Chairs

HOUSEKEEPERS ROOM BELOW STAIRS

A Roller Blind common mounting
A Carpet of Fine Carpeting
3 Eliptic roof Tent Bedsteads 4 feet wide
2 Setts printed Cotton Curtains bound with lace
2 Straw Palliaces
2 Hair Mattrasses
6 Blanketts
2 Binding do
2 Counterpanes
3 Feather Beds bolsters & Pillows
A Dressing Glass
4 Chairs
A Mahogany Washstand

SERVANTS HALL

A large deal Table
2 Benches

BUTLERS ROOM

A high Press Bedstead
2 Mattress
3 Blanketts
A Coverlet
A Bolster & Pillow
A Dressing Glass
A deal table
4 Chairs
A Bason stand

KITCHEN

4 Kitchen Chairs, A Meat Screen, A Plate Rack, A
    coal bucket

LAUNDRY

A large Ironing table, 4 Hardwood Chairs
2 Cloathes screens

49 pieces flock paper for Drawing Room
529 feet gold moulding

£1953 8s. 4d.

INTERNAL PAINT AND VARNISH

INTRODUCTION

Plastered walls and ceilings are painted to provide a decorative finish, to reflect or absorb light or to facilitate cleaning. Painting or varnishing timber seals it from dirt and protects it from surface deterioration and staining.

In the First New Town built before 1800, two types of paint were in general use for internal decoration. An oil paint based on white lead was used on woodwork usually in an off-white or stone colour. This type of paint was often used on plaster walls, being applied both above and below the chair rail. A water-borne paint, called distemper or size colour, was often used in areas unlikely to receive much wear.

Between 1810 and 1820 it became fashionable to use paint for graining, panelling and general joinery. Marbling, too, was practised but was generally reserved for the hall and staircase. These imitations were protected by a coat of an oil varnish.

For most people, period restoration means imitating an earlier colour scheme in modern emulsions, eggshells or gloss paints. For historical accuracy, however, and to allow the plaster to breathe, it is worth attempting to use traditional or near-traditional materials as well as appropriate colours. Lead-based paints, once the standard material, are now regarded as a health hazard, and may only be used on Grade A listed buildings, and then only with special permission from Historic Scotland.

The vast majority of Georgian schemes have been overpainted many times, so that even when paint looks old it may only date from the early years of this century. If relevant, an investigation by an experienced paint analyst will reveal both the number and type of earlier decorative coats.

PAINT

There are three recognized classes of paints: distempers, washable water paints and oil paints.

Distemper

Distemper has a dead flat finish and was used on plaster walls and ceilings. True distemper, known as soft distemper, is made from common whiting bound with animal glue size. The name is often wrongly applied to any form of water paint or water-thinned paint. The advantages of using distemper instead of a modern paint lie in its appearance, its permeability to water vapour, which enables the structure to breathe, and its solubility in water, which means that it can be easily washed off and is therefore unlikely to build up and obscure fine detail on cornices and other enrichments. Its disadvantages are that it cannot be recoated and has to be thoroughly washed off before re-application. In paste form it has a short shelf life and in powder form it is awkward to mix. It must be applied quickly and confidently in order to avoid heavy brushmarks.

Being soluble in water, distemper is easy to identify, although there are 'washable' distempers of a slightly different composition which have some resistance to water, but even these can be removed with a little wet rubbing. Soft distemper is still available and the ENTCC has the names of suppliers.

Washable water paints and emulsions

Soft distemper for walls was superseded by oil-bound water paints (sometimes wrongly called washable distempers) in the late nineteenth century, while ceilings continued to be painted with the non-washable material until recently. Although they are more durable, oil-bound water paints are not as permeable and are harder to remove than soft distemper. The vehicle is an oil-in-water emulsion that dries to a matt finish. To mix oil and water together requires the addition of an emulsifier such as soap, casein, glue or starch. This process is different from that for modern emulsion paints, which are not a suitable substitute.

Modern emulsion paints should more correctly be termed dispersions, as they consist of resin particles such as vinyl, acrylic and synthetic rubber, together with pigment particles dispersed in water. Modern emulsion paints can be applied over washable water paints provided that all powdery material is removed, and over old gloss paints if they are well washed down with detergent and abrasive paper.

GRAINED DOOR
HOPE TRUST, MORAY PLACE

216

Photo: Joe Rock

Oil paints

Oil paint was used on timber and, during the eighteenth and nineteenth centuries, on plaster walls and ceilings where cost was not a major concern. It was almost invariably based on white lead dispersed in a drying oil, such as linseed. The degree of sheen on the dried coat could be varied, a flat finish being fashionable for surfaces that did not receive much wear. The unadjusted finish would have had a mid-sheen. If a glossy finish was required a coat of oil-varnish would have been applied on top of the paint.

Modern gloss paints are unsuitable for authentic redecoration because they have a harder and more brilliant appearance than traditional oil paints. Even the more subdued semi-gloss and eggshell finishes are poor imitations, although there is little choice nowadays. A flat oil paint is perhaps a reasonable substitute for a flatted oil finish, although it should be remembered that this was not found universally.

CLEAR FINISHES

Oil varnishes made from various resins and oil are the traditional clear finish. The oil is responsible for the elasticity and durability of the film, the functions of the resins being to impart gloss and hardness. Oil varnishes are referred to as 'long-oil' or 'short-oil'; varnishes with high resin content give a hard, quick-drying glossy surface which is less durable than the 'long-oil' exterior varnishes.

Spirit varnish is made by dissolving a gum or resin in alcohol or other spirit. The solvent dries by evaporation, leaving the resin as a thin film on the surface. It dries rapidly and is used as a knotting for sealing knots in wood and as a spirit stain for floors, doors and mahogany handrails.

Varnishes need to be chosen with care. For interior use an alkyd varnish is to be preferred to the widely available polyurethane variety. The cheaper varnishes tend to give a thick, treacly finish which chips easily. The process of varnishing was taken very seriously in the past, many tips coming from the coachpainter. Each coat was polished carefully, and a high gloss was generally desired. A lower sheen was obtained by polishing the dried varnish with a very fine abrasive powder, such as rottenstone or finely ground pumice stone.

Well-buffed wax polishes, composed of beeswax and turpentine (carnauba wax appears to have been imported from about 1846) were often applied to floors, and these may still be used. For a wax finish, two coats of sealer, made up of 1:3 methylated spirits/white shellac, should be applied first. Wax is very difficult to remove if it penetrates timber and, when a paint finish is being restored after wax has been used on bare timber, copious washing with turpentine will be required. Spirit varnish can be used as a primer over traces of beeswax, but not over silicone wax. Any clear finish applied in place of paint should be reversible, so that a paint finish can be re-applied.

INSPECTION AND MAINTENANCE

The frequency of internal repainting depends largely on personal preference, but decorative plasterwork should be repainted as seldom as possible so that the detail does not become obscured (see PLASTERWORK). Surfaces should be examined prior to redecoration to ensure that the new finish will not be spoiled by poor adhesion of the existing film; new paintwork always shrinks slightly as it dries, and this may disturb weakly adhering paint. The application of new paint to a trial area in an inconspicuous part of a room will give some indication of the soundness of the existing surface.

Except for distemper (which will wash off), paint can be gently cleaned with warm water to which a small amount of de-ionized detergent has been added. Ordinary household detergent and abrasive cleaners are not suitable, as they will destroy the surface finish and make it more susceptible to dirt in the future. A test cleaning area should be made.

All timber surfaces should be cleaned and inspected annually for signs of wear. Spirit varnish finishes can be rubbed up with furniture polish, used sparingly so as not to dull the surface.

DEFECTS AND REMEDIES

Defects can be caused by water penetration, by salts present in walls or migrating with rising damp, by the unsuitable condition of new plasterwork, insufficient preparation before repainting, or the wrong choice of paint.

Persistently damp conditions indoors will encourage the growth of moulds, giving rise to black or coloured spots on the surface of the plaster. In order to prevent the recurrence of bad cases, it is necessary to remove the source of dampness and allow the surfaces to dry out thoroughly after scrubbing with a detergent and fungicide. Various types of wash are effective, including aqueous solutions of sodium pentachlorphenate or sodium orthophenylphenate, which can be obtained as proprietary brands. They should only be used in accordance with the manufacturer's instructions. It is wise to avoid painting during damp weather or in excessive heat, even for internal work. Good ventilation is always essential.

New plaster must be dry before painting; a few days' drying out will suffice for small patches, but several weeks are necessary for large areas of plaster. If there is any doubt, test with a hygrometer or a conductivity meter. Any new plaster, especially new lime plaster, should not be decorated for about a year except with a fully permeable paint such as a soft distemper.

Lack of adhesion is the most frequent defect in paintwork, manifested by blistering, peeling or flaking of the film. Areas of flaking paint should be thoroughly cleaned off and, though other paint on the same wall may appear sound, every effort should be made to remove it, because its weakness may become apparent after repainting. Where the area of blistering or flaking is small it may be rubbed down with an abrasive paper; water should not be used because it may penetrate the plaster and adversely affect sound paintwork. All residual loose and powdery material should be dusted off, but note that the dust from lead paints is very harmful to health.

Adequate preparation is essential to avoid defects in new paintwork. If an existing finish is in very poor condition, it may be necessary to strip it down to the plasterwork, but in most cases it will be sufficient to clean all surfaces thoroughly and prepare them in accordance with the paint manufacturer's requirements. Washable water-based paints and oil-bound water paints should be rubbed down dry to clean the surface, remove any loose material and improve the key for the new coat. Soft distemper should be washed off completely and the surface allowed to dry before applying a new finish.

Choosing a compatible paint will depend on identifying the existing finish, but there are various sealants available which act as barriers between finishes. Undercoats and finishing coats should be supplied by the same manufacturer.

REMOVAL OF PAINTS

Stripping doors, windows and panelling to expose the grain is now popular but not recommended; the surface of the timber may be damaged and it is historically inappropriate for most interiors in the New Town. However, it may be necessary to remove successive layers of clogged paint prior to repainting.

The two main classifications of paint removers are spirit and alkaline; spirit is safer and more effective but can be very inflammable. It consists of a solvent which softens the paint film, together with a thickening agent, usually wax, to retard evaporation and to hold it onto the surface. The solvent is left until the surface softens and then scraped off. All traces of wax must be removed completely by washing with white spirit.

Alkaline removers are strong caustic solutions which are not recommended for use on wood, plaster and other porous materials, because they are extremely difficult to remove completely or to neutralize; the caustic may affect subsequent finishes, destroy the structure of wood and dissolve glue. After the paint remover has been scraped off, the surface should be well rinsed with several changes of clean water, and a wash of weak acetic acid or vinegar used to neutralize the caustic. It is advisable to test the wet surface with litmus paper to see whether any alkalinity remains.

Emulsion paints are difficult to remove but may be cleaned off with an application of methylated spirits and steam, which is a long and laborious process. Using a paste of wallpaper stripper, old emulsion paint can be soaked until it is soft enough to scrape off (see PLASTERWORK).

Oil paint can be removed from joinery using a hot-air gun, which has many advantages over a blow-lamp; the temperature range of a hot-air paint stripper is lower, it gives a better surface finish and the risk of fire and of scorching the timber with a direct flame is reduced. Hot-air guns can be a health hazard if not used correctly. The chapter on FIRE SAFETY outlines some precautions needed when working in historic buildings with hot materials.

COLOUR

Choice of colour will depend largely on individual taste and current fashions, and whether one wishes to recreate an historical scheme. It is difficult to construct an historical colour chart because there was little standardization of paint colours and mixes and even contemporary references in documents and bills to the use of 'stone' or 'straw colour', 'pea green' or 'French grey' are open to different interpretations. A small range of the colours which would have been used c. 1780 may be seen at the office of the ENTCC.

In the eighteenth century the number of pigments suitable for use in paint was limited, but the mixing properties of each were exploited to produce as large a range of colours as possible. Paint could be bought semi-prepared from the early years of the century, and many pigments were available from the colourman ready-ground in oil. Families of colours were produced rather than specific shades or tints. In 1821, in his *Nomenclature of Colours*, Patrick Symes identified eight whites, seven blacks, ten blues, eleven purples, sixteen greens, fifteen yellows, six oranges, eighteen reds and eleven browns. It was not until the early twentieth century that an attempt was made to standardize paint colours.

In the Georgian era the most widely used colours were known as common colours. These were largely derived from earth pigments, red and yellow ochres, umbers, and from black – and produced a number of reddy browns, dull yellows, greys, browns and whites. When in an oil medium they could be used both internally and externally. The most expensive pigments were not always the most stable or colour-fast. Blue and green verditer and most of the organic yellow pigments could only be used in distemper.

Until the mid-eighteenth century, light colours such as grey, stone and white were preferred for plasterwork, and white was generally used for ceilings. It was never the pure, brilliant white of today's paints; there was usually a slight grey or ochre cast. After the 1760s there are increasing references to the use of pea green and light greens, and to reds and blues. Skirtings were often painted in a dark colour or in a darker version of the wall colour. Bright yellows were very difficult to achieve, and not widely used, although dull yellows were often found on common stairs. A mid-yellow or straw colour tended to be used for the walls of print rooms. Stone colour saw a great deal of use as did the drabs, of which there were a variety. Olive green was more expensive and is likely to have appeared on exterior doors.

By the beginning of the nineteenth century the use of colour in decoration schemes had become more elaborate. Gradually darker colours came to be used, and there was some influence from the recently discovered archae-ological sites of the ancient world. Terracottas, Tuscan and Pompeian colours in particular were added to the repertoire. From the 1820s a bright yellow became a possibility although it was very expensive, and if this was mixed with blue a range of cleaner greens could be created. The introduction of an artificial ultramarine also gave further options, adding a non-green blue to the range of colours at the house-painter's disposal.

Marbling and graining

A marbled or grained finish is achieved by building up a pattern of veins or grain on a ground colour, using a flat or mid-sheen oil-based paint. To achieve the best effect the technique should imitate one of the numerous patterns which occur in nature; this may be done using a wood colour oil glaze (e.g. oak, walnut, mahogany) applied over a suitably tinted ground colour, which is in turn brushed to partially reveal the ground colour and give the illusion of grain. There are a number of publications which describe the techniques of imitative painting.

Gilding

Gilding has always been an expensive internal finish used in the drawing rooms of important houses, and it was used lavishly in the New Town, c. 1820. There are two methods: water gilding was usually applied to furniture and oil gilding to architectural detail. Of the two, oil gilding is the simplest process and therefore somewhat cheaper. When it is only applied to the edges and tips of ornament it is called part gilding.

There are modern substitute finishes available but none produces as satisfactory or as lasting a finish as gold leaf. Gold leaf can discolour, though this mainly occurs with cheaper kinds of gold leaf containing copper and other alloys. Discolouration can sometimes be removed, but expert advice should be sought. Gold should be protected by the application of thin parchment size. In the eighteenth century lacquered silver was occasionally applied to furniture but seldom to architectural details. Untreated, it tends to tarnish.

WALLCOVERINGS

INTRODUCTION

By the time of the first New Town, panelling had been abandoned as a finish for fashionable rooms. Although it was common to hang the grandest drawing rooms in London or in country houses with silks, wallpapers were the norm in the New Town. Unfortunately, very little is known about the earliest New Town papers. At first, they were probably most commonly used in drawing rooms, and often they imitated more expensive textile wall-coverings. Dining rooms, by contrast, tended to be more simply finished and usually had painted walls. The most common choice for bedrooms were chintz papers related to the bed-hangings and, by the late eighteenth century, it was possible to have papers printed to match fabrics in the most expensive houses.

During the early nineteenth century, wallpaper became the standard finish for almost all rooms other than dining rooms, lobbies, staircases and passages. Their popularity is reflected in the first Scottish attempts to print papers in Edinburgh. The leading manufacturer here was William M'Crie, who was very ambitious and even went so far as to cut his own blocks instead of buying ready-cut patterns from specialist Southern firms. However, because wallpaper could be so easily imported into Scotland, he probably realized from the outset that he could not compete with the finest English papers, and his productions seem to have been confined to the middle of the market. There is considerable evidence to show that the very grandest French wallpapers, including panorama cycles, were hung in New Town houses during the 1820s and 1830s, but none have survived *in situ*. An example of a painted imitation of a panorama paper, with a dado balustrade between columns and a pastoral landscape beyond, was uncovered in a villa in Minto Street recently.

The use of wallpaper was attacked by D.R. Hay, the leading house-painter in Edinburgh, who condemned it as unhealthy in the damp Scottish climate because it was composed of decaying vegetable matter and held in place by putrifying animal glues. Hay sought to replace its use by stencilled effects, recapturing wallpaper patterns in paint. Many Edinburgh drawing rooms were therefore stencilled in gold size with repetitive patterns which were then gilded; imitation damask, which gave the effect of flock papers in raised textures by incorporating sand in the paint, was his favourite choice for dining rooms. Wallpapers, however, continued to be used extensively throughout the nineteenth and early twentieth centuries.

At the end of the nineteenth century there were experiments with bonding textiles to a backing of paper and several examples have been discovered in the New Town. These innovative techniques were characteristic of Edinburgh's Tynecastle Company led by William Scott Morton. Characteristically, when the 4th Marquess of Bute restored No. 5 Charlotte Square in 1905 to its putative 'Adam' appearance, he hung a silk lampas, but this was a far more luxurious finish than the New Town's original occupants would have indulged in.

It was always the best practice in decorating to strip off early wallpapers before a new wallpaper was hung. For this reason, any information about early papers is scant indeed, and even a stray fragment concealed by later alterations can be an important addition to knowledge. All interesting discoveries should be reported to ENTCC and it may be possible for the Royal Commission on the Ancient and Historical Monuments of Scotland to record the most important. Fragments should be carefully preserved. It is desirable to preserve fragments of earlier papers by concealing them behind mirrors, pictures or pieces of furniture. The open backs of bookshelves, too, can often preserve an earlier scheme without distracting from a modern one in the rest of the room.

There have been great advances in the reproduction of wallpapers in recent years. Before selecting a paper it is recommended that restorers should 'get their eye in' by examining some genuine samples of early papers (a small collection is held in the National Monuments Record of Scotland). Early papers were printed in brilliant clear colours which are very different from the subdued and antiqued effects favoured in many so-called reproductions. Reproductions are often screen-printed, but this technique gives a much less crisp effect than hand blocking which is the only method of creating the original quality of early papers. Wallpaper is now so expensive that it is worth an extra investment at the outset. It is also worth taking care of a costly paper, rather than making do with a cheap inferior imitation that may not wear so well.

DEFECTS AND REMEDIES

Damage to fabric and paper wallcoverings can result from human contact, fluctuations in relative humidity and exposure to daylight, direct sunlight and artificial light, but a great deal of damage can be prevented by good housekeeping practice. The extent of deterioration and fading will depend on the quality of the raw materials and also on the dyes and pigments used, some being more fugitive than others. High relative humidity can affect the fastness of some colours, and can create the damp conditions in which mould and fungi thrive on the size which was used to bind the paper together. Contact with iron, such as iron fixings, will cause fabric to rot. The Museums Association publishes useful leaflets on the control of lighting and humidity.

Wallcoverings can become detached from walls because of damp conditions, building movement, the poor condition of the wall surface or previous finish, or the deterioration of the backing material. Treatment in the last case will depend on the type of backing material used and it is best to consult a paper or textile conservator; the Scottish Conservation Bureau has names and addresses.

Dampness will cause mould and fungal growth, but the problem will not be solved by extra heating alone, as this may stimulate growth. Instead, ventilation needs to be improved and the source of dampness located and dried out.

A certain amount of dirt will not detract from the appearance of wallpaper and may be considered part of the character of a room. It may only be necessary to clean areas of concentrated dirt (around light switches, for example). Any cleaning method to be used should be tested on a small, inconspicuous area of wallpaper, to judge the effect, before embarking on more obvious areas.

Wallpaper can be damaged by acidity, either acquired during manufacture or on ageing, and it is also susceptible to attack by sulphur dioxide, which is present in the polluted atmosphere of urban areas.

Cleaning is an important part of textile and wallpaper conservation but should only be attempted by conservators who can establish the nature of the problem, the type of fabric or paper and the best methods to use. Information on conservators who undertake this sort of cleaning is available from the Scottish Conservation Bureau, the Conservation Unit of the Museums and Galleries Commission, or from the Institute of Paper Conservation.

DAMPNESS

Moisture is always present in buildings, and certain levels of natural moisture are necessary to keep the building fabric in a stable condition. The desirable moisture content varies widely with the type of material, ranging from one per cent for plaster to 18 per cent for timber. The term 'dampness' is reserved for conditions under which moisture is present in sufficient quantities to cause deterioration of the fabric, decoration or contents of the building.

Preventing or curing dampness is a major maintenance problem, but it is also important to remember that an essential characteristic of a traditional building is its permeability to water vapour and the ability of the materials to breathe; the introduction of unnecessary vapour barriers and damp-proof courses can do more harm than good.

INSPECTION

There are four main causes of dampness:

- Water penetration through walls or roof;
- Condensation;
- Rising damp;
- Other causes such as leaking pipes.

An inspection should identify the cause or causes, which may be far from obvious. It is advisable to employ an architect or a building defects consultant, who will take the following factors into account:

1 *Extent:* Is the damp widespread or does it occur in isolated patches?
2 *Pattern:* Does the damp relate to certain features, such as window openings or chimneys?
3 *Period:* Does the dampness occur only after long spells of rain, when the air is humid, or after a rise in temperature? Is it persistent or spasmodic? Is it related to previous building repairs?
4 *Use:* Does the dampness occur with a change of use of the house, a change in the lifestyle of the occupants or a change in the heat source, for example to paraffin or bottled gas heaters?

A good moisture meter may help the initial investigation, bearing in mind that a surface reading is not always a reliable guide to a source of damp; in fact, such readings are known to have led to unnecessary and expensive repairs. More reliable information is obtained by weighing samples of material taken from the wall or by using a carbide moisture meter. These techniques require skill and special equipment. The ENTCC can provide names of firms who can obtain and analyse samples.

The inspection should be systematic, beginning perhaps in the roof space and going down through all the rooms to the basement, followed by an external examination. Touch, sight and smell are often the best aids to diagnosis.

WATER PENETRATION

Penetration of water from an external source is often due to poor maintenance of the fabric so that rain or melting snow and ice can enter. The most common signs are:

- *Blocked or damaged gutters and downpipes,* resulting in damp patches appearing after rain. External stonework can be stained and there may be signs of vegetation. A faulty internal conductor from a valley gutter can remain undetected for some time;
- *Defective roof coverings,* resulting in damp patches after rain or snow. The source of dampness may not always be apparent, because moisture often runs down the rafters and drips off at some distance from the point of entry. Cracked, loose or missing slates can usually be replaced but sometimes, for example with a lead covering, it is preferable to replace a whole section rather than to patch and risk a repeat of the leak from an adjacent weak spot;
- *Defective flashings, soakers, mortar fillets* and so on, particularly if exposed to high winds;
- *Faulty stonework,* open vertical joints or cracks in horizontal projections are liable to conduct water into the wall. Stone cills, cornices and string courses are particularly vulnerable;
- *Faulty pointing* to window frames.

222

The cause of penetrating dampness must be remedied as soon as it is diagnosed; severe occasional penetration is often less dangerous than a slow drip which may go undetected for years. Once a leak has occurred it is essential to check all adjoining timbers for signs of decay. If the timbers are sound and well ventilated, they will dry out without further damage. The possibility of plaster ceilings collapsing some weeks, months or even years after severe water damage should not be forgotten (see PLASTERWORK).

CONDENSATION

Condensation occurs when warm, moist air comes into contact with a cooler surface and is invariably accompanied by mould growth, which can form on any cold surface, including the contents of cupboards and wardrobes. Dampness from condensation usually affects the whole area of walls, ceilings and floors, unlike the patchiness of penetrating or rising damp, but it can be confined to isolated pockets of cold, stagnant air (behind pictures, wardrobes and so on).

Much has been written on the subject of condensation, but in many cases the best and simplest remedy is natural ventilation with improved heating. Constant rather than intense intermittent heating is the most effective and economical method of keeping the structure warm and therefore reducing the risk of condensation. Natural ventilation can be supplemented by mechanical fans – the most effective being located at the source of the water vapour (usually the kitchen or bathroom) and controlled by a humidistat. The introduction of insulation may cause interstitial condensation unless a vapour barrier is included and it is virtually impossible to insert such a barrier into existing buildings of traditional construction. More information can be obtained from the BRE.

Condensation within a flue, bringing hygroscopic salts and tar to the surface, can cause damp patches on the chimney breast and, in some cases, on larger areas of wall. In mild cases, sealing the wall with aluminium paint, applied once the plaster has dried out, will hold back the stain and permit redecoration. Otherwise it may be necessary to strip and replaster the chimney breast and line the flue with an impervious lining. To avoid condensation and damage to chimneys, all flues serving gas fires and gas boilers should be lined with non-ferrous metal liners.

Disused flues should always be ventilated at the top and bottom and the ENTCC can supply information on suitable clay vent terminals for use with New Town chimneypots.

Portable bottled gas and paraffin heaters produce enormous quantities of water vapour and are not recommended; a gallon of paraffin, when burnt, produces a gallon of water.

RISING DAMP

Georgian buildings were not built with damp-proof courses and lower walls are subject to a degree of rising damp, often recognizable by a continuous band of dampness, discolouration and efflorescence on the plasterwork. However, these buildings have often performed well for two hundred years and rising damp should not cause undue concern. Unless it is severe, rising damp is not a structural problem, although the damage to plaster and decoration may be unacceptable. There is a possibility of damp causing dry rot in adjacent timber but with adequate ventilation this is slight.

When rising damp does constitute a problem, there are three possible solutions:

1   *Inserting a damp-proof course:* The construction - two skins of stonework with a rubble core – makes it impracticable, if not impossible, to insert a traditional felt or lead damp-proof course into external walls in the New Town. There are proprietary methods which claim to alleviate rising damp either by chemical fluid or grout injection or by electro-osmosis (neutralizing or reversing the positive potential of the wall relative to the earth using an electric charge). None of these methods can be recommended without reservation because installation may disrupt the integrity of the wall or damage stone arrises and, once carried out, the work is non-reversible. It may not be possible to ensure full penetration by chemical injection into rubble walls of composite construction containing voids and the chemicals themselves may break down under continuous damp conditions, while the claims for electro-osmotic systems have yet to be fully proven. A third system uses high-capillarity ceramic tubes; there is little evidence to suggest that this method works at all and it has been more or less abandoned.

While techniques are continually being developed and studied, it is wise to obtain impartial advice before specifying any one proprietary system of damp-proofing. Waterproofing agents can never provide a permanent barrier to the passage of moisture, and only hinder the natural drying-out process. To be effective, damp-proofing has to be thorough and must protect all floors and walls against penetrating moisture. Damp which is prevented by a damp-proof course from rising in basement walls will be diverted to unsealed floors or other weak points in the construction. Where basements are inhabited and properly ventilated and heated, rising damp is rarely a serious problem.

2   *Hindering the access of moisture to the wall:* This is often the most effective remedy and involves digging out earth which has been banked against a wall, and clearing away trees and shrubs which have been planted too near the building. A dry area along the base of the wall, formed by lowering the water table with a rubble drain, will assist in most cases. Paving adjacent to the building may need to be repointed or relaid to new falls.

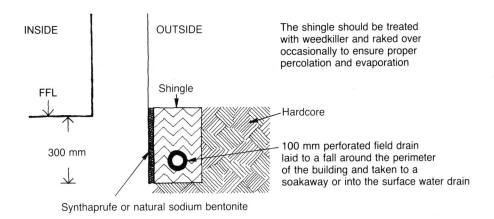

INSIDE

OUTSIDE

The shingle should be treated with weedkiller and raked over occasionally to ensure proper percolation and evaporation

FFL

Shingle

Hardcore

300 mm

100 mm perforated field drain laid to a fall around the perimeter of the building and taken to a soakaway or into the surface water drain

Synthaprufe or natural sodium bentonite

3   *Hiding the visual symptoms:* Concealment of damp areas is not recommended, but may be acceptable occasionally, if the enclosed space is well ventilated. A plasterboard lining on battens is the simplest form of concealment, but all battens must be pressure impregnated with a non-staining, inodorous preservative and both the wall surface and the back of the lining should be treated with fungicide. Alternatively a barrier, such as corrugated pitch or bitumen lathing, can be fixed to the stonework using stainless steel fixings and plastered over. It should be held approximately 50 mm away from the wall with good ventilation introduced at both the top and the bottom of the cavity. No timber strapping should be in direct contact with damp masonry.

Generally any method which introduces a waterproof material into the construction must be treated with great caution because it will inhibit the breathability upon which a traditional building depends. Where plaster is on the hard, proprietary waterproof cement paints or plaster will hold back considerable penetrating and rising dampness, provided that the background is well wetted before the plaster is applied so that it achieves good adhesion; most normal modern finishing plasters can be used on top of the waterproof undercoat. However, this method can force rising damp further up the treated wall, causing a tide mark above the level of the new plaster. It may also cause the breakdown of mortar and stonework if the stone is in a permanently wet condition, this only becoming evident at a much later date (see PLASTERWORK).

Once rising damp has been eradicated and the internal wall surface is dry, it should be replastered to a height of about 300 mm above the level of the old damp-affected plaster. If possible, replastering should be postponed for up to six months before renewal, to allow all the harmful soluble salts to be absorbed by the old plaster. Traditional lime plaster is most suitable for areas which are vulnerable to slight dampness.

EXTRANEOUS CAUSES OF DAMP

These can to be due to plumbing leaks, faulty radiator valves, washing machine floods and so on. The consequences can be as severe as with other sources of dampness and dry rot is just as likely to occur if the faults are not remedied quickly and correctly.

TIMBER DECAY

INTRODUCTION

This chapter is concerned with damage due to rot (decay) and insect attack. Its purpose is not to teach individuals how to eradicate decay but to guide building owners and their advisers in what can be a complicated and costly area of building maintenance.

Timbers vary, according to species, in their susceptibility to decay and insect attack. Many timbers, such as tropical hardwoods, are durable while home-grown softwoods are far less so. Whatever the species, timber is not normally at risk if it is kept dry, i.e. below about 20 per cent moisture content. As a result, it is very important that every building should be kept water-tight and well ventilated, both to prevent decay starting in the first instance and, once decay has occurred, to reduce its spread and allow the fabric to dry out as quickly as possible. Other factors which can contribute to timber deterioration include wear and tear, over-heating and chemical decomposition.

The initial survey and subsequent diagnosis of decay and dampness should be as accurate and detailed as practically possible in each situation. Although many specialist contractors will normally provide free inspections and estimates for visible decay, there may be a case for employing an independent professional consultant, possibly opening up some of the fabric in order to quantify any decay as accurately as possible. Because of the way that Georgian and Victorian houses were built, decay can often exist or spread to areas which are normally unseen. If eradication works are commenced without having a reasonably reliable quantification, there is often a risk of significant increases in costs.

The precise manner in which areas of decay are treated is likely to depend on a combination of circumstances. These are likely to include the nature and extent of the decay, the involvement of other parties (such as in flats, terraced or semi-detached buildings), the needs of the building occupier or owner and the original cause of the decay. The traditional methods of attending to decay involve expensive stripping out and replacement of timbers and adjacent material, together with treatment of surrounding areas. Alternative methods of treatment are now being advocated, but these may not be appropriate in every instance. Such alternative methods can involve a combination of assessing and repairing structural damage while retaining *in situ* as much timber as possible, curing the original cause of damp ingress and monitoring the extent of that dampness and drying out. These methods often also involve the use of fewer, and less hazardous, preservatives. Examples of these include borates (in solid or liquid form) and fungicides incorporated into water-based emulsions rather than hydrocarbon (petroleum) solvents. There are continual improvements in the types and effectiveness of fungicides and, as a result, the treatment is likely to be most effective if carried out by a company specializing in the use of such preservatives. Preservatives should be either an insecticide or a fungicide as required and not dual-purpose formulations.

When fungus is deprived of moisture it will die, and where this has happened it should not be necessary to remove any timber unless it has been structurally weakened. In historic buildings, the fabric should only be disturbed if the fungus is active, and the presumption that timbers are sound should be held until proved wrong. A great deal of unnecessary damage has been done because of exaggerated fears about the spread of rot, but it is important that the moisture content of areas where there has been active rot in the past should be carefully monitored.

Consultants should prepare detailed documents for competitive tenders. These documents should specify the extent and type of treatment and the quality of replacement materials.

Following treatments, guarantees are normally sought from the specialist company: it is in the owner's interest to seek an insurance-backed guarantee in case the company ceases to trade during the guarantee period. Alternatively, the chemical manufacturer may operate a guarantee scheme. Guarantees are always limited to the areas which have been treated; they should, of course, cover the cost of retreatment and not just the chemicals required. Contractors should not be selected just on the basis of back-up guarantees but on the basis of survey reports, specification, personal recommendation and so on.

Joinery and carpentry sub-contractors working for eradication companies can be employed to first-fix replacement structural timbers but proprietors are advised to appoint their own joiners to fit exposed and decorative finishes.

INSPECTION

Inspection for decay and insect infestation should normally commence on the outside of a building to ascertain signs of dampness or areas of risk. The initial internal inspection is also likely to be a visual one unless inspection panels or other access points are readily available. The most susceptible areas in an old building are:

- Roof timbers adjacent to wallhead or valley gutters;
- Panelling or lath and plaster linings to external walls, particularly around window and door openings;
- Timbers built into any external stone wall (for example, lintels over windows and doors, floor joists, dooks and bilgates);
- Timber floor construction in basements or rooms where there is inadequate damp-proofing;
- Timbers near to plumbing works, leaking pipes, showers, rainwater outlets, sanitary fittings and so on.

If an area of decay is suspected, it is advisable to carry out a more detailed inspection of the surrounding fabric either by opening up in the traditional way or by making use of modern technology in the form of fibre optics:

- To establish the type of decay; certain forms of decays can look fairly similar but may require different methods of treatment;
- To establish the extent of any decay;
- To help quantify the type, extent and cost of appropriate treatments.

IDENTIFICATION OF DECAY

Dry rot

Unlike most other fungi which destroy timber, those which produce strands have the ability to grow over inert or non-nutrient material for some distance. *Serpula lacrymans* (dry rot) produces strands that can grow vigorously, penetrating thick walls and mortar in order to reach timber, and it can frequently pass from skirting board to ceiling laths by growing upwards between plaster and masonry. Dry rot is commonly associated with damp masonry or ceiling plaster. This may be because the rot needs the calcium in mortar and plaster in order to neutralize the oxalic acid it produces when it breaks down the cellulose in timber. Dry rot can be identified as follows:

- In damp, humid conditions, the mycelia may well appear snowy white, rather like cotton wool. In dry environments, however, the mycelia may occur as thin silver-grey sheets;
- If it is in an advanced state, red rust spore dust may be detected signifying a sporophore (fruiting body) which may or may not be visible;
- There may be water-carrying strands or rhizomorphs at the edge of the mycelia, these may be as thick as 6 mm and are grey/mushroom coloured. There may also be a musty unpleasant smell, particularly when an area of dry rot is first opened up.

Wet rots

Wet rots are normally associated with saturated wood, the optimum moisture content for decay being in the region of 40-50 per cent. It most commonly occurs where timber is in close contact with damp masonry, for instance at the underside of wall plates, beneath window cills, at the base of external door frames, built-in joist ends and so on. Timber joists in contact with a damp solum are also prone to wet rot attack.

The most common wet rot fungus is *Coniophora puteana* (Cellar rot) although there are other wet rots, and correct diagnosis is important. Wet rots can be identified as follows:

- The wet rot fungus, *Coniophora puteana*, appears on timber as very fine, dark brown or black strands in a vein-like pattern. The affected timber suffers loss of strength and weight, discolouration to dark brown, and small cuboidal cracking;
- Another form of wet rot is the white rot fungus, *Phellinius contiguus*, which has the effect of bleaching the wood by the action of its cellulose and ligninase enzymes, so that the wood develops a stringy appearance, with an attendant loss of weight and strength;
- Mine fungus, *Fibroporia vaillantii*, has brilliant white strands;
- Ink Cap, *Coprinus domesticus*, has ginger strands and is found only in very wet situations on damp brickwork or plaster.

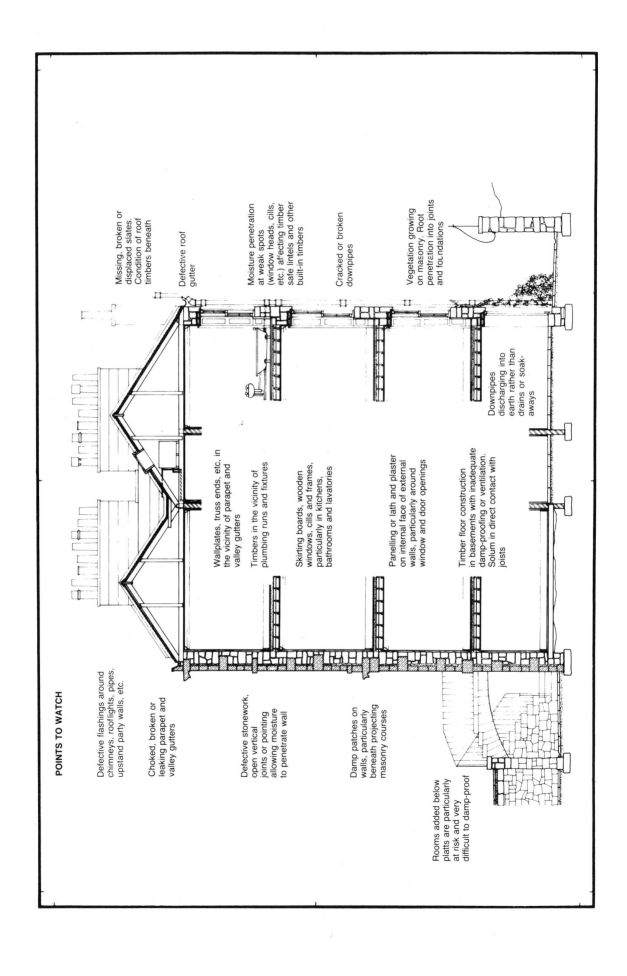

POINTS TO WATCH

Defective flashings around chimneys, rooflights, pipes, upstand party walls, etc.

Choked, broken or leaking parapet and valley gutters

Defective stonework, open vertical joints or pointing allowing moisture to penetrate wall

Damp patches on walls, particularly beneath projecting masonry courses

Rooms added below platts are particularly at risk and very difficult to damp-proof

Missing, broken or displaced slates. Condition of roof timbers beneath

Defective roof gutter

Moisture penetration at weak spots (window heads, cills, etc.) affecting timber safe lintels and other built-in timbers

Cracked or broken downpipes

Vegetation growing on masonry. Root penetration into joints and foundations

Wallplates, truss ends, etc, in the vicinity of parapet and valley gutters

Timbers in the vicinity of plumbing runs and fixtures

Skirting boards, wooden windows, cills and frames, particularly in kitchens, bathrooms and lavatories

Panelling or lath and plaster on internal face of external walls, particularly around window and door openings

Timber floor construction in basements with inadequate damp-proofing or ventilation. Solum in direct contact with joists

Downpipes discharging into earth rather than drains or soak-aways

Insect attack

Deathwatch beetle, the scourge of historic timbers south of the border, is almost unknown in Scotland. The principal timber-attacking insects encountered in Scotland are the common furniture beetle and wood-boring weevils. Other insect infestation is often noted but not all require treatment and so, in every instance, correct identification is essential. Evidence of insect infestation is often found in areas of fungal decay of timbers.

The common furniture beetle is the most frequent cause of insect damage to timber. The insect attacks not only furniture but also structural timbers, joinery and, in particular, decayed wood. Infestation also occurs where timber is often damp, e.g. timbers built into stonework, roof timbers, beneath cisterns or faulty coverings, rafter ends, floorboards near to external doors and windows.

Insect attack is normally confined to sapwood, especially at the edges of floorboards, although, if decayed wood becomes affected, infestation of the heartwood may occur. Attack by the common furniture beetle can be identified by a number of small holes approximately 1.6 mm diameter, although these may not be seen until the completion of at least one life cycle. Normally, the life cycle is three years and each adult female can lay 30-60 eggs, hence the need to carry out eradication treatment as quickly as possible.

TREATMENTS

Advice on specialist firms and contractors, guarantees and insurances can be obtained from the ENTCC.

Treatment of dry rot

Traditional treatments for dry rot are likely to involve the following broad course of action:

1 Remove the original cause of fungal attack by eliminating dampness and ensuring that the damp and decayed area is well ventilated.
2 Remove all structurally weakened timber, cutting away timber within a radius of approximately 500 mm from the last visible evidence of fungus or rot. This will involve removing laths, plaster, linings and so on.
3 Clean and treat the exposed surfaces of masonry with an appropriate biocide. The type of treatment (surface, toxic box or full irrigation) is likely to depend on the individual circumstances. Full irrigation of masonry walls is not as common as it used to be.
4 Replacement timber should be treated either by brushing with fungicide on site or under pressure off site. The treated timber should not be placed in contact with masonry or old timber which has been affected by damp or decay, nor should it be built into damp walls or wrapped in impervious felt. If possible, replacement timbers should be supported on joist hangers, or on exposed wallplates supported by stone corbels or non-ferrous brackets.
5 Treat all remaining sound timbers which are at risk with a minimum of two full brush coats of fungicide to reduce the likelihood of future infestation.
6 Provide permanent ventilation and damp-proof barriers to prevent a recurrence.

Alternative solutions for the treatment of dry rot are likely to involve, in broader terms, the following:

1 Making the building water-tight, and ensuring that it stays that way.
2 Improving ventilation to promote drying out of the existing structure.
3 Making the building structurally safe.
4 Carrying out more localized treatments and adopting a very regular and careful approach to monitoring levels of dampness and decay in the critical areas.

These alternative solutions, which minimize the use of fungicide, tend to require more specialist advice and input and may not be appropriate in every instance.

Treatment of wet rot

1 The source of dampness should be eliminated.
2 In mild cases, localized treatment of the rotting wood with fungicidal preservative will suffice to prevent further deterioration.
3 In more serious cases, remove all decayed timbers and replace with preservative-treated timber.
4 If the timbers adjacent to the affected area are very damp, they should be brushed with preservative to protect against decay during the drying-out period.
5 Permanent ventilation should be provided, together with damp-proof barriers, to prevent recurrence.

Treatment of common furniture beetle

Where infestation is severe, it may be necessary to replace infected timber and treat all new timbers with a suitable preservative prior to installation. However, in most cases *in situ* treatment by pressure spraying with insecticide is sufficient.

MAINTENANCE

The most effective safeguards against timber decay are the correct environment and appropriate techniques of timber preservation. Excessive moisture and dampness must be eliminated by ensuring that damp-proof courses, damp-proof membranes, vapour barriers, pipework and rainwater goods, flashings, roof coverings and falls, expansion joints, ventilation holes and so on, are correctly maintained and that timber and metalwork exposed to the weather are regularly painted.

Exceptional buildings may justify the permanent installation of a moisture monitoring system, such as at the Royal Pavilion, Brighton; this is particularly relevant if the building has been saturated, for example, by fire-fighting. However, less sophisticated moisture monitoring systems may be appropriate in houses where there is a higher than normal risk of water ingress into inaccessible or rarely inspected areas.

FIREPLACES

INTRODUCTION

The fireplace, burning wood or coal, was the focal point of the Georgian room and the only source of heat. Basically, the fireplace consists of a structural opening within which is placed the grate and in front of which sits the hearth stone. The opening is framed by a chimneypiece and occasionally an overmantle.

There are a variety of grates ranging from simple freestanding firedogs to built-in register grates or kitchen ranges. Chimneypieces also vary in range from simple stone to elaborately carved marble or wood surrounds.

Open fires have now been superseded in most buildings by central heating, gas fires or radiant heaters; Clean Air Zones are not really compatible with solid fuels. However, technologies come and go; gas and oil may run out and open fires could still be an essential part of our homes; the fireplace should not be thoughtlessly discarded.

CHIMNEYPIECES

The chimneypiece was the most important decorative element in the Georgian interior. The expense and elaboration of the chimneypiece was always related to the importance and function of the room, but it was very rare for an immense amount of money to be spent, unlike in London.

The most important chimneypiece in a New Town house was usually in the drawing room. In the best houses the chimneypiece would be designed by the architect himself, for example Robert Adam's chimneypieces at No. 8 Queen Street, built for Baron Orde in 1770 (now the Royal College of Physicians).

Adam, like many of his contemporaries, favoured white statuary marble for his elaborately ornamented architectural chimneypieces. Less important rooms would have had chimneypieces made of softwood such as yellow pine, painted and carved or enriched with *stucco* or *composition* (see PLASTERWORK). Even cheaper were chimneypieces with stone slips which were, of course, painted in imitation of marble. Mirrors were often placed over chimneypieces from the 1790s onwards.

Composition imitations of Adamesque chimneypieces were common in the New Town and a few at least seem to have been made in Edinburgh. Nothing is known about their manufacturer at present, but their distinctive Scottish character is betrayed by the few which depict Nelson's monument on Calton Hill on their central tablets. Nautical designs, depicting coral and shells, were also popular, such as those in Gayfield Square and Union Street.

In later Georgian houses in the New Town, chimneypieces followed a routine pattern, now conforming to the Grecian rather than the Roman orders of architecture. Both the front and the back drawing rooms had matching chimneypieces. The most popular forms were of marble; white Italian statuary marble, with yellow, sienna or verde antique inlays, was the most expensive, clouded white or other coloured marbles were cheaper. Black marble was favoured for dining rooms, and the variety of figuring and fossils is notable.

In the later New Town, composition chimneypieces were usually only found in bedrooms, and almost always painted to imitate marble. However, during the twentieth century many have been moved into drawing rooms because they are so pretty.

In the latter half of the nineteenth century iron chimneypieces were cast in one piece with the grates. After 1860, timber, iron and ceramic tiles were combined in a wide range of designs, to which brass trim and brass or copper smoke hoods were added.

GRATES

Firedogs

In its early form a fireplace consisted of stone cheeks and back, and a pair of cast iron firedogs or andirons which supported the ends of the burning logs above the stone hearth. A cast iron fireback, decorated with reeding or a low-relief design, was placed against the wall to reflect the heat into the room. In the eighteenth century when coal became the standard fuel, many New Town households retained at least one pair of firedogs for occasional wood-burning and these might have carried early coal-burning iron fire baskets, the forerunners of the elegant Georgian dog grate.

Register grate

Register grate

Dog gate

Hob gate

232

Photo: Inglis Stevens

Adam type chimneypiece
Marble with later dog grate *c.* 1796
The Georgian House, 7 Charlotte Square

Timber chimneypiece with cast iron
register gate *c.* 1810
Gayfield Square

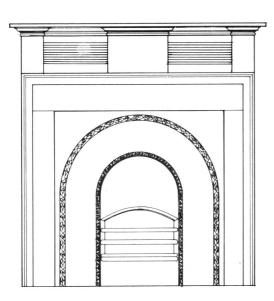

Typical bedroom fireplace
Stone surround with timber bolection moulding and mantel
shelf, and semi-circular cast iron register grate *c.* 1828
Dundonald Street

Dog grates

A dog grate is basically a free-standing firebasket with vestigial andirons as legs, standing in the hearth against a plain cast iron fire back. Dog grates were manufactured in wrought iron, cast iron or bright steel, sometimes with brass or bronze frets and mouldings. The Carron Foundry, near Falkirk, in which John Adam was a partner, designed and produced many fine dog grates which found their way into the hearths of early New Town houses. The Howarth Brothers were responsible for many of the Carron designs, including those for hob grates which came into use later in the eighteenth century.

Hob grates

Unlike the dog grate, the hob grate was built into the fireplace and it fitted neatly into the square recess of the opening, the tolerance being taken up with slate, stone or marble slips. Although the hob grate stretched right across the fireplace opening, the actual firebasket occupied only the middle third and was flanked on either side by metal flat-topped hobs. The cheeks of the fireplace above the hobs were usually plastered, although in some cases iron plates or Delftware tiles were used to line the recess. There are three common types of hob grate – the double semi-circle, the double ogee and the rectangular. These are named according to the shape of the ornamental front and were made of cast iron with bars of wrought iron. The fire, being raised high above the hearth, did not heat the air at floor level, and the wide flue opening created down-draughts which tended to blow smoke back into the room.

Register grates

The all-in-one register grate became popular during the later development of the New Town. The register grate completely filled the fireplace opening and its cast iron or bright-steel structure often formed the throating (with a device regulating draught flow) as well as the firebasket. Register grates had ornamental features or moulded edges in brightwork, polished with abrasives, such as sand, emery cloth or steel wool, to catch the light from the fire. Nineteenth century fireplaces often had decorative cast iron arched openings, and were black-leaded and polished daily. Technical improvements, devised by Count Rumford in 1797, were introduced gradually during the nineteenth century, and included the use of a non-conducting material such as firebrick which reflected heat forward into the room, although this did not come into general use until mid-century. There are numerous examples of Victorian grates in the New Town.

Kitchen ranges

Until the early nineteenth century, open-fire ranges, usually set into a plain stone chimneypiece with separate bread oven alongside, were installed in New Town kitchens. However, it was only a short step from the hob grate to the closed kitchen range patented in 1802. This began as a grate with an oven and grew in ingenuity and complexity through the nineteenth century. Usually it contained a small raised fire basket in the centre, with an ash pan beneath, an oven on one or both sides and a hot plate above. There might also be a small boiler for hot water.

Stoves

Used in Britain from the mid-eighteenth century, the slow-combustion stove was an early form of central heating which originated in northern Europe. It was made in cast iron, with a base containing the fire and a decorative urn or some sort of statuary on top. In the New Town a stove would be placed in the lobby, either freestanding at the bottom of a stairwell or in a specially designed niche, so that it would warm the core of a house. A later version, described in the inventory for 3 Moray Place, gave out hot air piped directly from the kitchen range. The stove was probably only used on special occasions such as when entertaining guests.

HEARTHS

Hearths were generally constructed of one large slab, up to 100 mm thick, of stone such as Caithness, Hailes or Arbroath, bedded on lime mortar mixed with ash and laid on boards over joists; they were set flush with, or just proud of, the floor boarding. Marble and slate hearths are generally later replacements.

Accessories

A number of accessories sit on the hearth. Fire-guards hooked to the top of the grate were widely used. Pierced steel or brass fenders came into use at the end of the eighteenth century, often similar in design to the grate. Until stands began to be made in the nineteenth century, fire-irons – the shovel, tongs and poker – were propped up against the side of the fire opening or against the fender. Coal and wood for burning was brought into the room by servants when needed.

If a fireplace was not in use a vase of flowers or a decorative chimney board might be placed in the empty hearth; very few chimney boards survive. Cast iron shutters or a hinged flap were also used to seal off the flue.

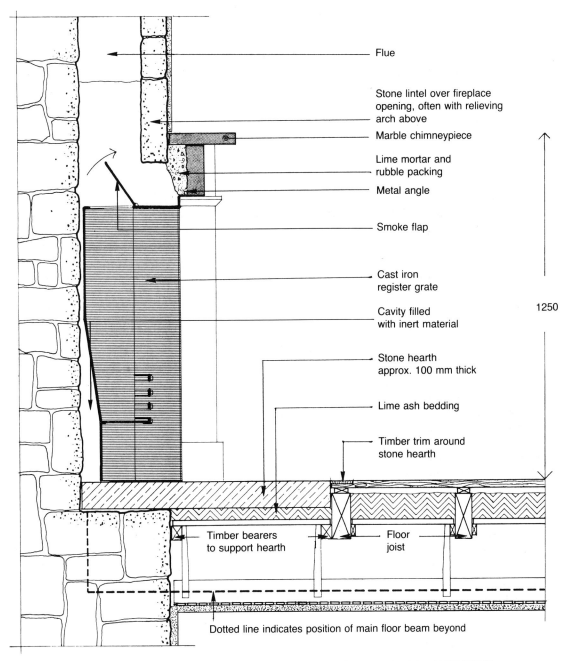

Flue

Stone lintel over fireplace
opening, often with relieving
arch above

Marble chimneypiece

Lime mortar and
rubble packing

Metal angle

Smoke flap

Cast iron
register grate

Cavity filled
with inert material

1250

Stone hearth
approx. 100 mm thick

Lime ash bedding

Timber trim around
stone hearth

Timber bearers
to support hearth

Floor
joist

Dotted line indicates position of main floor beam beyond

SECTION THROUGH FIREPLACE AT 27 CLARENCE STREET, SHOWING ORIGINAL CONSTRUCTION

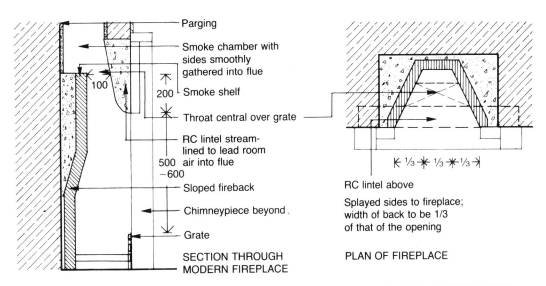

Parging

Smoke chamber with
sides smoothly
gathered into flue

Smoke shelf

Throat central over grate

RC lintel stream-
lined to lead room
air into flue

Sloped fireback

Chimneypiece beyond

Grate

100

200

500
−600

SECTION THROUGH
MODERN FIREPLACE

⅓ ⅓ ⅓

RC lintel above

Splayed sides to fireplace;
width of back to be 1/3
of that of the opening

PLAN OF FIREPLACE

COUNT RUMFORD'S PRINCIPLES, SETTING OUT THE IDEAL ARRANGEMENT TO AVOID SMOKING FLUES

DEFECTS AND REMEDIES

Smoke blowing back down the chimney into the room can be a problem with some old fireplaces with dog grates or fire baskets, large openings and wide flues. This defect was first studied by Count Rumford, who suggested that the fire would burn better if the width of the opening was reduced, the fireback brought forward, the sides splayed and the throating reduced in width; these principles are illustrated in the plan and section of a modern fireplace, which should be compared with a typical New Town fireplace. Other remedies are to raise the base of the fire, fit a canopy, raise the chimney or fit a higher can, and to improve the air supply by fitting ventilators above the door of the room or, if the floor is ventilated, in the floor boards near the hearth.

Marble chimneypieces were made of flat pieces of marble braced with marble off-cuts, set in plaster and built up to form a box section. The chimneypiece was held in position by steel wires set in the plaster and fixed to dooks (wooden plugs) in the wall, with two or three wires to each jamb or lintel. This has given rise to problems, because wire can rust and stain white or light-coloured marble, and in extreme cases the chimneypiece can become detached from the wall. Today the usual practice is to use copper wire with a bent or curled end to hold the chimneypiece more securely, or to use purpose-made brackets. The final coat of plaster would have been put on the wall around the chimneypiece once it was in position.

Repairs to timber or marble chimneypieces or to iron grates should always be undertaken by a specialist. A list of suitable firms is available from the office of the ENTCC. If any part of a chimneypiece breaks off, the pieces should be kept, so that they can be repaired or copied. Chimneypieces made of composition are likely to dissolve if an inappropriate paint-stripper is used on them.

Despite its appearance, marble is a fragile and porous material; it is easily stained by rust, verdigris, mould, wine, paints or cosmetics. It will distort in heat and fracture or star if heavy objects are dropped on it. Unpolished marble will become grimy and greasy from handling and sooty and tarry from smoke and heat. Acid, pumice, soda, alkaline soaps and abrasive cleaners are sometimes inadvisedly used to clean marble and can have a damaging effect. Atmospheric pollution and high relative humidity will also attack marble. Advice on the cleaning of marble varies a great deal, but because of its susceptibility to damage, caution is always advised. Do-it-yourself treatments are not generally recommended but, if attempted, a small inconspicuous area should be tested first. The ENTCC has further information on cleaning methods, and can provide a list of specialist contractors able to undertake marble cleaning and repair. Slate can be washed using a neutral pH soap.

Cracked fireclay bricks can be repaired with fireclay cement. A defective hearth poses a more serious problem, however. Corrugated metal may have been introduced to support a three-slab replacement hearth or to act as shuttering for *in situ* concrete. If the hearth is opened up the metal should be lifted to ensure that the underlying timbers are structurally sound and are not exposed to damage by open fires. Some modern high-rate solid fuels burn at too high a temperature for traditional fireplaces and may cause charring or even fire in the support timbers. In many cases the main timber beam passes beneath the hearthstone and is housed into the chimney breast; there is always a danger that heat from the hearth will ignite this beam and therefore, if for no other reason, the old raised grate or basket should still be used for open fires.

Too often, the character of a room is destroyed when a fireplace is blocked up and the chimneypiece removed during modernization. If a modern heating appliance is installed, it should be done in such a way that at a later date it is possible to reverse the situation. It should also be remembered that the fireplace is an essential source of ventilation which helps to prevent condensation.

SERVICES

INTRODUCTION

Originally, New Town houses had very basic services; open fire heating and cooking, candle or oil lamp lighting, and a cold water supply fed to a lead-lined timber or stone tank in the basement. In some cases, water would have been hand pumped from a well. Hot water from a copper pot or boiler in the kitchen was carried upstairs to the bedroom in jugs and buckets.

During the nineteenth century the standard of services gradually improved, lighting by gas and later by electricity was introduced, central heating systems were installed and hot water could be piped around the house. In the twentieth century solid fuel kitchen ranges have been superseded by gas and electric cookers, and open fires are now quite rare. Since 1994, the New Town has been controlled by the Clean Air Act, and only smokeless fuel is permitted.

RENOVATION AND RENEWAL OF SERVICES

Compliance with mandatory Building Regulations and the recommendations of the Chartered Institute of Building Services Engineers (CIBSE), the Institution of Electrical Engineers (IEE) and the relevant Code of Practice will satisfy safety standards and provide comfort for the occupiers. Modern services should be installed in such a way as not to damage the building fabric or detract from the historic character of the building.

Three-quarters of all buildings in the New Town remain in residential use, and the Planning Department has consistently discouraged further changes from residential to office use. Buildings in office or commercial use, chiefly in the First New Town and West End, have suffered because of alterations to satisfy modern business practice. Large firms have left prestigious addresses in Charlotte Square because flexible accommodation can be found only in new buildings, but future innovations in communications systems may allow some firms to remain without having to disturb the historic fabric. All new installations and modifications to existing services should be carried out by competent and experienced specialists, and all alterations should be discussed in advance with the Planning Department. The advice of a specialist architect or consulting engineer can avoid unnecessary damage and disruption.

Where the premises are being refurbished as office accommodation requiring provision for computer technology, the installation of a pre-wired voice/data network should be considered, using structured cabling, which is compatible with all major computer systems. Wiring for up to four terminal outlets can be contained in a single 20 mm conduit which can be concealed below floors and within walls, thereby precluding the need for skirting or dado trunking, which are incompatible with a Georgian interior.

Removal of services

Broad pine floor boards cannot easily be lifted and relaid without damage. They are such an essential feature of fine houses in the New Town that they should only be lifted if absolutely necessary and even then, only with great care. Replacement with new narrow flooring should be avoided. Lifting boards or 'traps' were often provided when previous services were installed and these should be located and used for access where possible. If old pipes cannot easily be removed without damaging finishes, they may be left in position but closed off to prevent access to vermin. They must also be disconnected from any live distribution pipe in order to avoid leaving a section of stagnant water in which bacteria could breed.

All redundant or exposed electrical wiring should be made safe and, if possible, removed. Ideally, all gas, water and drainage pipes made of lead should be replaced with copper or PVC as appropriate; the salvage value of the lead will help pay for the new materials.

DESIGN AND LAYOUT OF NEW SERVICES

When oil or gas lamps were used, timber ducts were sometimes provided in the ceiling as ventilation passages. If these are encountered they should not be blocked or removed because they may still ventilate underfloor spaces.

The removal of redundant services can create enough space, within typical New Town floor construction, to install new services without having to damage floor joists by further notching. Disturbance to stone-flagged kitchen floors is usually damaging and costly, and services should be routed elsewhere.

237

ST STEPHEN'S CHURCH 1828
WILLIAM PLAYFAIR ARCHITECT

Photo: John Johnson

The possibility of locating all services in an easily accessible and vented common service duct, situated in the common stair, should be investigated, particularly when a comprehensive repair scheme is being carried out. Bye-laws 48 and 49 require that pipes should be insulated and protected. This can result in bulky trunking which disfigures a listed building, particularly in the internal common stair of a tenement. Ingenuity and even an official waiver may be needed to overcome this problem.

Techniques have changed so often in this century that historic buildings can be protected only if all service installations are reversible without damage.

Water supplies

Grants administered by the District Council can help with the cost of providing a wholesome supply of drinking water at the kitchen sink. The following work is covered:

- Providing a direct supply where none exists, or replacing existing lead pipework between the house main and the kitchen sink;
- Replacing lead-lined, cold-water storage tanks within the property with galvanized steel or plastic tanks;
- Replacing or by-passing a lead main pipe between the house and the boundary of the property.

Any charges made by the water authority for upgrading supply pipework from the boundary of the property to the street main will be covered by this grant.

While many kitchens and bathrooms need to be brought up to current standards, consideration should be given to the retention of fine Victorian or Edwardian sanitary fittings, many of which are still functional. Leaks undetected over a long period can cause serious damage to structural timbers and foundations. Where possible, drainage connections under floors should be examined and, if necessary, renewed. If the existing WC (or WC position) is to be retained, the floor boards beneath should be checked for timber decay.

Ventilation

Current regulations require that all bath/shower rooms and most kitchens are provided with mechanical ventilation, whether or not natural ventilation is available. The purpose is to reduce condensation problems and, although the regulations are not retrospective, provision of mechanical ventilation should be considered even in existing WCs (which are often inadequately ventilated), bathrooms and kitchens, when extensive refurbishment is being carried out. Extract ducts must be carefully and unobtrusively planned and exhausted, if possible, at the rear of the building. Many bathrooms and kitchens had an open fire and flexible ventilation ducts can sometimes be inserted in the flues to avoid damage to decorative plasterwork. Where external grilles are necessary these should be kept to the rear of the building. Modern aluminium or plastic grilles may be visually acceptable if painted to match adjacent stone. Ventilation grilles on both front or rear elevations require Listed Building Consent.

Where there is no alternative to having ventilation openings on the front facade, modern styles of grilles would be inappropriate and some imaginative thought might be necessary to provide a discharge terminal which is either concealed from view or in keeping with the facade. For the relatively low air volumes associated with domestic kitchens and bathrooms, circular or rectangular traditional-styled cast iron grilles set into the stone can be quite effective.

Space heating

A central heating system should provide constant low background heat and be installed so that pipework is unobtrusive; microbore is recommended. If the heating is really constant then the boilers can be quite small; the fabric of a typical house or tenement in the New Town will retain heat for considerable periods and it is better to avoid intermittent systems which have to warm up from cold each time, thereby increasing the risk of interior decoration being attacked by condensation. Each radiator should be fitted with a thermostatically controlled valve.

A radiant focal point can be provided by an open fire, or by a gas, electric or solid fuel appliance, carefully fitted into the existing fireplace opening; damage to the chimneypiece must be avoided. If a new space heating system is installed using an existing flue, the flue should be lined; this will avoid structural damage caused by condensation and acid attack within the flue. The diameter of the flue must be checked against the minimum requirements of the fire or boiler to be installed. Although hot-air central heating is efficient and unobtrusive, it is not recommended for historic buildings, because the excessively dry atmosphere can cause damage to joinery and plasterwork. Excessive ambient or localized temperatures can damage the building fabric, particularly joinery, and so the location and output of heaters and boiler vents should be very carefully considered.

Compared with other appliances, open solid fuel fires are relatively inefficient. For the best results, the fire should have a deep flue bed and controllable air supply, and the fire should be maintained in good condition (see FIREPLACES).

Rooms containing appliances which burn gas, oil or solid fuel should be adequately ventilated to avoid the release of toxic gases, prevent oxygen starvation and ensure complete combustion. (See SASH WINDOWS for further guidance on ventilation.)

Gas supplies

In 1984-5 Scottish Gas carried out an extensive renewal programme of all gas mains and supply pipes in the New Town. The following guidelines were then issued by the Planning Department for installations in Conservation Areas and listed buildings:

1   A maximum of 450 mm of supply pipe to be visible on the front wall of buildings.
2   External pipes which are both horizontal and vertical to have the horizontal section within the basement areas where it will not be visible from the street. This is to ensure that horizontal pipes are kept as low as possible on the facades.
3   Meter boxes to be fitted in locations which are not visible from the public footpath.
4   Pipework and meter boxes to be painted in colours which match the adjacent stone.
    Adjacent stone paint specifications:
    Grey BS 10 A 11
    Normal BS 08 B 21 / 10 B 21 (50/50 mix)
    Dark BS 08 B 25 / 10 B 25 (50/50 mix)
5   All external holes to be marked prior to drilling, and the Planning Office notified of locations.
6   Stone ledges not to be cut unless unavoidable and the stone reinstated to match the original.
7   Holes not to be drilled through stone joints, especially 'V' jointing.
8   Cornices and other decorative plasterwork not to be cut unless unavoidable. Plasterwork to be reinstated to match the original.
9   All redundant surface-run pipework to be removed, the surfaces made good and painted to match existing materials and colour.
10 Setts which are damaged to be replaced by stones of similar size, colour and material (i.e. whin or granite). This must be completed within one year of the installation of new pipes.

Installation of cables for TV and communications requires similar guidelines if damage or disfigurement is to be avoided.

Electrical services

The location of switch gear, meters and incoming services should be unobtrusive but accessible. Meter and consumer units for electrical distribution should be contained in existing cupboards. It is very seldom visually satisfactory to locate these in hallways or corridors where a cupboard has to be constructed to conceal them. Existing cupboards often allow access to the floor above and below for wiring, without the risk of damaging skirting boards and cornices.

To meet the demand for more outlets in each room, mineral insulated copper cable (MICC) or PVC-sheathed twin and earth cabling should be used for power circuits and, where possible, the existing conduit system should be used for lighting circuits.

PVC-sheathed twin and earth cabling can often be installed without disturbing a wall. If MICC is installed behind lath and plaster walls, a PVC sheath should be used to give protection against chemical attack. Rewireable fuses should be replaced with miniature circuit breakers (MCB's) in order to provide greater electrical protection. The electrical installation can be further protected against earth faults by installing a residual current circuit device (RCCD) (previously known as earth leakage circuit breakers). It is recommended that split consumer units are used so that a fault on any of the power circuits will not automatically switch off the lights. RCCDs are available with different disconnection times; a 30 milli-amp-rated device will protect against both shock and fire. Earthing and bonding should be checked against the current recommendations of the IEE.

In many existing buildings the light switch may still be located within the bathroom. This should be replaced with a pull cord switch or repositioned outside the bathroom.

Television and communications

In the interests of visual amenity and to avoid damage to chimneys and roofs, consideration should be given to the replacement of individual television aerials by communal systems, especially when a comprehensive conservation scheme is being carried out. In most parts of the New Town, external aerials are an unnecessary liability; aerials situated in the roof space, or even set-mounted aerials are usually perfectly adequate. Where possible, television and telephone cabling should be installed as renovation work progresses, or alternatively, concealed conduit provided to an accessible vertical riser in the common stairwell.

Alarms

Burglar alarms systems are increasingly being installed in New Town houses, but they present their own problems both practically and aesthetically (see LISTED BUILDINGS AND CONSERVATION AREAS). Most systems have ugly alarm boxes with a flashing light mounted externally, usually adjacent to the front door.

Requirements for intruder alarm systems should be ascertained at an early stage so that wiring can be concealed prior to decoration. Wiring between movement detectors and the main panel can be difficult to conceal in old premises and for this reason the use of wireless systems should be considered. In any event the system should comply fully with the relevant British Standards and should be installed by a NACOSS (National Approved Council for Security Systems) approved installer.

Alarm boxes require Listed Building Consent and the Policy Guidelines published by the Planning Department advise against locating them on the front elevation but if this unavoidable then they should be inconspicuous, painted to match the background colour, and not normally located above ground floor level.

INSPECTION AND MAINTENANCE

Water supplies

Once a year, the storage tank and all water supply piping should be carefully examined for leaks and damaged or missing insulation, and the tank drained and cleaned. If the ball valve assembly is not functioning properly, the ball and washer should be replaced. Stopcocks should be located and tested to ensure that they can be shut off easily in an emergency; they should never be painted over. Tap washers should be replaced every five years. Overflow pipes from storage tanks should be at least two sizes larger than the ball valve nominal size.

When replacing or maintaining water storage, consideration should be given to upgrading the tank to 'potable water' standards (Bye-law 30). Similarly insulation, particularly on main supply pipes, should be well up to standard to avoid damage by freezing.

Space heating

Solid fuel stoves should be maintained at least annually. Stove doors, particularly those with glass fronts, and ash compartment doors should be checked to ensure that they are airtight. Refractory bricks should be repaired or replaced as necessary and flues regularly cleaned; the weighted-ball method causes damage to the flue and is not recommended.

Gas fires and *gas boilers* should be inspected and serviced once a year, preferably by a properly qualified engineer. Broken gas fire radiants should be replaced and particular attention paid to the correct functioning of thermostatic controls.

Oil-fired boilers should be serviced twice a year. Vaporizing, pot-type burners can be serviced by the occupier; atomizing-type burners should be serviced by a competent engineer. Thermostatic controls should be checked, oil storage tank filters cleaned and fire valve mechanisms tested. Boiler doors should be left open when the boiler is not in use.

Low pressure hot water systems associated with boilers should be inspected annually. All pipework and valves should be checked for leaks. The feed and expansion tanks, which should be covered to keep out dirt, and the insulation to the hot water cylinder, pipework and so on should be checked and renewed or repaired as necessary. There are special safety standards for the removal of asbestos insulation jackets.

Electric heating systems, including cables, terminals and earthings, should be inspected annually to ensure safety in use. Reflective surfaces on bar fires should be polished with a soft cloth and the interior of convector heaters cleaned. Off-peak storage heaters should be inspected and maintained by an electrician.

Electrical services

The main board and distribution wiring should be inspected every five years by a member of the Electrical Contractors' Association (ECA) of Scotland and an IEE Certificate obtained. If wiring is defective, it should be renewed to current IEE standards.

Record drawings

Detailed drawings and comprehensive operating and maintenance instructions should be provided by the engineer after any alteration or installation of services.

STEPS, PLATTS AND ARCHES

Most ground floors in the New Town are above pavement level, the common stair or main door being entered across a stone arched bridge carrying a stone platt and steps, with separate stone steps down from street level to the basement areas. Stone from Hailes Quarry, near Colinton in south-west Edinburgh, was commonly used for steps and paving. This stone is strongly stratified and is reddish-fawn, white, pink or blue in colour.

DEFECTS

Many steps and platts are badly worn and the soffit is sometimes softened and delaminated by water seeping through the joints. Rooms built as later alterations under the steps can suffer from chronic and incurable dampness. Arches and platts can be weakened by settlement of the area retaining wall (as at the west end of Fettes Row) or, less commonly, by movement in the front wall of the building. If the arch spreads, the joints between voussoir stones can open so that with increased erosion the structure becomes unstable. Rebuilding of the arch may be required; this is an expensive procedure requiring high quality masonry work and accurate setting out. Patch repairs to voussoirs are not recommended.

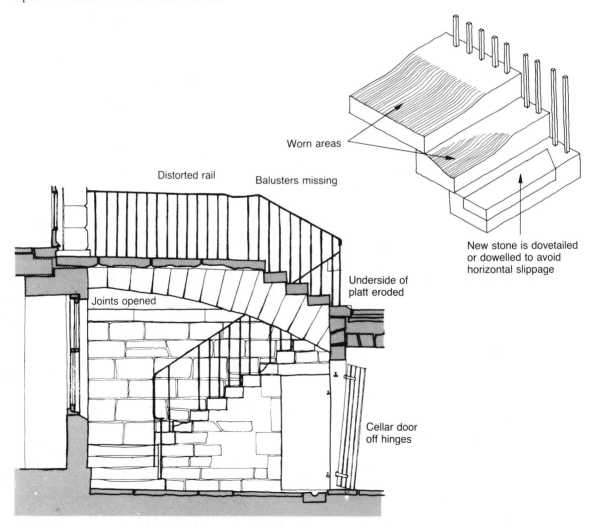

Worn areas

New stone is dovetailed or dowelled to avoid horizontal slippage

Distorted rail

Balusters missing

Joints opened

Underside of platt eroded

Cellar door off hinges

STEPS, PLATT AND ARCH DAMAGED BY MOVEMENT

REPAIRS

In old buildings, signs of wear are to be expected, and the greatest wear occurs at the entrances to common stairs. Minor damage is best left undisturbed, unless it affects the stability of the railings or masonry. Worn treads and damaged nosings can be ignored unless they constitute a hazard. Localized areas of wear or damage can be repaired by cutting back the stone to a sound base and inserting a new stone indent, bedded in lime mortar. The indent can either be dovetailed-in or secured with stainless steel dowels. All paving stones become worn and it may be necessary to dress and buff a new indented stone to match the adjacent worn profile.

More extensive damage can be remedied by renewing whole steps, although not all quarries are able to supply stones more than 2 m long. Matching the original stone colour is much less important than matching weathering characteristics; replacement steps and paving will very quickly go dark because of algal growth. Hailes Quarry stopped quarrying stone around 1931 and it is difficult to match even with stones of similar colour and texture from other quarries, but stone from Stainton or Dunhouse quarries, both in County Durham, or Clashach stone, from a quarry near Hopeman on the Moray Firth coast, should be considered.

The conspicuous position and frequent use of the steps and platts means that cheaper substitute materials are unsatisfactory; cement screeds, reinforced concrete, thin applied flagstones and any other such repairs are all unacceptable and are unlikely to obtain Listed Building Consent. The ends of stone steps often delaminate or they may be split by rusting cast iron balusters. If the stone around the balusters has spalled off it can be built up with a suitable mortar repair but it is better to cut the stone back to receive a stone indent fixed with stainless steel dowels. The baluster should be reset in a new hole filled with molten lead, and securely staved. On no account should the bases of the balusters be secured by a concrete haunching built up along the sides of the steps and platt.

Water penetration can cause considerable damage to rooms inserted below the platt, but it is impossible to waterproof butt joints between platt stones. Sealants can provide a temporary repair, and are best if used to limit the amount of seepage between each step. The majority of sealants now used are silicone based, replacing the polysulphides and butyl sealants. All sealants must be applied to clean, dry, dust-free joints. Joints must therefore be raked out and then primed in accordance with the sealant manufacturer's instructions.

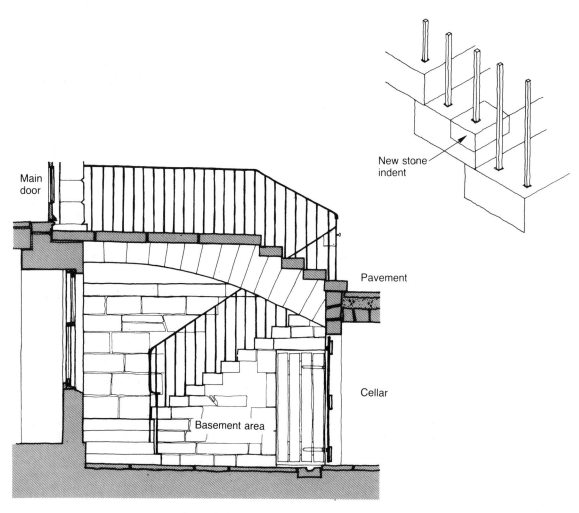

New stone indent

Main door

Pavement

Cellar

Basement area

STEPS, PLATT AND ARCH IN CORRECT POSTION

CELLARS AND BASEMENT AREAS

INTRODUCTION

Cellar floors and basement areas were usually built at the original ground level (see GEOLOGY AND TOP-OGRAPHY). Coal cellars provide useful, but often neglected and damp, external storage space at basement level in the barrel-vaulted construction which supports the pavement above. In the New Town, there are occasionally buildings with cellars at basement and sub-basement levels as, for example, at Clarence Street.

No timber should touch the damp masonry; for this reason, cellar doors were hung on crook-and-band hinges built directly into the wall, avoiding the need for a timber frame. Stone lintels are used to support the masonry above the door; timber was not used because it is too vulnerable in this location. Doors are braced and battened, with plenty of space top and bottom, and between the boards, to provide ventilation.

Basement areas are usually paved with flagstones about 600 mm × 400 mm × 75 mm thick, laid with the shorter side parallel to the street, to fall away from the building to a stone drainage channel which leads into the drainage system. The surface of the flagstones is often droved, to give a roughened surface for foot traffic. Most stones for paving and steps in the New Town were reddish-brown in colour and came from the Hailes Quarry, which is now closed.

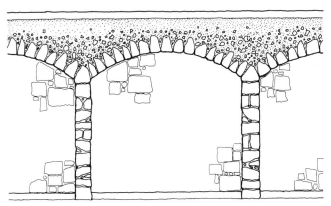

SECTION THROUGH CELLAR PARALLEL TO STREET FRONT

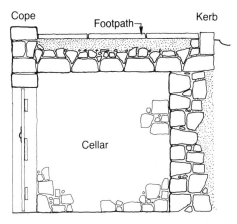

SECTION THROUGH CELLAR UNDER FOOTPATH

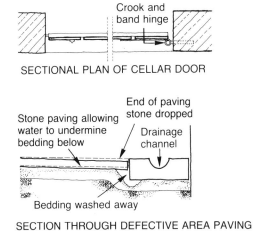

SECTIONAL PLAN OF CELLAR DOOR

SECTION THROUGH DEFECTIVE AREA PAVING

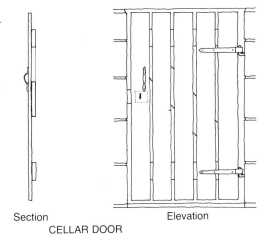

Section   Elevation

CELLAR DOOR

INSPECTION AND MAINTENANCE

Although maintenance of the pavement is the responsibility of the Regional Council's Highways Department, the owner is responsible for the structural stability of the cellar below. Cellars should be included in an architect's quinquennial inspection. Any signs of rising or penetrating damp, timber decay or structural movement should be noted. Cellars must be kept as clear and as well ventilated as possible.

Some basement area walls facing the facade, and undersides of platts, have been limewashed or painted over the years. This can be visually disruptive and, while it is permitted by the Planning Department, it is a practice that should not be encouraged because it materially alters the listed building. It should be noted that traditional limewash is the least harmful coating. For recoating old limewash, traditional limewash should be used, but if the stone has already been painted with modern masonry paint, an organic silicate-based paint is recommended. The ENTCC can give details of suppliers.

Stone slabs in shady areas tend to become covered with slippery algae and moss. This can be treated regularly with algicide (see CONTROL OF VEGETATION) although the long term effects of these chemicals are not known. Drains and gulleys should be cleared of leaves and other debris.

DEFECTS AND REMEDIES

Movement in the vaults is uncommon but lack of ties between the vault and area wall often leads to separation. This is important because collapse might endanger pedestrians above.

The original stone paving flags in the street were probably laid on puddled clay, as a waterproof membrane. This is likely to have been disturbed where flags have been replaced by precast concrete slabs or when services were introduced. The Regional Council disclaims any responsibility for water entering a cellar and so waterproofing must be carried out by the owners.

Seepage makes many cellars too damp to store anything; it is very difficult to make them waterproof and seldom worth the expense. A polythene lining will give temporary protection, but fixing is difficult because the barrier is punctured. Proprietary waterproof renders can sometimes help, but these are short-lived and break down, becoming detached from the stone to which they are applied and often causing decay in the stone at the same time. Some dampness in cellars is therefore to be expected, and as long as good ventilation is provided it will not do any real harm.

Pointing of basement area flagstones is often loose or missing and rainwater running through the open joints can cause the slabs to settle by washing away the base layer. Leaking joints may exacerbate dampness in sub-basement rooms and, in exceptional circumstances, may affect the foundations.

When relaying paving, the existing stones should be marked and their positions recorded on a plan before lifting; this allows them to be relaid in the same positions, correcting any misalignment that has been causing problems. It is rarely necessary to replace the hardcore base-layer but, if it has settled or has been disturbed, a thoroughly rolled and compacted 100 mm layer of crushed stones (maximum diameter 75 mm), mixed with sand, should be used as the new hardcore base. A 50 mm layer of washed and graded whin or builder's sand should be laid directly on both new and existing hardcore and thoroughly consolidated. The paving is then laid directly on the sand with a minimum fall of 1 in 80 away from the building. Original levels should be maintained where possible. The stones should have a large bearing area to avoid cracking. Joints should be well filled and flush pointed with a 1:3 putty lime and sand or 1:2:9 cement/non-hydraulic lime/sand mortar. Pointing with strong cement mortar is discouraged because it is liable to shrink and crack, increasing the likelihood of damage when the slabs are next lifted.

If stones need to be replaced, new or preferably secondhand stone flags should be used. If paving has to be patched with a substitute material, the original stones should be relaid together and not interspersed with the new material.

Particular care needs to be taken to avoid damage to flagstones under scaffolding, and during the repair or installation of services.

RAILINGS

INTRODUCTION

Cast iron railings are a distinctive feature throughout the New Town, serving as safety barriers at pavement level and at the sides of steps and platts around sunken basement areas. The simplest railings consist of cast iron uprights, or balusters, slotted through a wrought iron top coping rail, and set with lead into a stone plinth at the base. More elaborate railings may have double coping rails, ornamental finials, decorative husks, and dwarf balusters (dog bars). Different areas of the New Town have distinctive finial patterns such as the fleur de lys of Fettes Row and the anthemion of Manor Place. Examples of various finials can be seen at the ENTCC office. In many streets, lamp standards were either an integral part of the railings (as in Melville Street) or stood proud of them on the stone plinth (see LAMP STANDARDS).

The abundance of railings in Edinburgh was a result of the expansion of the city at a time when cast iron was relatively cheap. Even during the Second World War, when many ornamental railings around communal gardens were removed (but never actually used) for munitions use, the sunken basement still had to be protected, and consequently much of the original ironwork has survived.

MAINTENANCE

Maintenance is often limited to a new coat of paint over the old one, but successive coats can obliterate much of the decorative relief. Railings should be examined annually for signs of corrosion and fractures. They should be repainted every five years, or at the first signs of rust, whichever is the sooner. The old paint should be stripped or burned off and the metal primed with two coats of red lead or zinc phosphate before undercoating and finishing with black gloss. Guidance on the preparation and painting of ironwork is given in EXTERNAL PAINTWORK.

The Planning Department recommends that railings be painted black and Listed Building Consent and planning permission are required for changing to any other colour.

DEFECTS AND REMEDIES

Damage to railings can be due to weathering and corrosion, structural movement, injudicious repairs using incompatible materials, vandalism, or impact from vehicles. Iron will corrode in the presence of water and oxygen and requires protection by painting. As long as a paint film is unbroken, the protection will be effective. Rusting will usually begin in crevices and joints where it is difficult to paint. Although cast iron is less susceptible to rusting than wrought iron and mild steel, corrosion is exacerbated by poor maintenance and unsuitable design. The main principles of rust removal and prevention are set out in EXTERNAL PAINTWORK. More elaborate precautions against rusting may be advisable if the ironwork is unusually intricate or of particular architectural or historic importance.

It is essential to identify and remedy the cause of any structural movement which has caused damage to the railings themselves. Fractures most commonly occur at the joints between baluster and coping rail and between baluster and plinth. Ideally, a broken baluster or support upright should be replaced with a new casting, but a repair by electric or oxyacetylene welding may be satisfactory, provided that a constant temperature is maintained. All welds should be continuous and ground smooth in order to eliminate any gaps and crevices. If there is more than one fracture, a new casting may be cheaper than welded repairs.

Before replacing a baluster the lead must be chiselled out from the stonework at the plinth and the broken stump of the baluster removed. A slot cut in the stone on two adjoining sides of the balusters using a narrow cope chisel will enable the lead and the baluster to be removed together. Building up the plinth in concrete to strengthen the baluster is not recommended.

CAST IRON WORK AT ROYAL TERRACE 248 Photo: John Johnson

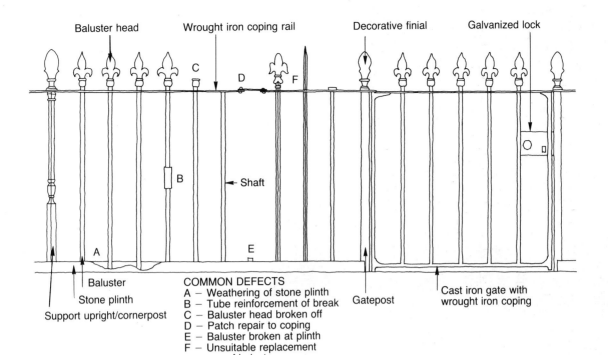

Baluster head — Wrought iron coping rail — Decorative finial — Galvanized lock

C   D   F

B   ← Shaft

A   E

Baluster
Stone plinth
Support upright/cornerpost

Gatepost

Cast iron gate with
wrought iron coping

COMMON DEFECTS
A – Weathering of stone plinth
B – Tube reinforcement of break
C – Baluster head broken off
D – Patch repair to coping
E – Baluster broken at plinth
F – Unsuitable replacement
      of balusters

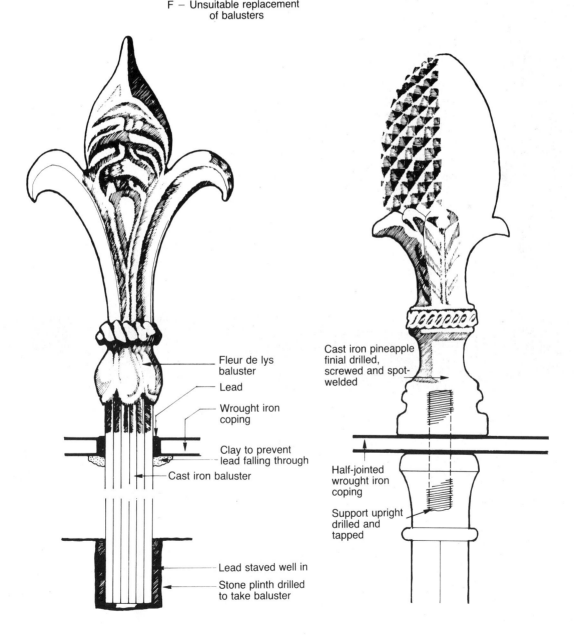

Fleur de lys
baluster

Lead

Wrought iron
coping

Clay to prevent
lead falling through

Cast iron baluster

Lead staved well in

Stone plinth drilled
to take baluster

Cast iron pineapple
finial drilled,
screwed and spot-
welded

Half-jointed
wrought iron
coping

Support upright
drilled and
tapped

Mild steel is not an acceptable material for replacement balusters, being more susceptible to rust and only available without the fluting which is a feature of many cast iron balusters. Ductile iron, a composite of steel and cast iron, can be used and is less likely to corrode than steel. It is more vandal-resistant because it is less brittle than cast iron but will, however, cost up to 15 per cent more. Ductile iron is also suitable for replacement coping rails.

Decorative finials should be fixed in the traditional manner, by tapping and screwing. Spot welding will prevent the finial becoming unscrewed.

Wrought iron copings can be repaired by welding or by replacing a complete length between support uprights. To avoid the costly and laborious process of removing each baluster from the plinth, they can be cut immediately below the baluster head and rewelded later. Baluster heads should be drilled and tapped, then welded and dressed up, otherwise they can be prone to falling off. True wrought iron is still available, but only in a limited range of sections from a few suppliers; the ENTCC can provide information on sources of supply. Mild steel is an acceptable substitute for wrought iron plain rectangular copings but is much more susceptible to damage by corrosion.

Handrails with moulded profiles, commonly found on steps leading down to basement areas, cannot be matched by mild steel sections and can only be repaired by welding lengths of salvaged wrought iron of similar design; short lengths of wrought iron can be specially made by specialist firms. Wrought iron handrails are fixed to the balusters by rivets, pins or screws set into countersunk holes in the handrail (see illustration). The same method is used to fix cast iron stair balusters to the wrought iron core of a timber handrail on an internal staircase. Alternatively, handrails can be fixed to balusters by electric welding.

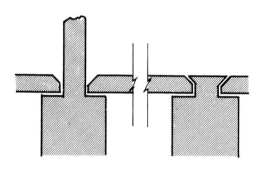

Pin is cast into baluster, sawn off above handrail and rivetted into countersunk hole in handrail

JOINT BETWEEN SIMPLE BALUSTER AND HANDRAIL

Balusters and support uprights should be fixed into the stone plinth in the traditional way, with staved lead set proud. The sinking should be dry and contain about a teaspoonful of powdered resin or a drop of machine oil to avoid a blow-out of molten lead. Repairs to the plinth itself, or to the steps and platt, are dealt with in STEPS, PLATTS AND ARCHES.

All railing repairs should be carried out promptly. Any replacement work should always match the original, and new castings can be made locally to match any pattern. The ENTCC can supply a list of foundries and suitable metalworking firms who are able to carry out repairs.

REPAIR TO THE MORAY ESTATE RAILINGS

More complex repairs are required for elaborate railings such as those in the Moray Estate, with double cope and ornamental husks, which are the most susceptible of all to rust damage. Water trapped between the shafts and husks will cause rust to build up. (Examples of this repair can be seen at 23 Moray Place and 3 Albyn Place.) The procedure is as follows:

1 Baluster sawn off above lower coping and set aside.
2 Wrought iron copings removed. Lower coping replaced with mild steel.
3 New husk positioned, the upper coping replaced and baluster heads inserted.
4 Lead poured into space between husk and shaft to seal this vulnerable joint against the weather. The lead must be sufficiently hot to ensure that it runs right through.
5 The husks, copes and balusters should be primed before reassembly.
6 Joints between husks, which are likely to be of varying size, must be thoroughly primed and filled with lead.
7 Primed with metal primer and two coats of paint.

Husks must follow the bends and rakes of the coping rails, and at least four different patterns will be required to suit a typical entrance on the Moray Estate. The top and bottom surfaces of the husks should be machined before fitting.

REPAIR TO MORAY ESTATE RAILINGS

BALCONIES

INTRODUCTION

Cast iron balconies, usually fitted to drawing room windows on the first floor, contribute greatly to the decorative quality of New Town facades. There are four basic types, each found in a variety of patterns:

- A long balcony running the length of a house or a street as at Regent Terrace and Carlton Street;
- A curved or rectilinear balcony which is the width of an individual window as at Moray Place and Royal Terrace;
- A shallow guard, the width of a window, often found on rear elevations and at nursery windows on upper floors;
- Balconies in front of ground floor shop windows as in Haddington Place and St Stephen Street.

Balconies may rest on the projecting band course at first floor level, or on the cornice at first or occasionally top floor level. In some cases a heavier balcony is supported on cast or wrought iron brackets. All balconies are fixed top and bottom to the stonework; at the bottom by lugs which are part of the cast iron floor of the balcony, and at the top by an extension of the wrought iron coping rail, both being staved into holes undercut in the masonry. Balconies running the length of a street are tied at the top to the stonework with wrought or cast iron bars. These ties sometimes take the form of decorative cast iron panels at the party walls between adjoining houses, as at Regent Terrace.

MORAY PLACE Anthemion and palmette

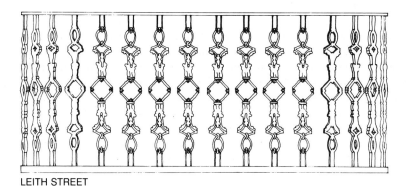

LEITH STREET

MAINTENANCE

Balconies are heavy and they may have to carry the load of people standing on them, so it is very important that they should be securely fixed to sound stonework and protected against rust. They should be thoroughly inspected every year, and repainted every five years or at the first signs of rust. Further guidance on the principles of rust prevention and painting is given in EXTERNAL PAINTWORK. The Planning Department recommends that all balconies be painted black; Listed Building Consent and planning permission are required for a change of this colour.

252

Great King Street

Walker Street

Melville Street

Northumberland Street

BALCONY BRACKETS

DEFECTS AND REMEDIES

Cast iron will fracture if subjected to tensile stresses or if a section is significantly reduced by corrosion. Fractures in balusters and coping rails can be repaired by welding *in situ* (see RAILINGS), and if an individual baluster or decorative panel cannot be repaired, a new one can be cast to match. Floor grid panels need to accommodate a certain degree of thermal and structural movement which may be restricted by welding *in situ*. An effective temporary repair can be achieved by bolting stainless steel plates above and below the fracture, but for permanent repairs these floor panels and broken or badly rusted balconies should be removed from the site for repair or recasting. Removal and re-erection require scaffolding but the cost is justified by the extended life of the balcony and the opportunity to examine the material at close quarters.

Wrought iron brackets are susceptible to rust and, where seriously weakened by corrosion, should be replaced by mild steel, preferably galvanized or, better, by stainless steel. Stainless steel lugs embedded in the wall and bolted to the brackets can avoid corrosion at this vulnerable position.

Water penetration may occur if fixings have been realigned on the wall face, leaving sinkings in the stonework. These should be made good with a lime mortar repair.

When restoring a missing balcony a suitable pattern may be found in the same street. The ENTCC can supply the names of the foundries able to make new castings, and blacksmiths to repair and restore existing balconies.

CASE STUDY RESTORATION OF BALCONY AT 8 REGENT TERRACE

Design

The cast iron balcony comprises three independent floor platforms and balustrade assemblies with lateral balustrade panels located on the property boundaries. Each floor platform is retained by three lugs built into the ashlar and connected to adjacent platforms by a simple lapped joint formed in the main edge members. The legs of the balustrade panels are located in sockets cast in the front member of the floor platforms and fixed by lead poured when it is molten.

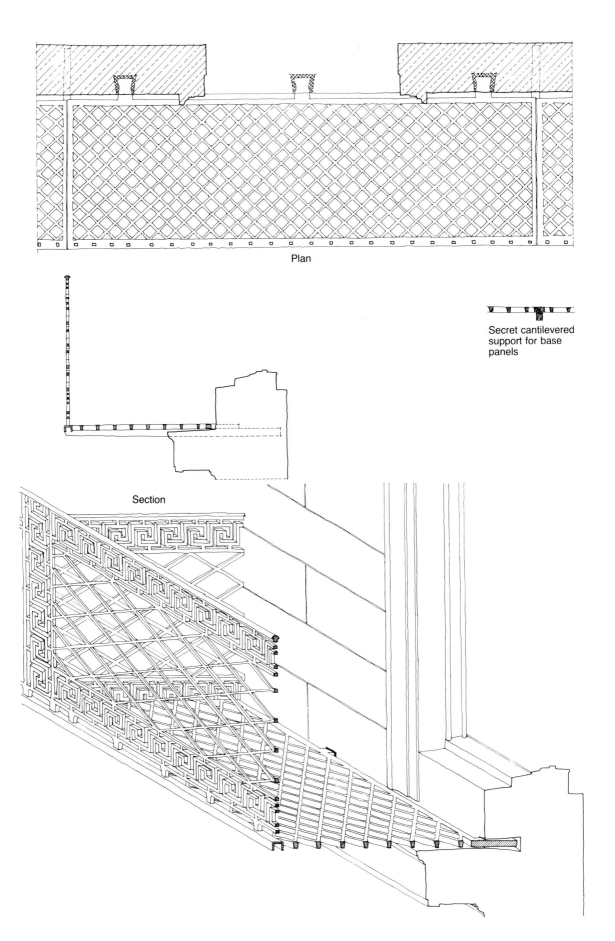

Plan

Secret cantilevered
support for base
panels

Section

Problem

Dry and wet rot was discovered in the floor timbers at first floor level; joist ends built into the external wall and the adjacent floor timbers were affected by damp penetrating the wall at balcony level.

The original detail at the junction of the cornice below the balcony and the first ashlar course above has shortcomings, for it was noted that there was no upstand to the cornice stone. This defect was compounded by the diamonds of the balcony floor panels having been infilled with mortar, which prevented rainwater draining away to the front of the cornice. It was also discovered that there was minimal cover (only 25 mm in some places) of the ashlar over the cornice masonry forming the aedicule.

Remedial action

The balcony assembly was removed with the exception of the wrought iron cantilever bracket located at mid-span, which was found to be in good condition. This enabled the metalwork to be shotblasted and repaired off site, before priming and painting in controlled conditions.

The problems of water penetration were solved by installing a lead cloak over the entire cornice with an upstand dressed into the ashlar. Pockets were formed in the leadwork to accommodate the integral lugs of the floor platforms.

The refurbished metalwork was reinstated and the joint between the ashlar and the rear floor member sealed with a proprietary two-part polysulphide sealant. Sealant was also applied to the joint between the coping rail and the balustrade panel to prevent water ingress.

LAMP STANDARDS

INTRODUCTION

The original Corporation oil lamps in the New Town were supported either on iron brackets projecting from the wall of a house or on simple wooden posts attached to the railings. None of these now survive.

At the beginning of the nineteenth century some proprietors put up their own lamp standards – wrought iron in Charlotte Square and York Place and cast iron of various designs in Heriot Row, Queen Street, Ann Street and Howard Place. Many of these private standards still survive, although some have lost their link horns used by the 'link boys' to extinguish their torches once they had escorted people to the house.

With the coming of gas, by the late 1820s most of the older posts were replaced by cast iron standards, mounted either on the railings or the pavement. The early four-sided lanterns were replaced by glass globes with chimneys, later in the nineteenth century by lanterns using incandescent mantles, and in the twentieth century by electric light.

The following extract by John Gifford, reproduced from *Edinburgh Lighting Vision* by kind permission of Morris and Steedman, Architects, gives a detailed insight into the history of street lighting in Edinburgh:

For much of the 16th and 17th centuries the provision of street lights in Edinburgh, still within its medieval boundaries, was the responsibility of private citizens. In 1533 the Town Council ordered householders to hang lanterns outside their dwellings at night, an order repeated by subsequent Council proclamations. It was not until 1688 that the Town Council took over the direct responsibility for lighting the main thoroughfares. In that year it erected lamps in the High Street and Cowgate, the Lord Provost having bought twenty-four lanthorns in London at a cost of £109 sterling. Fifteen lampposts were provided in 1734 and in 1742 the Council owned ninety-nine lamps for streets and closes and forty-six stoups (two of them spares) on which to set up lamps. The number of lamps in the Old Town continued to increase over the next fifty years so that by 1761 the Council was responsible for the maintenance of one hundred and forty and for three hundred and seven by 1786.

James Craig's plan for the First New Town on the north side of the North Loch valley was adopted in 1767, four years after the foundation stone of the North Bridge had been laid. By 1785, when much of the First New Town had been built, the Town Council had installed one hundred and sixteen lamps there, apparently placed either on wooden poles or, as was common, in the Old Town, on wall brackets,

Some New Town residents thought the wooden poles insufficiently elegant. In 1787 John Hunter, W.S., was given permission to replace (at his own expense) the pole outside his house in Queen Street with an iron lamp standard fixed to the railings, perhaps the first of the decorative cast and wrought iron standards put up by the residents of the grander First New Town houses at the end of the eighteenth century. Nevertheless in 1789 the Edinburgh Advertiser complained that 'while all strangers admire the beauty and regularity of the New Town, they are surprised at it being so badly lighted and wretched at night.'

Systematic replacement of the Town Council's wooden poles began in 1819 when the Committee of Commissioners of Police for the Lighting Department started to erect cast iron standards, each costing £1 6s. 0d., a cost reduced the next year to 14s. 6d. when they accepted a tender for their supply from John Anderson of the Leith Walk Foundry. These standards, like all the earlier brackets, poles and standards, were intended to carry oil lamps usually burning whale oil.

But 1819 was also the year in which gas lighting was introduced to Edinburgh by the erection of eighteen pillars for gas lamps on North Bridge. The following March the oil lamps in the High Street between Parliament Square and the Bridges were converted to gas which quickly began to oust oil as a street lighting fuel, the seventy-three oil lamps in Princes Street being replaced by fifty-three gas lamps in 1822.

By 1826 most of the New Town streets, as well as Leith Walk and Lothian Road, were gas-lit. The first gas lamp standards were very similar to the cast iron standards which had been adopted for oil lamps in 1819. Like them they were intended to be attached to railings although in practice they were almost always mounted on bollards (pall stones) standing on the pavement. In 1827 a design for a sturdier free-standing gas standard, a fluted column carrying the lantern, was adopted on the advice of the architect William Burn. Standards of this new type made their appearance in Princes Street in 1830 and quickly spread around most of Edinburgh, although never ousting the many brackets, originally designed for oil lamps but easily adapted to gas, from the Old Town.

The City's growing responsibility for street lighting did not mean that private lamps disappeared. They continued to be put up outside major buildings as advertisements for their importance. Notable surviving examples are the cast iron standards provided by the Shotts Foundry for the Royal Bank of Scotland at Dundas House in St.

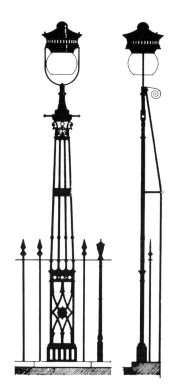

26 HERIOT ROW
Cast iron

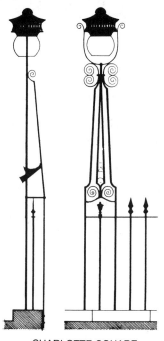

CHARLOTTE SQUARE
Wrought iron

NORTHUMBERLAND STREET
Cast iron

MELVILLE STREET
Cast iron

257

Andrew Square in 1828 and for the Royal College of Physicians at Queen Street in 1844, the doctors' lamps being adorned with miniature gilded cocks and cut glass bowls. The largest and grandest is the 45 foot high Portland stone standard designed by Sir Rowand Anderson to stand outside the McEwan Hall in 1897.

Electric street lighting was introduced to Edinburgh in 1881 when the Anglo-American Brush Electric Co. lit North Bridge, Princes Street and the Waverley Market but the company's power station proved inadequate for the purpose and the experiment had to be abandoned after three months. However, after the opening of the city's own power station at Dewar Place in 1895, electricity began to replace gas for lighting the principal thoroughfares, tall pole-and-bracket standards with decorative spandrels being provided by the Edinburgh firm of Mackenzie & Moncur. In 1922-3 T-shaped standards were placed down the centre of Princes Street, their double brackets carrying both lights and electric wires for trams. In 1955 the Town Council began a ten-year programme to replace all surviving gas lighting with electric light throughout the city. The 19th century cast iron gas standards were removed and concrete or steel poles put up, some of these new concrete standards being topped by 'Georgian-style' lanterns as a mark of respect for the New Town's architecture. As well as erecting standards of new design the Town Council reinstated or installed railing-mounted lamps in a few streets, the railing standards along the Mound and the south side of Princes Street being copied from examples of privately erected wrought iron oil standards of c. 1800 which had survived in Charlotte Square.

The example set by the Town Council has been followed since 1971 by several New Town street associations who, with the encouragement of the Historic Buildings Council for Scotland and the Edinburgh New Town Conservation Committee, have replaced concrete standards by cast iron railing standards, copied from a surviving example of c. 1820 in York Place. As a result, Northumberland Street, India Street, Fettes Row and Royal Crescent are now illuminated for the first time in their history by railing-mounted lamps. Wall-mounted flood-lights, first introduced to the Royal Mile in 1964-5 and to Princes Street in 1968, have now become the favoured means to light central Edinburgh's main traffic routes.

MAINTENANCE

Lamp standards should be painted every five years or at the first signs of rust. The principles of thorough painting and rust prevention are laid out in EXTERNAL PAINTWORK. The Planning Department requires that all railing-mounted lamp standards be painted black.

The Lighting Section of the Highways Department of Lothian Regional Council is responsible for the maintenance of public street lighting, including most of the railing-mounted lamp standards now converted to electricity. The few private lamps which are controlled from individual houses are maintained by the proprietor to whom they belong.

REPAIR AND RESTORATION

Cast iron is a robust material which should not need to be repaired unless it is accidentally damaged or badly rusted. Small sections can be repaired by welding, but standards in very poor condition may have to be replaced by new castings.

The ENTCC can provide a list of local foundries and blacksmiths, and of manufacturers who supply modern lanterns and bowls to fit the standards. The mounting ring on original private standards will have to be widened to take modern lanterns.

During the last twenty years some street associations and individual proprietors have erected new cast iron railing-mounted lamps, the first being those in Northumberland Street in 1971. The ENTCC now holds a good pattern from which further castings can be taken. This is based on an original private cast iron standard at 38 York Place, which has been the model for all of the new standards in Northumberland Street (1971), Fettes Row (1984), Royal Circus, India Street, Gloucester Place and Danube Street (1989), replacing kerbside concrete standards. The local residents subscribed to purchase the new standards, which were erected and wired by the Highways Department of Lothian Regional Council. In 1993 the Lighting Section replaced the large concrete standards in Great King Street, Drummond Place and London Street with new tubular steel columns regularly spaced along the kerb, which suit the scale and design of the street. More railing-mounted lamps will be fitted in Drummond Place, Nelson Street and Dundonald Street in the near future.

SHOPS AND SHOP FRONTS

INTRODUCTION

The First New Town was planned as a solely residential area and at first there were no purpose-built shops, so people continued to rely largely on the open air markets in the Old Town which sold fish, flesh and vegetables. However, by the end of the eighteenth century, basement and ground floor flats were being converted into shops by the addition of a door-piece and enlarged windows (for example in Thistle Street and Rose Street) and by the 1820s there were many purpose-built shops in the New Town selling a wide range of merchandise. These commercial premises were very small and the merchandise sold was very specialized. Typical inhabitants were bakers and confectioners, victual, tea and spirit dealers and bootmakers, but there was even a toy shop at 43 Hanover Street and a cutler and surgeons' instrument dealer at 34 Hanover Street. A permanent market was also established in Stockbridge in the 1820s, followed by Broughton Market in the 1840s.

GENERAL

In the second and third New Towns purpose-built shops are found at basement, half-basement and ground floor level, often grouped in threes at street corners, as in Dundas Street, Northumberland Street and Howe Street, or as parades of shops built along the entire length of minor streets, such as William Street and St Stephen Street (which has shops on two levels) and occasionally fronting onto major streets such as Haddington Place and Elm Row.

Georgian shops always had double doors flanked by one or two large, fixed astragalled windows. Ground floor shops are almost always accessed over a platt and, to allow customers closer scrutiny of window displays, were provided with cast iron or stone balconies running under the windows.

The degree of ornamentation of facades varies considerably, depending on the location of the shop and the period in which it was designed and built. Sometimes facades are very plain with only a band course dividing shop from first floor level, but more often there is a projecting stone cornice. Pilastered shop fronts as in William Street and Howe Street are common, but more elaborate fronts with detached fluted ionic columns flanking door and window openings can be found in Haddington Place.

The Department of Planning encourages the repair and retention of original shop fronts and has Design Guide leaflets which set out their general policy on *Shop Fronts, Dutch Canopies, Shop Front Security, Advertisement and Signage,* and *External Alarm Boxes.* Two, more specific, design guides have also been written: *A Policy for Shop Frontage: 20-60 St Stephen Street, 1 March 1983* and *Conservation Design Guidelines: 33-51 William Street,* which are intended to encourage the restoration and retention of the original character and appearance of these groups of shops. They cover painting colour schemes, location, size, type and colour of advertisement and signage, the street numbering system, fixing details for hanging and applied signage, and location of floodlighting.

DESIGN AND CONSTRUCTION

Structural openings

Window and door openings for shops were formed using long, flat, stone lintels, often with stone relieving arches behind, or sometimes, in the case of longer spans, a timber bonding beam was built into the wall above. Bonding beams are susceptible to rot, while cracking is a common problem in the long, thin stone lintels and these defects can go undetected when hidden behind timber fascias.

Windows

To achieve the largest possible area for displaying goods and to maximize daylighting within the shop, shop windows were made as large as possible by reducing the wall to a series of piers, and for this reason shop windows, unlike domestic window openings, do not have splayed ingoes. Shop window frames are not fixed behind stone reveals but are simply wedged within the stone opening. The junction between stone and timber is covered by a timber facing which generally also forms a frame for timber shutters which are fixed over the windows at night.

SHOPS IN WILLIAM STREET 260

Unlike in houses, astragal mouldings face to the outside and windows are glazed from the inside. At first, windows were divided into nine or more panes of very thin 2 mm glass, supported on thin astragals. Later, when plate glass became available, windows were commonly divided horizontally into two or three glazed sections, supported by heavier section astragals. Earlier glass and astragals were then removed, as in William Street.

Loose shutters were put over windows at night for security, each timber-framed and panelled shutter being fixed at the top by two iron pins located into holes in the frame and at the base by a long, iron bolt passing from the inside horizontally through the window frame and screwed into a threaded plate bedded in the shutter. It is likely that two or three shutters covered one window and, once installed, these were sometimes held by an iron bar with slots at each end put across the shutters and padlocked through iron rings set in the masonry. Shutters are no longer used, but holes for pins, bolt fixings and iron rings are still often visible unless they have been removed, blocked up or lost under layers of paint. Any surviving shutters and fixings should be retained.

Shop window frames are susceptible to rot because they are set close to the building face and are therefore particularly exposed to weathering. At cill level the facing on which the shutter rests forms a horizontal ledge and water gathering on or below this ledge can result in wet rot. Frequently rotten timber has been cut out and replaced with a cill board, losing the flush simplicity of the original detail and this should be avoided. Where major (workshop) repair work is necessary, it is a relatively simple matter to remove shop window frames. For further information on repairs see SASH WINDOWS.

Doors

All original shop doors in the New Town were timber-framed double doors, panelled at the base and divided horizontally into between four and eight glazed panes above. Diminished stile (gun-stock stile) doors are common, this being where stiles are reduced in width at the glazed section in order to maximize the area of glass. The glazing bars are recessed back about 25-30 mm from the face of the door, forming a frame to receive timber panelled shutters, which are secured with pins and bolts in the same way as for window shutters.

Each shutter is divided into two or three panel sections and strengthened at each corner with a triangular iron plate simply screwed into place. A cast iron plate of approximately 75 mm × 35 mm with two key-type holes is normally screw-fixed in the centre of the bottom rail of each shutter. This was used to lift off the shutters which are otherwise completely flush with the plane of the door. Doors were secured with a big cast iron rim lock and often a cast iron bar placed across the meeting stiles and padlocked.

Doors can be damaged in several ways. As with windows, the original glass and glazing bars have often been removed and replaced with one sheet of glass, or have been completely boarded over. Sometimes shutters have been painted over in position leaving them stuck in place. The proportions of shop doors are such that door knobs were fixed at a relatively low level on the locking rail and may have been moved or replaced at a higher position on the locking stile, damaging the stile, rail and rim lock. For further information see DOORS.

Balconies

Almost every shop entered over a platt had a cast iron or stone balcony under the windows and generally these are still in place. Cast iron balconies normally have an open diagonal grid base and are usually rectangular or with a curved end. Cast iron railings attached to the main base frame range from a very simple square section to an ornately cast pattern, such as in Haddington Place. There is evidence that a great number of balconies had simple wrought iron gates between platt and balcony, lockable with a padlock, but few remain. Balconies are normally supported on simple wrought iron brackets, fixed into the stonework (see BALCONIES) but occasionally they are formed using stone slabs cantilevered out from the wall, with cast iron supports. The joints between slabs are half-checked together. Examples can be seen in Haddington Place, St Stephen Street and Dundas Street.

Often fixings into stone have worked loose or rusted causing stone decay, and additional inappropriate supports have been introduced. Small structural movement or local failures in the elevation can affect stability and safety, particularly of stone balconies.

Cornices, fascias, canopies and signage

Dealers advertised their wares by sign-writing on the fascia or frieze above the window lintel, over which there is normally a projecting stone cornice. Cornices which have been wrongly bedded or are chronically damp may be decaying and the best remedy, after suitable stone repair, is to dress a lead apron flashing over the top of the cornice, and tuck it into the joint between cornice stone and wall (see FLASHINGS). The unifying effect of a cornice is destroyed when it is painted in bands of different colours along its length. When this has happened, paint should either be stripped off or the cornice should be painted in the same colour for its entire length (see REMOVAL OF PAINT FROM STONEWORK).

Where Dutch canopies or large fascia boards have been added, these not only alter the original proportions of the shop front and are unsightly, but can also result in the stone being damaged by the fixings. Planning permission and Listed Building Consent will be required before fitting new canopies and fascia signs which, in general, will not be allowed on finely detailed elevations. They are prohibited in certain streets in the New Town which are listed in the Design Guide: *Dutch Canopies* published by the Department of Planning. Traditional flat sun blinds are more

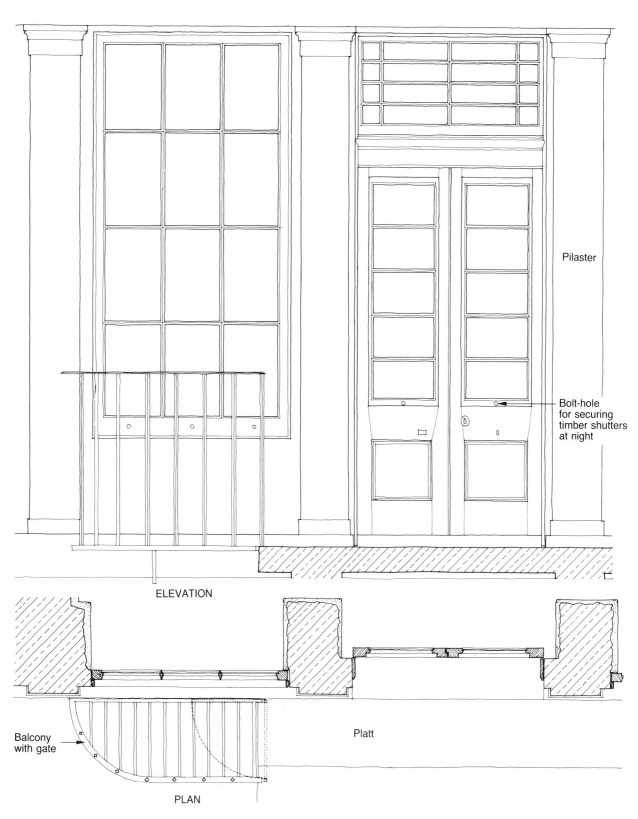

Pilaster

Bolt-hole
for securing
timber shutters
at night

ELEVATION

Balcony
with gate

Platt

PLAN

TYPICAL NEW TOWN SHOP FRONT   262

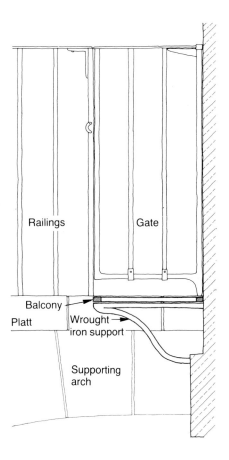

TYPICAL GATE TO A BALCONY

appropriate than the curved Dutch type of canopy. They should, if possible, be fitted within the window reveal rather than being fixed directly to the window surrounds and should be retractable. The awning should be of a canvas type material, and both canvas and canopy housing should be integrated into the general colour scheme.

Signs should be an integral part of the shop front design. Standard designs of multiple stores may have to be modified to avoid detracting from the building character, and sponsored signs (for example by soft drinks and cigarette manufacturers) are generally not acceptable. Fascia panels must be made of a good quality traditional material. The lettering should be well designed and executed and should not exceed two-thirds of the depth of the fascia up to a maximum height of 450 mm. Hanging signs are not traditional in the New Town and fascia signage is preferable. Where introduced, there should generally only be one per shop, made in timber and wrought iron. It should be a maximum of 0.5 m², be 2.25 m clear of the pavement level (or as agreed with the Department of Planning) and should project a maximum of 1 m from the building or to a maximum of half the width of the pavement, whichever is the smaller. When drilling stone for hanging and applied signage, fixing holes must not be drilled closer than 75 mm from stone arrises and should, generally, be kept to a minimum. Alternatively, new signs can be located inside the shop window where they are nearer eye level, can be more effective, less damaging to the fabric and do not require Listed Building Consent.

Paint colour and lighting

Unpainted masonry must not be painted because of the damage that this impervious layer can do to the stone (see REMOVAL OF PAINT FROM STONEWORK). Where a shop front has already been painted, colour schemes should be well integrated and muted, in keeping with Georgian paint colours, and all proposals must be discussed with the Department of Planning. The paint colour for astragal mouldings should be considered as part of the overall scheme and preferably be dark because it is easier to see through a window to the display behind when astragals are dark than where the eye is distracted by a grid of white lines.

Shops can be lit by floodlights facing the building from the basement retaining wall or, better, by lighting the shop display from inside because this attracts the eye into the shop. Light fittings should not be attached to the outside wall of the shop or to the fascia. External lighting will require planning permission and Listed Building Consent.

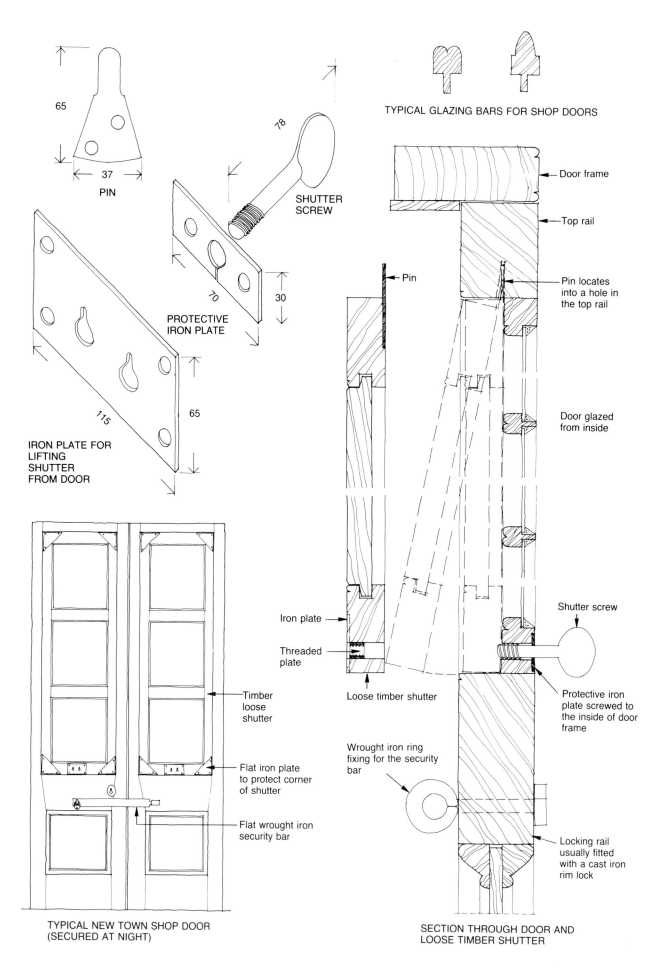

65

37

PIN

SHUTTER
SCREW

78

70

30

PROTECTIVE
IRON PLATE

115

65

IRON PLATE FOR
LIFTING
SHUTTER
FROM DOOR

TYPICAL GLAZING BARS FOR SHOP DOORS

Door frame

Top rail

Pin

Pin locates
into a hole in
the top rail

Door glazed
from inside

Iron plate

Threaded
plate

Loose timber shutter

Shutter screw

Protective iron
plate screwed to
the inside of door
frame

Wrought iron ring
fixing for the security
bar

Timber
loose
shutter

Flat iron plate
to protect corner
of shutter

Flat wrought iron
security bar

Locking rail
usually fitted
with a cast iron
rim lock

TYPICAL NEW TOWN SHOP DOOR
(SECURED AT NIGHT)

SECTION THROUGH DOOR AND
LOOSE TIMBER SHUTTER

Security

It is recommended that the original timber shutters should be used. A possible alternative is removable mesh grilles fitted over windows and doors because these are much less damaging and visually disruptive than fixed roller shutter systems. The design, material and colour of the grilles must be to the satisfaction of the Department of Planning.

USEFUL ADDRESSES

Architectural Heritage Society of Scotland
The Glasite Meeting House
33 Barony Street, Edinburgh, EH3 6NX

British Chemical Dampcourse Association
16a Whitchurch Road
Pangbourne, Reading, RG8 7BP

British Geological Survey
Murchison House
West Mains Road, Edinburgh, EH9 3LA

Building Research Station
Garston, Watford, WD2 7JR

The City of Edinburgh, District Council
City Archives
City Chambers, Edinburgh, EH1 1YL

The City of Edinburgh, District Council
Environmental Health Department
Johnston Terrace, Edinburgh, EH1 2PT

The City of Edinburgh, District Council
Housing Department
23-25 Waterloo Place, Edinburgh, EH1 3BH

The City of Edinburgh, District Council
Planning Department
1 Cockburn Street, Edinburgh, EH1 1BJ

The City of Edinburgh, District Council
Department of Property Services
329 High Street, Edinburgh, EH1 1PN

Civic Trust
17 Carlton House Terrace, London, SW1Y 5AW

Cockburn Association (The Edinburgh, Civic Trust)
Trunk's Close
55 High Street, Edinburgh, EH1 1SR

Council for the Care of Churches
83 London Wall, London, EC2M 5NA

Edinburgh, New Town Conservation Committee
13a Dundas Street, Edinburgh, EH3 6QG

English Heritage
Fortress House
23 Savile Row, London, W1X 2HE

The Georgian Group
37 Spital Square, London, E1 6DY

Historic Scotland (and the Historic Buildings Council
for Scotland), Longmore House
Salisbury Place, Edinburgh, EH9 1SH

Historic Scotland – Scottish Conservation Bureau
3 Stenhouse Mill Lane, Edinburgh, EH11 3LR

Institute of Paper Conservation
Leigh Lodge, Leigh, WR6 5LB

Lead Sheet Association
St Johns Road, Tunbridge Wells, Kent, TN4 9XA

Lothian and Borders Fire Brigade
Brigade Headquarters
Lauriston Place, Edinburgh, EH3 9DE

Lothian and Edinburgh Enterprise Ltd.
Apex House
99 Haymarket Terrace, Edinburgh, EH12 5HD

Lothian Regional Council
Highways Department
18-19 Market Street, Edinburgh, EH1 1BL

Museums Association
42 Clerkenwell Close, London, EC1R 0PA

The Conservation Unit
Museums & Galleries Commission
16 Queen Anne's Gate, London, SW1H 9AA

National Monuments Record of Scotland
John Sinclair House
16 Bernard Terrace, Edinburgh, EH8 9NX

Royal Commission on the Ancient
and Historic Monuments of Scotland
John Sinclair House
16 Bernard Terrace, Edinburgh, EH8 9NX

Royal Incorporation of Architects in Scotland
15 Rutland Square, Edinburgh, EH1 2BE

Royal Institute of British Architects
66 Portland Place, London, W1N 4AD

Royal Institution of Chartered Surveyors
9 Manor Place, Edinburgh, EH3 7DN

SALVO, PO Box 1295, Bath, BA1 3TJ

Scientific and Educational Services Ltd.
Romsley Hill Cottage
Farley Lane
Romsley, Halesowen, West Midlands, B62 0LW

Scottish Office Environment Department
Building Directorate,
New St Andrew's House, St James Centre
Edinburgh, EH1 3DG

Scottish Federation of Housing Associations
42 York Place, Edinburgh, EH1 3HU

Scottish Record Office
Register House, 2 Princes Street,
Edinburgh, EH1 3YY

Scottish Society for Conservation and Research
The Glasite Meeting House
33 Barony Street, Edinburgh, EH3 6NX

Society for the Protection of Ancient Buildings
37 Spital Square, London, E1 6DY

Stone Federation
82 New Cavendish Street, London, W1M 8AD

Timber Research and Development Association
Stocking Lane, Hughenden Valley, High Wycombe,
Bucks, HP14 4ND

REFERENCES

1 GENERAL
2 WALLS, OPENINGS AND BUILDING STONES
3 ROOFS
4 INTERNAL DETAILS AND TIMBER TREATMENT
5 EXTERNAL DETAILS

1 GENERAL

Architectural Heritage Society of Scotland, *Scottish Pioneers of the Greek Revival*. Architectural Heritage Society of Scotland, The Glasite Meeting House, 33 Barony Street, Edinburgh, EH3 6NX.

Ashurst, John and Ashurst, Nicola, English Heritage Technical Handbook *Practical Building Conservation* (5 vols). Gower Technical Press, 1988.

Ayres, James, *Shell Book of the Home in Britain 1500-1850*. Faber, 1981.

Begg, Ian, *Survey Techniques*. Unpublished Conservation Course paper. ENTCC Library, 1973.

Bigham, D Alistair, *The Law and Administration Relating to Protection of the Environment*. Oyez Publishing, London, 1973. Supplement 1974.

Bolton, Arthur T, *The Architecture of Robert and James Adam*. George Newnes Ltd, 1972.

Bowyer, Jack, *Small Works Contract Documentation and How to Administer It*. Architectural Press, 1976.

Bowyer, Jack, *Guide to Domestic Building Surveys*. Architectural Press, 1977.

Bowyer, Jack, *Property Maintenance and Management*. Orion Books, Eastbourne, 1984.

British Standards Institution: *Sectional List of British Standards and Codes of Practice*.

Building Research Establishment: *Building Research Digests*.

Bullock, Ralph, *A Scottish Building Glossary: Survival of Craft Terms*. Mimeograph Sheet in ENTCC Library from *The Builder*, 20 June 1952.

Carlsson, Ingrid and Holmström, Ingmar, *Care of Old Buildings, an Annotated Bibliography: Part 1*. National Swedish Research Institute for Building Research, Stockholm, 1975.

Chitham, R, *Measured Drawing for Architects*. Butterworth, 1991.

City of Edinburgh District Council Planning Department:
Development Control Handbook 1994.
Alterations to Listed Buildings − a design guide.
Automatic Teller Machines.
Balanced Flues on Historic Buildings.
Burglar Alarm Boxes.
Dutch Canopies - planning guidelines.
Exterior Painting of Buildings.
Flagpoles, Flags and Banners.
Gas Pipes and Gas Meter Boxes.
Painting of Listed Buildings.
Parking in Front Gardens − a design guide.
Planning Policies for the first New Town.
Rear Stairs on Listed Buildings − a design guide.
Roller Shutters.
Roof Policies − a design guide.
Shopfronts − planning guide-lines.
Solar Panels.
Stockbridge Colonies − a maintenance and design guide.
Sub-dividing Older Property, March 1986.
Window Alterations − a design guide.

Clark, Lord Kenneth, *Civilisation − A Personal View*. BBC Publications, London, 1969.

Council for the Care of Churches, *How to Look after your Church*. Church Information Office, London, 1970.

Cockburn, Lord, *Memorials of his Time*. 1856.

Cox, Michael, *Exploring Scottish History*. Scottish Library Association and Scottish Local History Forum, 1992.

Cunningham, P, *Care for Old Houses*. Prism Alpha, 1984.

Curl, J S, *English Architecture, An Illustrated Glossary*. David & Charles, 1977.

Curl, Prof. James Stevens, *Encyclopaedia of Architectural Terms*. Donhead, 1992.

Dunbar, J G, *Historic Analysis: Sources of Information*. Unpublished Conservation Course paper. ENTCC Library, 1973.

268

Eldridge, H J, *Common Defects in Buildings*. HMSO, London, 1976.

English Heritage, *The Directory of Public Sources of Grants for the Repair and Conservation of Historic Buildings*, 1990.

English Heritage, *Theft of Architectural Features*. (Leaflet), English Heritage, London Division, Historic Buildings and Monuments Commission for England.

Feilden, Sir Bernard, *Conservation of Buildings*. Butterworth, 1982.

Fleming, J Honour, H, and Pevsner, N, *Penguin Dictionary of Architecture*. Penguin, 1984.

Freeth, Evelyn, and Davey, Peter, (editors), *AJ Legal Handbook*. Architectural Press, 1978.

Gifford, J, McWilliam, C, and Walker, D, *Edinburgh (The Buildings of Scotland)*. Penguin Books, 1984.

Gilbert, John, and Flint, Ann, *The Tenement Handbook*. RIAS, 1992.

Gill, William H, *Rimmer's – The Law Relating to the Architect*. Stevens and Son, London, 1964.

Gloag, W M, and Henderson, R C, *Introduction to the Laws of Scotland*, 9th edition (ed. Wilkinson and Wilson), Edinburgh, 1987.

Harrison, Patrick, *Civilising the City*. Nic Allen Publishing, 1990.

Henderson, Ross C, *Planning in Scotland: A Basic Guide*, 1985.

Historic Buildings Council for Scotland *Annual Report 1991-92*.

Historic Scotland *Annual Report 1992-93. Scotland's Listed Buildings, a guide to their protection*, 1993.

Holmström, Ingmar and Sandström, Christina, *Maintenance of Old Buildings: Preservation from the Technical and Antiquarian Standpoint*. National Swedish Institute for Building Research, Stockholm, 1975.

ICOMOS *Guide to Recording Historic Buildings*. Butterworth Architecture, 1990.

Insall, D W, *The Care of Old Buildings Today*. Architectural Press, London, 1972.

Institute of Advanced Architectural Studies, *Preserving the Character of Conservation Areas*. Research Section, IAAS Publications, 1981-3.

Kelsall, Moultrie, and Harris, Stuart. *A Future for the Past*. Oliver & Boyd, Edinburgh, 1961.

King, D, *The Complete Works of Robert and James Adam*. Butterworth, 1991.

Lindsay, Ian G, *Georgian Edinburgh*. Oliver & Boyd, Edinburgh, 1948. Revised by David Walker, Scottish Academic Press, Edinburgh, 1973.

Linn, Björn, *Conservation of Old Buildings*. National Swedish Institute for Building Research, Stockholm, 1975.

Lothian and Borders Fire Brigade, *Fire Defence of Historic Buidings*. (Leaflet).

MacGregor, A G, *The Mineral Resources of the Lothians*. Geological Survey of Great Britain, 1945.

McKay, W B, *Building Construction vols 1–4*. Longmans, Green and Co., London, 1948.

McWilliam, Colin, *Edinburgh New Town Guide: The Story of the Georgian New Town*. 1984.

McWilliam, Colin, *Scottish Townscape*. Collins, London, 1975.

Martin, D C, *Maintenance and Repair of Stone Buildings*. Church Information Office, London, 1970.

Matthew, Sir Robert, Reid, John and Lindsay, Maurice (editors), *The Conservation of Georgian Edinburgh*. Edinburgh University Press, Edinburgh, 1972.

Melville, Ian A, and Gordon, Ian A, *The Repair and Maintenance of Houses*. The Estates Gazette, London, 1973.

Mills, Edward D, (editor), *Building Maintenance and Preservation*. Butterworth, 1980.

Oresko, Robert (editor), *The Works in Architecture of Robert and James Adam*. Academy Editions, London, 1975.

Plenderleith, H J, and Werner, A E A, *The Conservation of Antiquities and Works of Art*. Oxford, reprint, 1979.

Powys, A R, *Repair of Ancient Buildings*. SPAB, J M Dent, 1929.

Pride, Glen L, *Glossary of Scottish Building*. Fameham Publishers Ltd, Scotland, 1975.

Ramsey, Stanley, C, and Harvey, J D M, *Small Georgian Houses and Their Details 1750-1820*. Butterworth Architecture, 1989.

Richardson, A E, *An Introduction to Georgian Architecture*. Art and Technics Press, 1950.

Richardson, B, *Remedial Treatment of Buildings*. Constructional Press, 1980.

Robinson, Peter, *The Tenement House*. National Trust for Scotland, 1986.

Robinson, Peter, *Tenements: The Industrial Legacy, 1750-1918. Review of Scottish Culture No. 2*. John Donald, 1986.

Robinson, Peter, *Tenements – A Scottish Urban Tradition*. Scottish Vernacular Buildings Working Group, Annual Conference, 1980. Unpublished paper, ENTCC Library.

Salvo, *Directory of British Dealers*. 1994.

Sandwith, H, and Stainton, S, *National Trust Manual of Housekeeping*. Viking, 1991 revision.

Scott John, S, *A Dictionary of Building*. Penguin Books, 1974.

Scottish Civic Trust, *The Architectural Heritage: A Maintenance Crisis*. Conference Report, 1980.

Scottish Development Agency, *Scottish Architects in Conservation*. Published by the Bureau of the SDA and RIAS, Edinburgh, 1985.

Scottish Society for Conservation and Restoration, *Resin in Conservation* (eds Tate, J O, Townsend, J, and Tennent, N) SSCR, 1982.

Scottish Vernacular Building Working Group pamphlet, *Building Construction in Scotland*. 1976.

Scottish Office Environment Department:
Buildings of Special Architectural or Historic Interest. Descriptive list.

Historic Scotland: *Buildings at Risk Bulletin*. Scottish Civic Trust, 1992.

Historic Scotland: *Memorandum of Guidance on Listed Buildings and Conservation Areas 1993*.

Historic Scotland: *Scotland's Listed Buildings*.

Improve your Home with a Grant. Amended 1986.

Simpson, J W, and Horrobin, P J, *The Weathering and Performance of Building Materials*. Medical and Technical Publishing Co. Ltd, 1970.

Stillman, Damie, *The Decorative Work of Robert Adam*. Academy Editing, London/St Martin's Press N.Y. 1973.

Studdards, Roger W, *Listed Buildings: the Law and Practice*. Sweet & Maxwell, 1982.

Thomson, G, *The Museum Environment*.

Ward, P (editor). *Conservation and Development in Historic Towns and Cities*. Oriel Press, 1968.

Woodforde, J, *Georgian Houses for All*. Routledge and Kegan Paul, 1978.

Worsdall, F, *The Tenement – A Way of Life*. Chambers, 1979.

Yerbury, F R, *Georgian Details of Domestic Architecture*. Benn, 1985.

Youngson, Professor A J, *The Making of Classical Edinburgh*. Edinburgh University Press, 1966.

2 WALLS, OPENINGS AND BUILDING STONES

Arnold, Edward, and Howe, Allen J, *The Geology of Building Stones*. 1910.

Ashurst, John, *Cleaning Stone and Brick*. SPAB, 1977, rev. 1986.

Ashurst, John, *The Cleaning of Natural Stone Buildings*. Unpublished manuscript, 1972. ENTCC Library.

Ashurst, John, *Mortars, Plasters and Renders in Conservation*. Ecclesiastical Architects and Surveyors Association, 1983.

Ashurst, John, and Ashurst, Nicola, *English Heritage Technical Handbook, Practical Building Conservation, vol. 1, Stone Masonry*. Gower Technical Press, 1988.

Ashurst, John, and Ashurst, Nicola, *Cleaning Stone and Brick*. SPAB, Technical Pamphlet 4, 1991.

Ashurst, J, and Dimes, F G, *Stone in Building: Its Use and Potential Today*. Architectural Press, London, 1977.

Ashurst, John, and Dimes, Francis G, *Conservation of Building and Decorative Stone*. 2 vols, Butterworth-Heinemann, 1990.

HMSO Publications:
BRE, *The Building Sandstones of the British Isles*. 1986.

BRE Digest 139, *Control of Lichens, Moulds and Similar Growths*.

BRE Digest 245, *Rising Damp in Walls: Diagnosis and Treatment*.

BRE Digest 297, *Surface Condensation and Mould Growth in Traditionally-Built Buildings*.

Bunyan, I T, and Fairhurst, J A, *Building Stones of Edinburgh*. Edinburgh Geological Survey, 1987.

Burnell, G R, *Limes, Cements and Mortars*, Weale, 1857.

Caroe, A D R, and M B, *Stonework: Maintenance and Surface Repairs*. C10 Publishing, 1984.

Craig, George, *Building Stones used in Edinburgh: Their Geological Sources, Relative Durability and Other Characteristics*. Vol. VI, Transactions of Edinburgh Geological Society, 1892.

Georgian Group, The, *The Georgian Group Guides, No. 1 Windows* (Booklet).

Georgian Group, The, *The Georgian Group Guides, No. 3 Doors* (Booklet).

Hendry, Prof. A W, *Tests on Masonry Walls built of Craigleith Stone*. Unpublished manuscript for the ENTCC, 1976.

Historic Scotland *Technical Advice Note 1: Preparation and Use of Lime Mortars*. HMSO, revised 1993.

Historic Scotland *Technical Advice Note 3: Performance Standards for Timber Sash and Case Windows*. HMSO, 1993.

Historic Scotland/Scottish Enterprise, *Practitioner's Guide to Sandstone Cleaning*. 1994.

Historic Scotland/Scottish Enterprise, *Stone Cleaning in Scotland: Literature Review*. 1992.

Historic Scotland/Scottish Enterprise, *Stone Cleaning in Scotland: Research Summary*. 1991.

Historic Scotland/Scottish Enterprise, *Stone Cleaning and the Nature, Soiling and Decay Mechanisms of Stone*. Proceedings of international conference 14-16 April 1992, edited by Professor G M Webster. Donhead Publishing, 1992.

Howe, J Allen, *The Geology of Building Stones*. Edward Arnold, 1910.

Hughes, Philip, *The Need for Old Buildings to Breathe*. SPAB Information sheet 4, 1987.

Induni, Bruce and Induni, Liz, *Using Building Limes*. Induni, 1990.

Jones, D, and Wilson, M J, *Chemical Activity of Lichens on Mineral Surface – a Review*. International Biodeterioration, vol. 21, No. 2, 1985.

Leary, E *The Building Sandstones of the British Isles*. BRE, 1986.

MacGregor, John E M, *Outward Leaning Walls*. SPAB, Technical Pamphlet 1, 1985.

Millar, W, *Plasterwork, Plain and Decorative*. Batsford, 1897. (Information on lime.)

Price, C A, *Decay and Preservation of Natural Building Stones*. BRE, 1974.

Price, C A, *Testing Porous Building Stones*. BRE, 1975.

Report of Commissioners Respecting Stone to be used in Building the New Houses of Parliament. London, 16 March 1839.

Richardson, B A, *Control of Biological Growths*. Reprinted from Stone Industries. March/April 1973.

Richardson, B A, *Selecting Porous Building Stone*.

Schaffer, R J , *The Weathering of Natural Building Stones*. BRE, reprint, 1972.

Schofield, Jane, *Basic Lime Wash*. SPAB Information Sheet 1, 1985.

Stone Industries, *Natural Stone Directory*. Stone Industries, 1994/5.

Thomas, Andrew R, Williams, Gilbert, and Ashurst, Nicola. *The Control of Damp in Old Buildings*. SPAB, Technical Pamphlet 8.

University of Aberdeen, Department of Chemistry, *Report on the Durability of Moray Stone*. 1978.

University of Newcastle-upon-Tyne, Department of Geology, 1978, *Tests on Sandstone*. (Samples supplied by Natural Stone Quarries Ltd.) ENTCC.

Walkden, Gordon M, and Kathleen M, University of Aberdeen, *Report on Durability Tests on Greenbrae Stone* for Alexander Hall & Son (Builders) Ltd. Report No. 78/2, March 1978. ENTCC Library.

Walkden, Gordon M, and Kathleen M, University of Aberdeen, *Report on Durability Tests on Clashach Stone* for Alexander Hall & Son (Builders) Ltd. Report No. 78/3, March 1978. ENTCC Library.

Walkden, Gordon M, and Kathleen M, University of Aberdeen, *Report on Durability Tests on Craigleith Stone* for Edinburgh New Town Conservation Committee. Report No. 78/4, March 1978. ENTCC Library.

Warland, E G, *Modern Practical Masonry*. Stone Federation, 1929, 2nd edition, 1953.

Warnes, Arthur R, *Building Stones*. Ernest Benn, London, 1926.

Warnes, A R, *Building Stone*. Benn, 1926.

Watson, John, *Building Stones*. Cambridge University Press, 1911.

Williams, G B A, *Pointing Stone and Brick Walling*. SPAB, Technical Pamphlet 5, 1991.

Wingate, Michael, *An Introduction to Building Limes*. SPAB, Information sheet 9.

Wright, Adela, *Removing Paint from Old Buildings*. SPAB, Information Sheet 5, 1989.

3   ROOFS AND CHIMNEYS

Bickerdike Allen Partners, *Edinburgh New Town Wall-head Gutters*. Unpublished report for ENTCC on remedial work.

BRE Digest 60: *Domestic Chimneys for Independent Boilers*.

Williams, G B A, *Chimneys in Old Buildings*, SPAB Technical Pamphlet 3. 1990.

4   INTERNAL DETAILS AND TIMBER TREATMENT

Ashurst, John and Ashurst, Nicola, *English Heritage Technical Handbook, Practical Building Conservation, vol. 5 Wood, Glass and Resins*. Gower Technical Press, 1988.

Bamford, Francis, *A Dictionary of Edinburgh Wrights and Furniture Makers, 1660-1840*. Furniture History Society, 1983.

Beard, Geoffrey, *Decorative Plasterwork in Great Britain*. Phaidon, 1975.

Bravery, A F, Berry, R W, Carey, J K, and Cooper, D E, *Recognising Wood Rot and Insect Damage in Buildings*. BRE, 2nd edition 1992.

BRE Digest 197, *Painting Walls, Part 1: Choice of Paint*. 1982.

BRE Digest 201, *Wood Preservatives: Application Methods*.

BRE Digest 230, *Timber Fire Doors.*

BRE Digest 354, *Painting Woodwork.*

Bristow, I B, *The Architect's Handmaid: Paint Colour in the 18th Century.*

Carron Company 1759-1938, A History of the Company. London Press Exchange, 1938.

Cartwright, K, *Prevention.* Department of Scientific and Industrial Research. HMSO, 1946.

Colefax, J, and Fowler, J, *English Decoration in the 18th Century.*

Cornforth, J, *English Interiors, 1790-1848. The Quest for Comfort.* Barrie and Jenkins, 1978.

Fowler, J, and Cornforth, J, *English Decoration in the Eighteenth Century.* Barrie and Jenkins, 2nd edition 1978.

Georgian Group, The, *The Georgian Group Guides No. 4, Paint Colour.* (Booklet).

Georgian Group, The, *The Georgian Group Guides No. 6, Wallpaper.* (Booklet).

Georgian Group, The, *The Georgian Group Guides No. 7, Mouldings.* (Booklet).

Georgian Group, The, *The Georgian Group Guides No. 9, Fireplaces.* (Booklet).

Goodier, J H, *Dictionary of Painting and Decorating.*

Gow, I, Undercoats of Edinburgh, *Country Life,* 20 April 1989.

Gow, I, *The Scottish Interior.* Edinburgh University Press, 1992.

Health and Safety Executive, *Remedial Timber Treatment in Buildings.* HMSO, 1991.

Historic Scotland *Technical Advice Note 2: Conservation of Plasterwork.* 1993.

Hughes, Philip, *Patching Old Floorboards.* SPAB. 1988.

Hunt Alistair, *Electrical Installations in Old Buildings.* SPAB Technical Pamphlet 9, Rev. 1989.

Hurst, A E, and Goodier, J H, *Painting and Decorating.* 9th edition, Charles Griffin, 1980.

Innes, Jocasta, *Paint Magic.* Frances Lincoln, 1981.

Institute of Advanced Architectural Studies, *The Use of Colour in Historic Buildings.* Research Sections, IASS Publications, 1978.

Jenson & Nicholson Ltd., (compilers and editors), *Paint and its Part in Architecture.* 1930.

Kelly, A, *The Book of English Fireplaces.* Country Life Books. London, 1968.

Lothian and Borders Fire Brigade, Fire Prevention Department, Standard No. 21, *Upgrading of Doors using Intumescent Treatment.*

Lothian and Borders Fire Brigade, Fire Prevention Department, Appendix FP10, *Methods of Improving Fire Resistance of Existing Doors and Frames.*

Lothian and Borders Fire Brigade, Fire Prevention Department, Appendix FP11, *Construction of Fire Check Doors: Half-hour Type.*

McDonald, Roxana, *The Fireplace Book.* Architectural Press, 1984.

MacGregor, John, E M, *Strengthening Timber Floors.* SPAB Technical Pamphlet 2, 1991.

Marwick, T P, *History and Construction of Staircases,* J & J Grey, Edinburgh, 1888.

Munn, Harry, *Joinery for Repair and Restoration Contracts.* Orion Books, Eastbourne, 1984.

Nicholson, Peter, *New Carpenter's Guide.* Jones & Co., London, 1825.

Richardson, B, *Wood in Building.* Constructional Press, 1980.

Richardson, B, *Wood Preservation.* Constructional Press, 1978.

Ridout, Brian, V, *An Introduction to Timber Decay and its Treatment,* Scientific and Educational Services, Ltd, 1992.

Stagg and Masters, *Decorative Plasterwork.* Orion Books, Eastbourne, 1984.

Sugden, A V, and Edmonson J L, *A History of English Wallpaper.*

Timber Research and Development Association, *Information Sheets.*

Tomlin, Maurice, *Catalogue of Adam Period Furniture.* V & A, 1972.

Victorian Society, The, *Care for Victorian Houses, No. 1: Doors.* (Leaflet).

West, Trudy, *The Fireplace in the Home.* David & Charles, 1976.

Weston, R, Paper Trace, *Traditional Homes* No. 2, November 1984.

5 EXTERNAL DETAILS

Morris and Steedman/Lighting Design Partnership, *Edinburgh Lighting Vision,* for Lothian and Edinburgh Enterprise Ltd.

English Historic Towns Forum, *Shops and Shop Fronts in Historic Towns,* 1991.

Thoms, A W, *Edinburgh New Town Railings: A Supplementary Study.* Unpublished papers, ENTCC Library.

Tindall, B V, *Edinburgh's Cast Iron Railings.* Unpublished manuscript, ENTCC Library.

ACKNOWLEDGEMENTS

The first edition of this book was sponsored by the Edinburgh New Town Conservation Committee with the financial assistance of the Manpower Services Commission, and it combined experience gained in the first seven years of the Committee's work with wider knowledge of building conservation elsewhere.

The draft sheets were critically examined and discussed by a voluntary panel of architects, engineers and planners. Draft sheets and comments were circulated to research organizations and conservation agencies, and the Building Research Establishment was commissioned to write a report on technical aspects.

For the fourth edition the authors again discussed the text with a group of experts including:

John Addison BSc (Eng) FFB CEng MICE, Peter Stephen and Partners. Patrick Dignan BArch (Hons) RIBA ARIAS, Dignan, Read and Dewar. Peter Donaldson, Principal Architect, Historic Scotland. Pat Gibbons, Area Architect (Central), Properties in Care, Historic Scotland. Ian Gow Hon FRIAS, National Monuments Record of Scotland. Leonard Grandison, Leonard Grandison and Sons, Plasterers. Peter Lynn BArch ARIAS RIBA. A D McAdam BSc CGeol, British Geological Survey. J M MacDonald, Stirling Stone (Restoration) Ltd. Ingval Maxwell, DA RIBA FRIAS FSA Scot, Director of Technical Conservation, Research and Education, Historic Scotland. David Narro BSc CEng MICE MIStructE, David Narro Associates. Thomas Pollock BArch (Hons) RIBA FRIAS, William A Cadell Architects. Alistair Rae, D Blake and Co. Ltd. Douglas Read BArch (Hons) RIBA ARIAS, Dignan Read and Dewar. Ted Ruddock BA MAI MSc (Eng) MICE. R Gerard Walmesley ARIBA, FRIAS, Walmesley and Savage. J D Wren BSc CEng FICE MConsE, Wren and Bell.

Detailed comments and advice were also received from:

Iain Abbot, Architect. Neil Adams, Burgess Adams Architects. Ian Davidson, David Adamson and Partners. Mike Armstrong, Principal Planner, Planning Department, City of Edinburgh District Council. Geoffrey Bailey, former Deputy Director, Technical Services, City of Edinburgh District Council. Sarah Ball, SPAB Scholar 1993. Ballantine Bo'ness Iron Co. Ltd. Maurice Batten, Law and Dunbar-Nasmith. Dorothy Bell, Conservation Unit, Department of Architecture, Heriot Watt University. David Beveridge, Conservation and Projects Manager, Planning Department, City of Edinburgh District Council. H A Brechin and Co. Neil Cameron, The Royal Commission on the Ancient and Historical Monuments of Scotland. James H Clark, ENTCC. Emma Crawford, Architectural Heritage Society of Scotland. J Currie, Conservator, Historic Scotland. John Daly, Department of Housing, City of Edinburgh District Council. Anthony Dixon, Stewart Tod and Partners. Martin Easson, Planning Department, City of Edinburgh District Council. Neil Barrass, Edinburgh Architectural Salvage Yard. Edinburgh Stained Glass House. Peter Elliott, Elliott and Company. Richard Emerson, Historic Scotland. Richard Ewing, Rowand Anderson Partnership. John Evans, Stonemason/carver. Donald Faulkner, Property Services Department, CEDC. John Foster, Building Directorate, Scottish Office. Jack Gillon, Planning Department, City of Edinburgh District Council. Peter Gillon, The Cast Iron Railing Company. Martin Haddlington, Architect. Iain Harthill, W H Banks and Son Ltd. Christopher Hartley, National Trust for Scotland. John Higgitt, Department of Fine Art, University of Edinburgh. D B Hodgson, Dorothea Restorations Ltd. John C Hope, Architect. Ian Ramsey, Hurd Rolland Partnership. Meryl Huxtable, The Wallpaper History Society. David Hyde, Remtox (Chemicals) Ltd. Bruce Induni, Lime Specialist. Nancy Jamieson, Senior Planner, Planning Department, City of Edinburgh District Council. Nigel Somner, formerly of J and F Johnson Limited. John Knight, Area Architect (South), Properties in Care, Historic Scotland. Charles Laing and Sons Ltd., Brassfounders. Alan Lisle, Stonemason/carver. The London Crown Glass Company. Bruce Lonie, Heritage Policy Division, Historic Scotland. James McCormack, Clerk of Works, ENTCC. Andrew McErlain, Silvermills Forge Ltd. Michael Mclenaghan, Chimney Sweep, A Auld Reekie. Graham McLennan, Alexander McLennnan Blacksmith. M Andrew, McLeod and Aitken. SDO Mackinlay, Fire Safety Department, Lothian and Borders Fire Brigade. Ian MacPhail, Lothian Regional Council Department of Highways. Dr Eric W Marchant, Unit of Fire Safety Engineering, University of Edinburgh. John Mengham, Edinburgh Old Town Renewal Trust. James Morris, Morris and Steedman. David Morrison, Principal Building Control Officer, City of Edinburgh District Council. Paul Morrod, PAR Scaffolding Ltd. Douglas Mure, Chartered Quantity Surveyor. Raymond Muszynski, Ian Begg Architects. T and J W Neilson Ltd. Steve Newsom, Simpson and Brown. J O'Brien, Scottish Office Environment Department. Keith Parker, IAAS, University of York. R Paterson, Lighting Section, Lothian Regional Council, Department of Highways. J Paterson and Son (Plumbers) Ltd. A Peacock, Building Control Division, Property Services Department, City of Edinburgh District Council. John Perreur-Lloyd, The Michael Laird Partnership. Michael Pontin, Law and Dunbar-Nasmith. Graham Reid, Historic Scotland. Richardson and Starling. Ian Riddell, Architect. Debbie Robertson, Planning Department, City of Edinburgh District Council. Joe Rock, Photographer. The Royal Incorporation of Architects in Scotland. The Royal Institution of Chartered Surveyors in Scotland. A G Sheel, Health and Safety Executive. James Simpson, Simpson and Brown. J Skilling, Health and Safety Executive. Rab Snowden, Principal Conservator, Historic Scotland, Stenhouse Conservation Centre. Alex Stark, W and J R Watson. Henrietta Stephen, Whytock and Reid. A D O Stevenson, Fire Safety Department, Lothians and Borders Fire Brigade. Charles Taylor, Woodwork and Design Ltd. Sebastian Tombs, Deputy Secretary, RIAS. Harry Turnbull, Stirling Stone. Iain Lawrence, Walfords, Chartered Quantity Surveyors. Hunter Walker Associates. The Thomas Walmesley Forge, Blists Hill,

Ironbridge, Shropshire. Donna Watts, BRE Bookshop. Dick Wilson, National Trust for Scotland. Maureen Young, School of Surveying, Robert Gordon's University. Rory Young, Limeworker.

The Committee would also like to thank A D McAdam BSc CGeol, British Geological Survey, who has rewritten the chapter on **Geology**; Patrick Baty, Papers and Paints Ltd, for his technical advice on **Internal Paint and Varnish**; Rachel Dodd BA (Hons), Dip Arch, for the new chapter on **Shops and Shop Fronts**; Professor Max Fordham MA FEng FCIBSE MConsE Hon FRI-BA, Raymond L Doctor, MAP, and Ian Low I Eng MIEIE ACIBSE, for their advice on **Services**; Ian Gow Hon FRIAS, National Monuments Record of Scotland, for his revision of **Interior Decoration** and **Wallcoverings**; Andrew Kerr WS for his amendments to **Legal Agreements**; Lori McElroy BSc MA MCIBSE, EDAS Consultant, for her advice and work on **Insulation**; Colin Mitchell-Rose, Craig and Rose plc, who advised on **External Paintwork**; James Simpson BArch FRIAS RIBA who wrote the **Introduction**; John H Robertson BSc MBA MCIOB CTIS, Apex Property Care, for his contribution towards **Timber Decay** on which Dr Brian Ridout, Ridout Associates, was also consulted.

The Committee thanks Gillian Bishop MA BArch ARIAS for her generous help with the final draft; Elizabeth Nicholson for her many hours of typing; and Gillian Stewart BA (Hons) for her enthusiastic help with surveys of doors, door furniture and salvaged materials in the New Town.

New photographs were specially taken for the fourth edition by Alan Forbes, Stewart Guthrie, Desmond Hodges, Ian Riddell, the Royal Commission on the Ancient and Historical Monuments of Scotland, André Sypestyn and Hunter Walker.

Extra drawing equipment was kindly lent by Alan Marshall, Gray Marshall and Associates.

New and revised drawings were prepared by Adrian Boot BA (Hons) Dip Arch, Rachel Dodd and Fiona Rankin. The drawings for **Flashings** were prepared by Michael Pontin, Law and Dunbar-Nasmith.

All of this information was collated and the text further revised by Rachel Dodd BA (Hons) Dip Arch and Fiona Rankin MA Dip Arch.

Edinburgh, February 1995

Andy Davey
Bob Heath
Desmond Hodges
Mandy Ketchin
Roy Milne

INDEX

Sodium pentachlorophate, 69
Sodium sulphate Chrystallisation Test, 55
Solder, for flat roofs, 140
Solid fuel fires/stoves:
    efficiency, 239
    inspection and maintenance, 241
    old fireplaces, 236
Solvent-based paint removers, 132
Solvent-thinned primers, 132
Sound insulation, 173
Sources of information, 40
Spalling of stonework, 71, 75–6, 244
Spanish slates, 137
Spirit paint removers, 218
Spirit varnishes, 217
Sporophore (fruiting body) of dry rot, 227
Spot welding, 250
Springwell stone, 55
Square-snecked rubble, 60
Squared rubble, 60
Stacks, chimney, 163–7
Stainless steel:
    clips, 140
    corbel plate, 76
    fixings, 76, 78, 89–90, 135, 169, 177, 199, 253
    lathes, 142
    nails, 139
    screws, for roof fixings, 159
Stainton stone, 55, 244
Stair door cranks, 104
Stairs:
    cleaning, 185
    common, 184–7
    handrail, 183–7
    house, 183
    lighting, 187
    pen-checked, 80, 184–7
    repairs to, 185, 187
    structural defects, 80, 185
Stairwells, 184
Stancliffe stone, 55
Standard Improvement Grants, 9–10
Stanton Moor stone, 55
Statutory Notice for repairs, 7–8, 22–3
Steel angles sash window repair, 116
Steel beams, for floor strengthening, 181–2
Steps, xxv
    basement, 242–4
    delamination of, 244
    internal, 184
    repair of, 244
Stone:
    Acid Immersion Test for, 54
    Arbroath, for hearths, 234
    ashlar quarrying, 61
    bedding planes in, 56
    Binney, 54
    Blaxter, 55
    Caithness, for hearths, 234
    Cat Castle, 55
    Clashach, 55
    Craigleith, xx, 45, 53–5, 69
    Crosland Moor, 55
    Cullalo, 54
    Dalgety, 54
    Darney, 55
    decay and defects, 56–7
    Doddington, 55
    Dunhouse, 55
    durability of, 54

durability tests for, 54–5
    Greenbrae, 55
    Hailes, 54, 109, 184, 244–5
        for hearths, 234
    local history, 53–5
    matching, 55
    Murrayfield, 54
    pore size distribution of, 54
    porosity of, 54
    quarrying of, 53–5
    Ravelston, 54
    Redhall, 54
    for replacement, 89
    salvaged, 16, 55
    saturation coefficient of, 54
    Sodium sulphate Crystallisation Test, 54
    Springwell, 55
    Stainton, 55, 244
    Stancliffe, 55
    Stanton Moor, 55
    substitute materials for, 92
    tests on, 54–5
    water absorption of, 54
    Woodkirk, 55
Stone flags, external, 244
Stone paints, removal of, 73–4
Stone-flagged kitchen floors, 237
Stonework:
    abrasive blasting for paint removal, 73
    algae growth on, 69
    arches, 61, 65, 76, 242–4
    architraves, 65
    ashlar, see Ashlar stonework
    balustrades, 168
    band courses, 65
    bedding of, 56–7, 64–5, 81–6
    blocking courses, 63, 147–52, 168
    bonding of, 59–66, 175–7
    chemical attack on, 57
    chimneys, 77, 80, 162–6
    cills, 65, 75, 114
    cleaning, xx, 67, 73–4
    columns, 92
    construction, xx
    contour scaling of, 88
    copings, 162, 164, 168
    cornices, 63
    crystallization damage, 57
    cutting out, 89–90
    decay, xx
    disintegration in chimneys, 162
    door openings, 75–6
    dummy windows, 130
    edge bedding of, 57, 64–5
    exfoliation of, 56
    face bedding of, 57, 64, 66, 130
    fixings for, 76, 90, 168
    friezes, 63
    frost action in chimneys, 162
    frost action on, 56, 168
    indenting, 88–90, 92, 164, 244
    inspection, 88
    inspection of, 58
    jambs, 60
    jointing and pointing, xx, 81–6
    joints, see Joints; Mortars
    lamination/delamination, 65, 88, 162
    lichen growth on, 69, 71
    lintels, 65, 75–6
    mortars for, see Mortars

Upper Dean Street, 111

V-jointed ashlar, 61, 65
Valley gutters, 134
    organic growth on, 71
Vapour barriers, 222
    floors, 171
Varnishes:
    alkyd, 217
    oil based, 217
    spirit, 217
Vegetation:
    growth of, 69–72
    removal of, xx, 69–72
Vented common services ducts, 239
Ventilation:
    of building fabric, 139, 171–2
    with cupolas, 156
    of flues, 162, 165
    for gas appliances, 170, 240
    and heat loss, 170
    mechanical, 118, 239
    for oil appliances, 240
    for open fires, 240
    of rooms, 116
Vermiculated finish, on ashlar, 63
Vesda system, fire detection, 21
Virginia creeper, 72
Voussoir, 62

Wainscotting, 188
Walker Street, 112
Wall plates, 135
Wallcoverings, xxv
    bedrooms, 220
    chintz papers, 220
    cleaning, 221
    damage to, 220–1
    dining rooms, 220
    drawing rooms, 220
    fading, 220–1
    gilded gold size, 220
    panelling, 220
    paper backed textiles, 220
    silk, 220
    stencilled, 220
    stencilled effects, 220
    wallpapers, see Wallpapers
Wallhead gutters, 147
    organic growth on, 71
Wallheads, 147
Wallpapers, 220
    acidity attack, 221
    early, analysis of, 202, 204
    French, 220
    pollution attack, 221
Wallplates, 135
Walls:
    acoustic insulation of, 173
    ashlar, see Ashlar stonework
    basement area, 76
    basement area painting, 73
    blocking course, 65, 168
    bonding of, 59–66, 175–7
    buckling of, 77, 175–7
    bulging, 77
    bulging of, 77, 175–7
    cavity filling, 59
    colour, 219
    dampness in, see Dampness

external, 59–66, 77
    gable, 80
    heat loss through, 170
    insulation of, 170–1
    internal, 59, 77, 175–7
    loading, 175–7
    parapet, 168
    plastering, see Plasterwork
    retaining, 76
    rubble, 59–60, 75–7, 175
    stone decay defects, 56
    strapping of, 77, 171
    strength of, 77
    structural defects, 75–80, 177
    thermal expansion, 77
    ties, 77
    timber stud partitions, 175–7
    'U' values, 170–1
Water, in mortar, 84
Water absorption by stone, 54
Water gilding, 219
Water of Leith, 45
Water penetration:
    around windows, 75, 222
    in cellars, 76
    from gutters and drainpipes, 147, 222
    in roofs, 92, 222
    in stonework, 69–72, 75, 222
    through cracks, 75–7, 92
    through platts, 76
    walls, 222
Water pipes:
    insulation for, 241
    removal of, 237
Water storage tanks, 239
Water supplies, 9–10
    inspection and maintenance, 241
Water thinned primers, 132
Water washing of stonework, 67
Watergates flashings, 144
Waterproof agents, 72
Waterproof renders, 225
Waverly Market, 258
WCs:
    inspection beneath old, 239
    ventilation for, 239
Weathering, stonework, 88
Welding:
    balconies, 253
    railings, 250
Welsh slates, 137
Wemyss Place, 92
Wet rot:
    Coniophora puteana (cellar rot), 227
    Coprinus domesticus (ink cap), 227
    Fibroporia vaillanti (mine fungus), 227
    identification, 227
    inspection for, 227
    Phellinius contiguus (white rot fungus), 227
    and plasterwork, 199, 227
    treatment, 229
    vulnerable areas, 114–20
    wetting and drying cycles of stone, 57
Wet-blasting stonework, 67
White rot fungus, 227
William Street, 259, 260
Window tax, 124
Windows:
    in ashlar walls, 62
    astragals, 114, 124–6

290